Contributions to Statistics

V. Fedorov/W.G. Müller/I. N. Vuchkov
(Eds.)
Model-Oriented Data Analysis,
XII/248 pages, 1992

J. Antoch (Ed.)
Computational Aspects of Model Choice,
VII/285 pages, 1993

W. G. Müller/H. P. Wynn/A. A. Zhigljavsky
(Eds.)
Model-Oriented Data Analysis,
XIII/287 pages, 1993

P. Mandl/M. Hušková (Eds.)
Asymptotic Statistics,
X/474 pages, 1994

P. Dirschedl/R. Ostermann (Eds.)
Computational Statistics,
VII/553 pages, 1994

C. P. Kitsos/W. G. Müller (Eds.)
MODA 4 – Advances in Model-Oriented
Data Analysis,
XIV/297 pages, 1995

H. Schmidli
Reduced Rank Regression,
X/179 pages, 1995

W. Härdle/M. G. Schimek (Eds.)
Statistical Theory and Computational
Aspects of Smoothing,
VIII/265 pages, 1996

S. Klinke
Data Structures for Computational Statistics,
VIII/284 pages, 1997

A. C. Atkinson/L. Pronzato/H. P. Wynn
(Eds.)
MODA 5 – Advances in Model-Oriented
Data
Analysis and Experimental Design
XIV/300 pages, 1998

M. Moryson
Testing for Random Walk
Coefficients in Regression
and State Space Models
XV/317 pages, 1998

S. Biffignandi (Ed.)
Micro- and Macrodata of Firms
XII/776 pages, 1999

W. Härdle/H. Liang/J. Gao
Partially Linear Models
X/203 pages, 2000

W. G. Müller
Collecting Spatial Data
XII/196 pages, 2001

Carlo Lauro · Jaromir Antoch
Vincenzo Esposito Vinzi
Gilbert Saporta (Editors)

Multivariate Total Quality Control

Foundation and Recent Advances

With 49 Figures and 34 Tables

Physica-Verlag
A Springer-Verlag Company

Series Editors
Werner A. Müller
Martina Bihn

Editors

Professor Carlo Lauro
Università degli Studi Napoli
Frederico II
Dipartimento di Matematica e Statistica
Via Cintia – Monte Sant'Angelo
80126 Napoli
Italy
carlo.lauro@unina.it

Dr. Vincenzo Esposito Vinzi
Università degli Studi Napoli
Frederico II
Dipartimento di Matematica e Statistica
Via Cintia – Monte Sant'Angelo
80126 Napoli
Italy
binci@unina.it

Professor Jaromir Antoch
Charles University
Faculty of Mathematics
Department of Statistics
Sokolovská 83
18675 Praha 8 – Karlin
Czech Republic
jaromir.antoch@karlin.mff.cuni.cz

Professor Gilbert Saporta
Conservatoire National
des Arts et Métiers
292, rue Saint Martin
75141 Paris Cedex 03
France
saporta@cnam.fr

ISSN 1431-1968
ISBN 978-3-7908-1383-8

Cataloging-in-Publication Data applied for
Die Deutsche Bibliothek – CIP-Einheitsaufnahme
Multivariate total quality control: foundation and recent control; with 34 tables / Carlo Lauro ... (ed.). – Heidelberg; New York: Physica-Verl., 2002
(Contributions to statistics)
ISBN 978-3-7908-1383-8 ISBN 978-3-642-48710-1 (eBook)
DOI 10.1007/978-3-642-48710-1

Physica-Verlag Heidelberg New York
a member of BertelsmannSpringer Science+Business Media GmbH

Softcover design: Erich Kirchner, Heidelberg

SPIN 10796564 88/2202-5 4 3 2 1 0 – Printed on acid-free paper

Preface

In the last decades, the production of goods and the offer of services have become quite complex activities mostly because of the markets globalisation, of the continuous push to the innovation and of the constant requests from more and more demanding markets.

The main objective of a company system has become the achievement of the quality for the business management cycle. This cycle goes from the design (*Plan*) to the production (*Do*), from the control (*Check*) to the management (*Action*), as well as to the marketing and distribution. Nowadays, the *Total Quality* of the company system is evaluated, according to the ISO 9000 regulations, in terms of its capacity to adjust the design and the production to the needs expressed (explicitly or implictly) by the final users of a product/service. In this process, the use of statistical techniques is essential not only in the classical approach of *Quality Control* of a product but also, and most importantly, in the Quality Design oriented to the satisfaction of customers. Thus, *Total Quality* refers to the global capacity of a company to fit its system to the real needs of its customers by designing products which are able to match the customers' taste and by implementing a statistical control of both the product and the *Customer Satisfaction*.

In such a process of design and evaluation, several statistical variables are involved and with a different nature (numerical, categorical, ordinal). These variables are linked by dynamic relationships which can be of different types (aleatory-statistical, logic and taxonomic) and, therefore, can give rise to complex data structures for the analysis. The data related to both the *Customer Satisfaction* and the characteristics of the product may be structured as classical two or three-way matrices (units by variables by occasions), as homogeneous or mixed data, or as matrices of symbolic data (units by variables by relationships). These structures of data demand for innovative statistical methods and techniques which allow to handle the complexity without losing interpretabily of results and a user-friendly usage.

Within this framework, this book represents a homogeneous collection of the major contributions from three European schools (Naples, Paris and Prague) whose researchers have faced new problems in the filed of *Multivariate Statistics for Total Quality*. The papers will gather results which, up to now, have been published in several international journals but also will show innovative results with a high interest for real applications whose data are naturally multivariate.

A process is known to be affected by causes of common variation and other causes called of specific or assignable variation. The common causes are associated with the overall behaviour of the system when it is in a (hypothetical) stable state of control (which still needs to be monitored). The specific causes are, instead, associated with deviations from the mentioned stable state so that possible evidences need to be flagged. They occur at specific points in time and usually require immediate corrective actions, while common causes are an opportunity for improvements with long-term payoffs. If these improvements are pursued, we are in presence of a proactive style of managing quality.

A proactive approach is distinguished from a reactive one because it designs for quality rather than just verifying quality, it identifies and removes causes of potential non-quality instead of discovering causes of occurred non-quality, it understands damages and gives a feedback to the whole process design rather than just quantifying damages, it finally deals with problems which are hard to detect but inexpensive to correct.

In a proactive quality control framework, two sets of variables are usually identified: the process variables and the quality characteristics. The process variables may consist in hundreds of very precise measurements routinely collected every minute, while the quality characteristics are rather noisy and generally less frequently collected because their collection is expensive and time consuming. Moreover, they should be measured on-line but are most often measured off-line in a laboratory and may be substituted by indirect (proxy) variables. All these discrepancies lead to data sets which are ill-conditioned for statistical modelling because of highly-correlated variables, missing data and measurement errors. In this framework, the book does not concentrate only on the standard situation of quality control when the errors accompanying the observed process are normally distributed. The more general situations that call for the use of the robust and non-parametric approaches are also described in details. Within these approaches, the use of recent methods of the multivariate analysis in the *Total Quality Control (TQC)* is enhanced with particular reference to the customer satisfaction area, the monitoring of quality characteristics and process variables, as well as the comparison of patterns generated from multioccasion observations.

The multivariate approach will be used in the following facets of TQC:

- improvement of the quality design by means of explicative methods;
- diagnosis of multivariate out-of-control situations for quality variables;
- identification of causes in terms of process variables;
- building of charts for controlling;
- visualization of relationships between variables for monitoring;
- comparison of standard (or expected) situations with respect to the achieved (or perceived) ones and explanation of eventual gaps.

The discussion covers both the theoretical and the practical computational aspects of the problem as well as necessary mathematical background, taking into account the requirements of the TQC. Several applications on real and simulated data help in giving a deeper insight on the proposed methods and the related interpretation rules.

First part of the contribution by J. Antoch, M. Hušková and D. Jarušková on off-line quality control deals with tests on the stability of statistical models. The problem is formulated in terms of testing the null hypothesis H against the alternative hypothesis A. The null hypothesis H claims that the model remains the same during the whole observational period, usually it means that the parameters of the model do not change. The alternative hypothesis A claims that, at an unknown time point, the model changes, which means that some of the parameters of the model are subject to a change. In case we reject the null hypothesis H, i.e. when we decide that there is a change in the model, we concentrate on a number of questions that arise, i.e. when has the model changed, is there just one change or are there more changes and what is the total number of changes etc. Aside testing for a change, the interest is to estimate change point(s) in different models. The least squares, M-, R- and MOSUM estimators are introduced and studied. Of course, the authors also estimate other parameters of the model(s), offer approximations to the distributions of the change point estimators and show that the estimators of the change points are usually closely related to some of the test statistics treated in the first part. Further, three types of confidence intervals are developed, one based on the limit distribution of the (point) estimators of the change point and two based on the bootstrap methods. All three methods are suitable for local changes while only the bootstrap constructions apply also to fixed changes. The test statistics described are typically certain functionals of partial sums of independent, identically distributed variables and their distribution is very complex. Therefore, selected limit results are presented that form the basis for establishing the limit distribution of considered test statistics, functionals of partial sums and change point estimators.

Next four contributions are closely connected with the on-line process control. At first, the paper of J. Antoch and D. Jarušková provides a critical overview of classical methods for the detection of a change in a sequence of observations. Sequential approach is presented. Attention is payed to the Shewhart, CUSUM, EWMA and Girshick-Rubin-Shiryayev procedures. Numerous applications to the normal distribution suitable for statistical process control are considered. Instead of detailed proofs only their main ideas are presented.

Contributions of L. Jaupi and N. Niang to multivariate process control both refer to the theory of robustness but in different frameworks. N. Niang gives first a survey of some charts for multidimensional data, including Hotelling's, EWMA and UWMA charts based on the Mahalanobis distance to the target mean vector. She then shows that the classical T^2-chart suffers from severe limitations in the presence of outliers during the preliminary study where provisional control limits are estimated. These limits are unreliable. The solution consists in the use of robust M-estimators of the mean for grouped data, and of the mean and covariance matrix for individual observations.

L. Jaupi, in his contribution, uses the duality between robustness and outlier detection: detecting an out of control situation is a particular case of outlier detection. Influence functions are thus a tool for outlier detection and form the basis of a new type of control charts for the process mean, the process variability, and the process orientation (correlation structure). The case of a large number of quality characteristics split in two sets (important ones and less important ones) raises to the use of nonsymmetric PCA (or Rao's PCA with instrumental variables) for dimension reduction, combined with influence functions of the eigenvalues.

The paper by G. Scepi considers the situations where control charts for monitoring individual quality characteristics may be not adequate for detecting changes in the overall quality of products. It compares classical and parametric multivariate approaches with new and non parametric ones. In particular, control schemes both of simple data as well as batch and time dependent data are considered. Furthermore, the problem of stability common to parametric schemes and non parametric ones is discussed. Finally, some new developments on the control of complex and structured data, like symbolic data, are introduced.

The paper by V. Esposito proposes different strategies for comparative analyses which are based on a joint use of *Non Symmetrical Multidimensional Data Analysis* and *Procrustean rotations*. As a general purpose, this paper aims at comparing and visualizing the behaviour of paired elements (products/services) under different observational conditions with respect to the dependence structure between the quality/sensory characteristics and the process variables. This is a general problem in *Total Quality Measurement* when one wants to compare the really observed quality characteristics with the in-control situation and explain the eventual differences with respect to the process variables. The paper proposes also a strategy which may be very helpful in the field of *Customer Satisfaction* where one generally aims at measuring the gap between perceived and expected quality by the customers of a product/service with respect to a pre-defined set of scenarios.

Finally, the paper by G. Giordano focuses on *Multiattribute Preference Data*, which refer to preference judgements expressed with respect to a set of stimuli described by any relevant attributes, in order to perform a conjoint analysis. An explorative strategy of multidimensional data analysis, which allows to take into account the complexity of the multiattribute data structure, is discussed. This strategy allows to represent on a two dimensional space the relationships among the judges, the attribute-levels and the stimuli. Each factorial axis is a synthesis of the preference variables and is useful in describing the preference of homogeneous subsets of respondents towards the attribute levels.

It must be stressed that this book can be used for teaching purposes in courses related to quality control and industrial statistics but also as the reference book for the real-world practitioners in both engineering (quality control and production departments) and in business (customers department) during the phases of planning and checking for quality. Some perspectives of research are also outlined for those who wish to invest in this very promising and highly demanding research field.

Napoli April 2001,	*Natale Carlo Lauro*
Praha April 2001,	*Jaromír Antoch*
Napoli April 2001,	*Vincenzo Esposito Vinzi*
Paris April 2001,	*Gilbert Saporta*

Contents

Contributors

Jaromír Antoch

Charles University of Prague, Department of Statistics, Sokolovská 83, CZ – 186 75 Praha 8, Czech Republic; jaromir.antoch@karlin.mff.cuni.cz

Vincenzo Esposito Vinzi

University of Naples Federico II, Faculty of Economics, Department of Mathematics and Statistics, Via Cintia – Complesso Monte Sant'Angelo, I – 80 126 Naples, Italy; binci@unina.it

Giuseppe Giordano

University of Naples Federico II, Faculty of Economics, Department of Mathematics and Statistics, Via Cintia – Complesso Monte Sant'Angelo, I – 80 126 Naples, Italy; giordano@dms.unina.it

Marie Hušková

Charles University of Prague, Department of Statistics, Sokolovská 83, CZ – 186 75 Praha 8, Czech Republic; jaromir.antoch@karlin.mff.cuni.cz

Daniela Jarušková

Czech Technical University, Department of Mathematics, Thákurova 7, CZ – 166 29 Praha 6, Czech Republic; daniela.jaruskova@fsv.cvut.cz

Luan Jaupi

Département de Mathématiques, Conservatoire National des Arts et Métiers, 292 Rue Saint Martin, F – 751 41 PARIS CEDEX 03, France; jaupi@cnam.fr

Ndèye Niang

Département de Mathématiques, Conservatoire National des Arts et Métiers, 292 Rue Saint Martin, F – 751 41 PARIS CEDEX 03, France; niang@cnam.fr

Germana Scepi

University of Naples Federico II, Faculty of Economics, Department of Mathematics and Statistics, Via Cintia – Complesso Monte Sant'Angelo, I – 80 126 Naples, Italy; scepi@unina.it

OFF-LINE STATISTICAL PROCESS CONTROL

Jaromír Antoch, Marie Hušková and Daniela Jarušková

Charles University of Prague, Department of Statistics, Sokolovská 83, CZ-186 75 Praha 8, Czech Republic; jaromir.antoch@karlin.mff.cuni.cz. marie.huskova@karlin.mff.cuni.cz
Czech Technical University, Department of Mathematics, Thákurova 7, CZ-166 29 Praha 6, Czech Republic; daniela.jaruskova@fsv.cvut.cz

Abstract: First part of this paper deals with tests on the stability of statistical models. The problem is formulated in terms of testing the null hypothesis H against the alternative hypothesis A. The null hypothesis H claims that the model remains the same during the whole observational period, usually it means that the parameters of the model do not change. The alternative hypothesis A claims that, at an unknown time point, the model changes, which means that some of the parameters of the model are subject to a change. In case we reject the null hypothesis H, i.e. when we decide that there is a change in the model, we concentrate on a number of questions that arise:

- when has the model changed;
- is there just one change or are there more changes;
- what is the total number of changes etc.

The time moment when the model has changed is usually called *change point*. Aside testing for a change, our interest is to estimate change point(s) in different models. The least squares, M-, R- and MOSUM estimators are introduced and studied. Of course, we also estimate other parameters of the model(s), show approximations to the distributions of the change point estimators and show that the estimators of the change points are usually closely related to some of the test statistics treated in the first part.

Three types of confidence intervals are developed, one based on the limit distribution of the (point) estimators of m and two based on the bootstrap methods. All three methods are suitable for local changes while only the bootstrap constructions apply also to fixed changes.

The test statistics described below are typically certain functionals of partial sums of independent, identically distributed variables and their distribution is very complex. Therefore, we present selected limit results that form the basis for establishing the limit distribution of considered test statistics, functionals of partial sums and change point estimators.

Several Matlab codes, that implement selected methods described below and include detailed description and links to the previous sections, are presented to illustrate the possibilities of the studied methods. Complete Matlab codes are available from the authors on request.

2000 *Mathematics Subject Classification.* 62-01, 62F03, 62F05, 62F10, 62F12, 62F15, 62F35, 62F40.

Key words and phrases. Change point problem; $L_1-, L_2-, ML-, M-$ and R-estimators; Bayesian, maximumum-type and trimmed maximum-type statistics; MOSUM; location model, linear regression, abrupt and gradual changes, multiple changes; AR and ARMA processes; hypotheses testing, approximate critical values, Bonferroni inequality; Brownian bridge, Gaussian and Wiener processes; bootstrap.

Acknowledgement. The work was partially supported by grants GAČR 201/00/0769, MSM 113200008 and MSM210000001. We are grateful for valuable comments we got from Prof. J. Steinebach.

CONTENTS

Part I. Testing

1. Introduction

In the scope of mathematical statistics the decision whether observed series remained stationary or whether a change of a specific kind occurred is usually based on hypotheses testing. *The null hypothesis claims that the process is stationary while the alternative hypothesis claims that the process is non-stationary and the stationarity was violated in a specific way.*

We start with the simplest situation that arises if we assume that a certain characteristic (e.g. that of a manufacturing process) varies around a certain constant μ_0 given by the production design. We suppose that at the beginning the process is *in control.* However, it can happen that due to a failure of the production device, e.g., the observed characteristic suddenly starts to vary around another *out-of control* constant $\mu_1 \neq \mu_0$. The same sudden failure may cause a change of the variance as well, but it is also possible that the variance remains the same. Moreover, sometimes we can even suppose, because of our long experience with the production process, that the variance σ^2 is known.

In this simple case, with starting value μ_0 and variance σ^2 known both, one can standardize the observations to obtain the standardized variables Y_i, $i = 1, \dots, n$, which have at the beginning a zero mean and a unit variance, and to test the following null hypothesis H against the alternative A, i.e.,

$$
\begin{aligned}
&H : Y_i = e_i, && i = 1, \dots, n, \qquad \text{(I.1)}\\
&A : \exists \mathsf{m} \in \{0, \dots, n-1\} \quad \text{such that} && \\
&\qquad Y_i = e_i, && i = 1, \dots, \mathsf{m},\\
&\qquad Y_i = \mu + e_i, && i = \mathsf{m}+1, \dots, n,
\end{aligned}
$$

where $\mu \neq 0$ and e_i are independent identically distributed (iid) random variables (errors). *The quantity* m *is called change point.*

2. Methods for deriving test statistics

The decision rule for rejecting the null hypothesis H is based on test statistics. Two basic methods may be applied to derive them, namely, the *maximum likelihood method* and the *pseudo-Bayes method.* We demonstrate on the testing problem (I.1) how both these approaches can be applied. For simplicity we suppose that e_i are independent and distributed according to the standard normal distribution $N(0,1)$ with the density $\phi(x)$ and the distribution function $\Phi(x)$.

2.1. Likelihood ratio method

♣ Let us suppose for a moment that the change point m is known and put $\mathsf{m} = k$. If $\mu \neq 0$, the log-likelihood ratio for testing H against A is

$$\Lambda_k = \sup_{\mu} \log \frac{\prod_{i=1}^n f_A(Y_i)}{\prod_{i=1}^n f_H(Y_i)} = \sup_{\mu} \log \frac{\prod_{i=1}^k \phi(Y_i) \prod_{i=k+1}^n \phi(Y_i - \mu)}{\prod_{i=1}^n \phi(Y_i)}$$

$$= \sup_{\mu} \left\{ -\frac{1}{2} \sum_{i=k+1}^{n} (Y_i - \mu)^2 + \frac{1}{2} \sum_{i=k+1}^{n} Y_i^2 \right\} = \frac{1}{2(n-k)} \left(\sum_{i=k+1}^{n} Y_i \right)^2 .$$

The null hypothesis H is rejected when $\Lambda_k > C_\alpha$, which can be equivalently expressed as

$$\left| \frac{1}{\sqrt{n-k}} \sum_{i=k+1}^{n} Y_i \right| > \sqrt{2C_\alpha},$$

where C_α is a constant chosen so as to correspond to the fixed significance level α. In other words, the log-likelihood ratio is a function of the average of the second part of the series of observations $\{Y_i\}$.

To simplify the notation, we put

$$\overline{Y}_k = \frac{1}{k} \sum_{i=1}^{k} Y_i \quad \text{and} \quad \overline{Y}_k^{o} = \frac{1}{n-k} \sum_{i=k+1}^{n} Y_i . \tag{I.2}$$

Notice, that $\overline{Y}_k^{o}$ is the least squares estimator of the unknown constant μ and $\sqrt{n-k}\ \overline{Y}_k^{o}$ has a standard normal distribution $N(0,1)$.

- For the one-sided alternative with $\mu > 0$, we obtain the test statistic

$$\frac{1}{\sqrt{n-k}} \sum_{i=k+1}^{n} Y_i,$$

while for the two-sided alternative with $\mu \neq 0$, we apply its absolute value

$$\left| \frac{1}{\sqrt{n-k}} \sum_{i=k+1}^{n} Y_i \right| .$$

♣ When the change point m is unknown (so that both μ and m are unknown), we have to take the supremum of the log-likelihood ratio with respect to both of them, i.e.,

$$\max_{0\le k\le n-1} \sup_{\mu} \log \frac{\prod_{i=1}^{k} \phi(Y_i)\prod_{i=k+1}^{n} \phi(Y_i-\mu)}{\prod_{i=1}^{n} \phi(Y_i)}$$
$$= \max_{0\le k\le n-1} \frac{1}{2(n-k)} \left(\sum_{i=k+1}^{n} Y_i \right)^2,$$

and the test statistics usually applied for the case with the unknown change point m are of the form

$$\max_{0\le k\le n-1} \left\{ \frac{1}{\sqrt{n-k}} \sum_{i=k+1}^{n} Y_i \right\}, \tag{I.3}$$

and

$$\max_{0\le k\le n-1} \left\{ \left| \frac{1}{\sqrt{n-k}} \sum_{i=k+1}^{n} Y_i \right| \right\}. \tag{I.4}$$

These statistics belong to the so-called *maximum-type statistics.*

• If we consider the two-sided alternative with $\mu \neq 0$, the null hypothesis H is rejected if for a suitably chosen constant $C_{1\alpha}$

$$\max_{0\le k\le n-1} \left\{ \left| \frac{1}{\sqrt{n-k}} \sum_{i=k+1}^{n} Y_i \right| \right\} > C_{1\alpha},$$

where $C_{1\alpha}$ is a constant chosen so as to correspond to the fixed significance level α. This rule is reasonable because it rejects H if for at least one k, $0 \le k \le n-1$,

$$\left| \frac{1}{\sqrt{n-k}} \sum_{i=k+1}^{n} Y_i \right| > C_{1\alpha}.$$

• Similarly, for the one-sided alternative with $\mu > 0$, the null hypothesis is rejected if

$$\max_{0\le k\le n-1} \left\{ \frac{1}{\sqrt{n-k}} \sum_{i=k+1}^{n} Y_i \right\} > C_{2\alpha},$$

where $C_{2\alpha}$ is again an appropriately chosen constant.

2.2. Pseudo-Bayesian method

The method is based on the assumption that the unknown change point m and unknown magnitude of shift $\boldsymbol{\mu}$ are random variables such that the prior distribution of m is uniform, i.e. $P(\mathsf{m} = i) = 1/n, i = 0, \ldots, n-1$, and $\boldsymbol{\mu}$ is distributed according to $N(0, \gamma^2)$. Since the density of $Y_1, \ldots, Y_n$ given $\boldsymbol{\mu} = \mu$ and $\mathsf{m} = k$ is normal, i.e.,

$$f\Big(y_1, \ldots, y_n \mid \boldsymbol{\mu} = \mu, \mathsf{m} = k\Big) = \prod_{i=1}^{k} \phi(y_i) \prod_{i=k+1}^{n} \phi(y_i - \mu),$$

the unconditional density can be expressed as

$$f(y_1,\dots,y_n)=\sum_{k=1}^{n}\frac{1}{n}\int_{-\infty}^{\infty}\prod_{i=1}^{k}\phi(y_i)\prod_{i=k+1}^{n}\phi(y_i-\mu)\frac{1}{\gamma\sqrt{2\pi}}\exp\Big\{-\mu^2/2\gamma^2\Big\}\,d\mu=$$

$$=\prod_{i=1}^{n}\phi(y_i)\left(\sum_{k=1}^{n}\frac{1}{n}\int_{-\infty}^{\infty}\frac{1}{\gamma\sqrt{2\pi}}\exp\left\{-\frac{1}{2}\left(-2\mu\sum_{i=k+1}^{n}y_i+(n-k)\mu^2+\frac{\mu^2}{\gamma^2}\right)\right\}d\mu\right)$$

and the corresponding likelihood ratio has the form

$$\widetilde{\Lambda}=\frac{f_A(Y_1,\dots,Y_n)}{f_H(Y_1,\dots,Y_n)}$$

$$=\sum_{k=1}^{n}\frac{1}{n}\sqrt{\frac{1}{1+(n-k)\gamma^2}}\exp\left\{\frac{\gamma^2}{2(1+(n-k)\gamma)^2}\Big(\sum\nolimits_{i=k+1}^{n}Y_i\Big)^2\right\}.$$

• Letting $\gamma\to 0$ and applying Taylor expansion, the likelihood ratio $\widetilde{\Lambda}$ is, for the two-sided alternative, equivalent to the test statistic

$$\sum_{k=1}^{n}\frac{1}{n}\left(\frac{1}{\sqrt{n}}\sum_{i=k+1}^{n}Y_i\right)^2. \tag{I.5}$$

The obtained statistic belongs to the so called *sum-type* test statistics. For details see Gardner (1969).

• For the one-sided alternative with $\mu>0$ (μ is assumed to follow a 50% truncated normal distribution) we may analogously derive the sum-type test statistic

$$\sum_{k=1}^{n}\frac{1}{n}\left(\frac{1}{\sqrt{n}}\sum_{i=k+1}^{n}Y_i\right). \tag{I.6}$$

For details see Chernoff and Zacks (1964) and Kander and Zacks (1966).

2.3. Critical values

For the decision about rejection of the null hypothesis H we need to know critical values of the suggested test statistics. It means to know their distributions under H.

• Let us start with the test statistic

$$\max_{0\le k\le n-1}\left\{\left|\frac{1}{\sqrt{n-k}}\sum_{i=k+1}^{n}Y_i\right|\right\}. \tag{I.7}$$

Assuming that $\{Y_i\}$ are iid with the standard normal distribution $N(0,1)$, then all statistics

$$\frac{1}{\sqrt{n-k}}\sum_{i=k+1}^{n}Y_i,\quad k=0,\dots,n-1, \tag{I.8}$$

have a $N(0,1)$ distribution. If m is known and equal to k, one would reject H at the significance level α if

$$\left| \frac{1}{\sqrt{n-k}} \sum_{i=k+1}^{n} Y_i \right| > u_{1-\alpha/2},$$

where $u_{1-\alpha/2}$ is the $100(1-\alpha/2)\%$ quantile of $N(0,1)$. Clearly, the statistic (I.7) is a stochastically larger variable than any of the absolute values of statistics (I.8), i.e., $\forall x \in \mathcal{R}^1$ and $0 \le k \le n-1$

$$P\left(\max_{0 \le k \le n-1} \left\{ \left| \frac{1}{\sqrt{n-k}} \sum_{i=k+1}^{n} Y_i \right| \right\} > x \right) \ge P\left(\left| \frac{1}{\sqrt{n-k}} \sum_{i=k+1}^{n} Y_i \right| > x \right).$$

Therefore, the $100\,\alpha\%$-quantile of the distribution of statistic (I.7) is larger than $u_{1-\alpha/2}$. Analyzing the same data set, we reject the null hypothesis in the case when the change point is known much more often than in the case of an unknown change point.

• To find the exact distribution of (I.4) means to find the distribution of the maximum of absolute values of standardized normal variables that are (unfortunately) correlated. The correlation coefficients are

$$corr\left(\frac{1}{\sqrt{n-k}} \sum_{i=k+1}^{n} Y_i, \frac{1}{\sqrt{n-l}} \sum_{i=l+1}^{n} Y_i \right) = \sqrt{\frac{n-l}{n-k}}, \quad k \le l.$$

Theoretically, it should not be a problem to find the distribution of (I.4). However, in practice the distribution is so complex, that its quantiles (desired critical values) may be computed only for small values of n, see Hawkins (1977).

• Sometimes, the approximate critical values may be quite satisfactory for practical use. To find approximate critical values we can use a very simple idea by applying the Bonferroni inequality as follows:

$$P\left(\max_{0 \le k \le n-1} \left\{ \left| \frac{1}{\sqrt{n-k}} \sum_{i=k+1}^{n} Y_i \right| \right\} > C \right) = P\left(\bigcup_{k=0}^{n-1} \left\{ \left| \frac{1}{\sqrt{n-k}} \sum_{i=k+1}^{n} Y_i \right| > C \right\} \right)$$
$$\le \sum_{k=0}^{n-1} P\left\{ \left| \frac{1}{\sqrt{n-k}} \sum_{i=k+1}^{n} Y_i \right| > C \right\} = n\,P\left(\left| \frac{1}{\sqrt{n}} \sum_{i=1}^{n} Y_i \right| > C \right).$$

Hence, the $100(1-\alpha/(2n))\%$-quantile of the standard normal distribution $N(0,1)$ may serve as an upper estimate of the critical value at the significance level α for the problem (I.1) applying the test statistic (I.4). The approximate critical values obtained in this way are good enough for small samples (for small values of n), but they are too conservative for n large.

• Therefore, for n large, the asymptotic behavior of the studied test statistic (I.4) is of interest. It can be proved, applying the law of iterated logarithm, that

$$\max_{0\le k\le n-1}\left\{\left|\frac{1}{\sqrt{n-k}}\sum\nolimits_{i=k+1}^{n} Y_i\right|\right\} \longrightarrow \infty \quad \text{almost surely as} \quad n\longrightarrow\infty.$$

It follows that the limit distribution of (I.4) does not exist and that critical values increase to infinity as $n \to \infty$. The problem is caused by the behavior of the sequence $\{(n-k)^{-1/2}\sum_{i=k+1}^{n} Y_i, i=1,\dots,n-1\}$ near to its end. Here the averages, whose departures from zero are studied, are calculated only for "a small" number of observations and it can happen, with a large probability, that at least one of them attains a rather large value.

Therefore, some authors suggest to use, instead of the statistics (I.3) and (I.4), the *trimmed maximum-type* test statistics

$$\max_{0\le k\le \lfloor(1-\beta)n\rfloor}\left\{\frac{1}{\sqrt{n-k}}\sum\nolimits_{i=k+1}^{n} Y_i\right\} \tag{I.9}$$

and

$$\max_{0\le k\le \lfloor(1-\beta)n\rfloor}\left\{\left|\frac{1}{\sqrt{n-k}}\sum\nolimits_{i=k+1}^{n} Y_i\right|\right\}, \tag{I.10}$$

where β is a small positive constant less than one and $\lfloor x\rfloor$ denotes the integer part of x. The advantage of the statistics (I.9) and (I.10) is that they are bounded in probability. The *trimming off* a $100\beta\%$ portion of the sample (upper time points) means that one assumes that the change did not occur during this time period. Notice that, typically, $\beta\in[0.01, 0.1]$. The decision "*How much to trim off?*" depends on the subjective decision of the statistician and his/her a priori knowledge of the problem. If the statistician decides to trim off only a very small portion of the time points or no time points (observations) at all, he/she pays for it by a loss of the power of his/her test as the critical values depend rather strongly on the value of β.

Tables 1. – 6. below contain critical values for statistics (I.3) – (I.4) and (I.9) – (I.10) obtained by simulations.

	significance level				
	10%	5%	2.5%	1%	0.5%
over-all	2.297	2.604	2.874	3.192	3.412
$\beta=0.01$	2.269	2.581	2.854	3.177	3.398
$\beta=0.05$	2.190	2.510	2.792	3.118	3.347
$\beta=0.10$	2.119	2.446	2.731	3.068	3.302

Table 1. Simulated critical values of the over-all maximum-type test statistic (I.3) and the corresponding trimmed maximum-type test statistic (I.9) for different trimming portions β, $n=100$.

	significance level				
	10%	5%	2.5%	1%	0.5%
over-all	2.603	2.874	3.117	3.410	3.613
$\beta = 0.01$	2.581	2.853	3.102	3.395	3.597
$\beta = 0.05$	2.510	2.792	3.040	3.344	3.555
$\beta = 0.10$	2.447	2.732	2.991	3.299	3.512

Table 2. Simulated critical values of the over-all maximum-type test statistic (I.4) and the corresponding trimmed maximum-type test statistic (I.10) for different trimming portions β, $n = 100$.

	significance level				
	10%	5%	2.5%	1%	0.5%
over-all	2.455	2.754	3.018	3.328	3.542
$\beta = 0.01$	2.387	2.696	2.969	3.286	3.508
$\beta = 0.05$	2.270	2.589	2.867	3.189	3.418
$\beta = 0.10$	2.186	2.515	2.801	3.130	3.363

Table 3. Simulated critical values of the over-all maximum-type test statistic (I.3) and the corresponding trimmed maximum-type test statistic (I.9) for different trimming portions β, $n = 500$.

	significance level				
	10%	5%	2.5%	1%	0.5%
over-all	2.753	3.016	3.255	3.540	3.737
$\beta = 0.01$	2.694	2.967	3.211	3.503	3.707
$\beta = 0.05$	2.589	2.869	3.117	3.420	3.629
$\beta = 0.10$	2.515	2.800	3.055	3.362	3.573

Table 4. Simulated critical values of the over-all maximum-type test statistic (I.4) and the corresponding trimmed maximum-type test statistic (I.10) for different trimming portions β, $n = 500$.

	significance level				
	10%	5%	2.5%	1%	0.5%
over-all	2.510	2.805	3.066	3.373	3.585
$\beta = 0.01$	2.416	2.721	2.991	3.311	3.530
$\beta = 0.05$	2.289	2.610	2.886	3.214	3.442
$\beta = 0.10$	2.205	2.533	2.817	3.149	3.373

Table 5. Simulated critical values of the over-all maximum-type test statistic (I.3) and the corresponding trimmed maximum-type test statistic (I.9) for different trimming portions β, $n = 1\,000$.

	significance level				
	10%	5%	2.5%	1%	0.5%
over-all	2.804	3.066	3.301	3.586	3.785
$\beta = 0.01$	2.723	2.994	3.238	3.531	3.735
$\beta = 0.05$	2.609	2.887	3.142	3.446	3.651
$\beta = 0.10$	2.531	2.815	3.071	3.374	3.590

Table 6. Simulated critical values of the over-all maximum-type test statistic (I.4) and the corresponding trimmed maximum-type test statistic (I.10) for different trimming portions β, $n = 1\,000$.

• For large n, the approximate critical values can be calculated from the asymptotic behavior of the probabilities under H, because we have $\forall x \in \mathcal{R}^1$:

$$P\left(\max_{0\le k\le n-1}\left\{\left|\frac{1}{\sqrt{n-k}}\sum_{i=k+1}^{n} Y_i\right|\right\} > \frac{x+b_n}{a_n}\right) \approx 1-\exp\left\{-e^{-x}\right\}, \quad \text{(I.11)}$$

where

$$a_n = \sqrt{2\log\log n} \quad \text{and} \quad b_n = 2\log\log n + \frac{1}{2}\log\log\log n - \frac{1}{2}\log\pi,$$

and

$$P\left(\max_{0\le k\le \lfloor(1-\beta)n\rfloor}\left\{\left|\frac{1}{\sqrt{n-k}}\sum_{i=k+1}^{n} Y_i\right|\right\} > x\right) \approx 2(1-\Phi(x)) + x\phi(x)\log\frac{1}{\beta}. \quad \text{(I.12)}$$

The approximation (I.11) was derived by Darling and Erdős (1956).

Both formulas (I.11) and (I.12) were derived from the approximation of the maximum of the sequence

$$\left\{ \left| \frac{1}{\sqrt{n-k}} \sum_{i=k+1}^{n} Y_i \right| \right\}$$

by the maximum of the random process

$$\left\{ \frac{|\, W(1-t)\, |}{\sqrt{1-t}} \right\},$$

where $\{W(t), t \geq 0\}$ denotes the Wiener process, see Part III Sections 11 and 12.

Notice that the distribution of

$$\max_{0 \leq t \leq 1-\beta} \frac{|\, W(1-t)\, |}{\sqrt{1-t}} \quad \text{is the same as the distribution of} \quad \max_{\beta \leq t \leq 1} \frac{|\, W(t)\, |}{\sqrt{t}}.$$

The approximation (I.12), as well as the approximations of the same type stated later on, are derived for x large and therefore they can be useful for finding critical values. For small x their validity fails.

The critical values obtained from (I.11) are not very good because they are too conservative. The approximation (I.12) gives more accurate critical values but the choice of β affects them significantly. The critical values obtained from (I.12) get better as β gets larger.

• For the one-sided alternatives with $\mu > 0$, the approximations (I.11) and (I.12) have to be adapted as follows, i.e., $\forall x \in \mathcal{R}^1$

$$P\left(\max_{0 \leq k \leq n-1} \left\{ \frac{1}{\sqrt{n-k}} \sum_{i=k+1}^{n} Y_i \right\} > \frac{x+b_n}{a_n} \right) \approx 1 - \exp\left\{ -\frac{1}{2} e^{-x} \right\}, \quad \text{(I.13)}$$

and

$$P\left(\max_{0 \leq k \leq \lfloor (1-\beta) n \rfloor} \frac{1}{\sqrt{n-k}} \sum_{i=k+1}^{n} Y_i > x \right) \approx (1 - \Phi(x)) + \frac{1}{2} x \phi(x) \log \frac{1}{\beta}. \quad \text{(I.14)}$$

• Now we come to the sum-type statistics obtained from the pseudo-Bayes method. For n large, the approximate critical values of (I.5) derived for the two-sided alternative can be calculated from the convergence

$$\sum_{k=1}^{n} \frac{1}{n} \left(\frac{1}{\sqrt{n}} \sum_{i=k+1}^{n} Y_i \right)^2 \xrightarrow{\mathcal{D}} \int_0^1 W^2(t)\, dt = \sum_{j=1}^{\infty} \frac{4}{(2j+1)^2 \pi^2}\, \chi_j^2(1), \quad \text{(I.15)}$$

where $\{\chi_j^2(1)\}$ are independent variables distributed according to the χ^2 distribution with one degree of freedom. The distribution of $\int_0^1 W^2(t)\, dt$ was studied by MacNeill (1978) who obtained the following selected $100(1-\alpha)\%$ quantiles

$$P\left(\int_0^1 W^2(t)\,dt < x\right) = 1 - \alpha,$$

which can serve as the $100\alpha\%$ critical values presented in the following table

α	0.1	0.05	0.025	0.01
x	1.196	1.656	2.134	2.788

Table 7. $100\alpha\%$ critical values for the sum-type statistic calculated according to (I.15).

• The calculation of critical values of (I.6) derived for the one-sided alternatives is very simple as the sum-type test statistic

$$\sum_{k=1}^{n} \frac{1}{n}\left(\frac{1}{\sqrt{n}} \sum_{i=k+1}^{n} Y_i\right) = \frac{1}{n}\frac{1}{\sqrt{n}} \sum_{k=1}^{n} (k-1) Y_k$$

is normally distributed as

$$N\left(0, \frac{1}{3} - \frac{1}{2n} + \frac{1}{6n^2}\right).$$

3. Detection of changes in the parameters of a normal distribution

In this section we will, in more details, concentrate on the detection of changes in the parameters of a normal distribution. More general situations will be considered in Section 4.

3.1. Change in mean with unknown starting value

It (very) often happens that, even before the change point, the mean (the starting value) is unknown. In this case we test the following null hypothesis H against the alternative A:

$$\begin{aligned}
&H: Y_i = \mu + e_i, && i = 1, \ldots, n, && \text{(I.16)}\\
&A: \exists \mathsf{m} \in \{1, \ldots, n-1\} \ \text{ such that} && &&\\
&\quad Y_i = \mu + e_i, && i = 1, \ldots, \mathsf{m}, &&\\
&\quad Y_i = \mu + \delta + e_i, && i = \mathsf{m}+1, \ldots, n,\ \delta \neq 0. &&
\end{aligned}$$

We suppose again that the variables e_i are iid with the normal distribution $N(0, \sigma^2)$, σ^2 known. Then the maximum-type test statistics have the form

$$\max_{1 \le k \le n-1} \left\{ \frac{1}{\sigma} \left| \sqrt{\frac{n}{k(n-k)}} \sum_{i=1}^{k} \left(Y_i - \overline{Y}_n\right) \right| \right\} \qquad \text{(I.17)}$$

and

$$\max_{\lfloor \beta n \rfloor \le k \le \lfloor (1-\beta) n \rfloor} \left\{ \frac{1}{\sigma} \left| \sqrt{\frac{n}{k(n-k)}} \sum_{i=1}^{k} \left(Y_i - \overline{Y}_n\right) \right| \right\}, \qquad \text{(I.18)}$$

while the sum-type test statistic has the form

$$\frac{1}{\sigma^2}\sum_{k=1}^{n}\frac{1}{n}\left(\frac{1}{\sqrt{n}}\sum_{i=1}^{k}\left(Y_i-\overline{Y}_n\right)\right)^2. \tag{I.19}$$

Notice that $\sum_{i=1}^{k}\left(Y_i-\overline{Y}_n\right) = -\sum_{i=k+1}^{n}\left(Y_i-\overline{Y}_n\right)$ and that we deal with the partial sums of residuals $S_k=\sum_{i=1}^{k}\left(Y_i-\overline{Y}_n\right)$ under H instead of with partial sums of observations.

• To find approximate critical values, we may proceed in the same way as in the case with the known starting value. For n small we can use the Bonferroni inequality, and for n large we may apply the limit behavior of the studied probabilities, i.e.,

$$P\left(\max_{1\le k\le n-1}\left\{\frac{1}{\sigma}\left|\sqrt{\frac{n}{k(n-k)}}\sum_{i=1}^{k}\left(Y_i-\overline{Y}_n\right)\right|\right\}>\frac{x+b_n}{a_n}\right)\approx$$
$$\approx 1-\exp\left\{-2e^{-x}\right\},\quad x\in\mathcal{R}^1, \tag{I.20}$$

$$P\left(\max_{\lfloor\beta n\rfloor\le k\le\lfloor(1-\beta)n\rfloor}\left\{\frac{1}{\sigma}\left|\sqrt{\frac{n}{k(n-k)}}\sum_{i=1}^{k}\left(Y_i-\overline{Y}_n\right)\right|\right\}>x\right)\approx$$
$$\approx 2\left(1-\Phi(x)\right)+2x\phi(x)\log\frac{1-\beta}{\beta}. \tag{I.21}$$

The formulas (I.20) and (I.21) are derived from the approximation of the sequence

$$\left\{\frac{1}{\sigma}\left|\sqrt{\frac{n}{k(n-k)}}\sum_{i=1}^{k}\left(Y_i-\overline{Y}_n\right)\right|\right\}\quad\text{by the process}\quad\left\{\frac{|B(t)|}{\sqrt{t(1-t)}}\right\},$$

see Part III Sections 11 and 12. The approximation (I.20) can be found in Yao and Davis (1986).

For normally distributed random variables, a better approximation suggested by James et al. (1987) may be used:

$$P\left(\max_{\lfloor\beta n\rfloor\le k\le\lfloor(1-\beta)n\rfloor}\left\{\left|\frac{1}{\sigma}\sqrt{\frac{n}{k(n-k)}}\sum_{i=1}^{k}\left(Y_i-\overline{Y}_n\right)\right|\right\}>x\right)\approx$$
$$\approx 2\left(1-\Phi(x)\right)+2x\,\phi(x)\int_{x\sqrt{\beta/(n(1-\beta))}}^{x\sqrt{(1-\beta)/(n\beta)}}\frac{1}{y}\,\nu\left(y+\frac{x^2}{n\,y}\right)dy, \tag{I.22}$$

where

$$\nu(y)=\frac{2}{y^2}\exp\left\{-2\sum_{n=1}^{\infty}\frac{1}{n}\Phi\left(-\frac{1}{2}y\sqrt{n}\right)\right\}\approx\exp\left\{-0.583\,y\right\},\quad y>0. \tag{I.23}$$

Tables 8.–11. below contain critical values for statistics (I.17)–(I.18) obtained by simulations when σ is known and equal to one. Tables 8'.–11'.

contain asymptotic critical values calculated using (I.20) – (I.21) for comparison purposes.

n	(I.17)	(I.18)		
		$\beta = 0.01$	$\beta = 0.05$	$\beta = 0.1$
50	2.709	2.709	2.639	2.558
100	2.809	2.783	2.703	2.627
200	2.892	2.855	2.763	2.682
300	2.931	2.884	2.780	2.694
500	2.973	2.916	2.804	2.714

Table 8. Simulated 10 % critical values for statistics (I.17) and (I.18).

(I.20)	$n = 100$	$n = 300$	$n = 500$
	3.226	3.285	3.310
(I.21)	$\beta = 0.01$	$\beta = 0.05$	$\beta = 0.1$
	3.082	2.920	2.810

Table 8'. Asymptotic 10 % critical values for statistics (I.20) and (I.21).

n	(I.17)	(I.18)		
		$\beta = 0.01$	$\beta = 0.05$	$\beta = 0.1$
50	2.960	2.960	2.900	2.823
100	3.065	3.040	2.965	2.900
200	3.143	3.110	3.030	2.950
300	3.176	3.135	3.042	2.967
500	3.218	3.169	3.068	2.983

Table 9. Simulated 5 % critical values for statistics (I.17) and (I.18).

(I.20)	$n = 100$	$n = 300$	$n = 500$
	3.637	3.671	3.686
(I.21)	$\beta = 0.01$	$\beta = 0.05$	$\beta = 0.1$
	3.320	3.173	3.074

Table 9'. Asymptotic 5 % critical values for statistics (I.20) and (I.21).

n	(I.17)	(I.18)		
		$\beta = 0.01$	$\beta = 0.05$	$\beta = 0.1$
50	3.200	3.200	3.144	3.070
100	3.294	3.275	3.203	3.146
200	3.371	3.340	3.262	3.194
300	3.410	3.370	3.286	3.212
500	3.440	3.397	3.301	3.229

Table 10. Simulated 2.5 % critical values for statistics (I.17) and (I.18).

(I.20)	$n = 100$	$n = 300$	$n = 500$
	4.041	4.049	4.056
(I.21)	$\beta = 0.01$	$\beta = 0.05$	$\beta = 0.1$
	3.541	3.404	3.312

Table 10'. Asymptotic 2.5 % critical values for statistics (I.20) and (I.21).

n	(I.17)	(I.18)		
		$\beta = 0.01$	$\beta = 0.05$	$\beta = 0.1$
50	3.486	3.486	3.441	3.370
100	3.563	3.546	3.490	3.436
200	3.649	3.623	3.548	3.481
300	3.684	3.649	3.581	3.510
500	3.703	3.664	3.587	3.518

Table 11. Simulated 1 % critical values for statistics (I.17) and (I.18).

(I.20)	$n = 100$	$n = 300$	$n = 500$
	4.570	4.545	4.539
(I.21)	$\beta = 0.01$	$\beta = 0.05$	$\beta = 0.1$
	3.809	3.684	3.600

Table 11'. Asymptotic 1 % critical values for statistics (I.20) and (I.21).

The approximate critical values for the sum-type test statistic may be obtained from the convergence

$$\frac{1}{\sigma^2}\sum_{k=1}^{n}\frac{1}{n}\left(\frac{1}{\sqrt{n}}\sum_{i=1}^{k}(Y_i-\overline{Y}_n)\right)^2 \xrightarrow{\mathcal{D}} \int_0^1 B^2(t)\,dt = \sum_{j=1}^{\infty}\frac{1}{j^2\pi^2}\,\chi_j^2(1), \quad \text{(I.24)}$$

where $\{B(t), 0 \le t \le 1\}$ is the Brownian bridge and $\{\chi_j^2(1)\}$ are independent variables distributed according to the χ^2 distribution with one degree of freedom, see, e.g., Csörgő and Horváth (1997). It follows from Theorem 2.1 of Anderson and Darling (1952), see also Anderson and Darling (1954) and Kiefer (1960), that $\forall x \in \mathcal{R}^1$

$$P\left(\int_0^1 B^2(t)\,dt > x\right) = \quad \text{(I.25)}$$

$$1-\frac{\sqrt{2}}{\pi^{3/2}\sqrt{x}}\sum_{j=0}^{\infty}\frac{\Gamma(j+1/2)}{\Gamma(j+1)}\sqrt{2j+\frac{1}{2}}\,\exp\Big\{-\frac{(4j+1)^2}{16x}\Big\}K_{1/4}\left(\frac{(4j+1)^2}{16x}\right),$$

where $K_{1/4}(\cdot)$ denotes a modified Bessel function of the second type (see, e.g., function `besselk(nu,z)` in Matlab or function `BesselK` in Mathematica).

Now, let us turn to the case when the variance is unknown. In all statistics (I.17), (I.18) and (I.19) one can replace the unknown σ^2 by its estimate

$$\frac{1}{n}\sum_{i=1}^{n}\left(Y_i-\overline{Y}_n\right)^2 \quad \text{(I.26)}$$

and the limit distribution remains the same under H. However, in the case of the maximum-type test statistics, one gets a more powerful test if one applies the test statistics

$$\max_{1\le k\le n-1}\left\{\left|\frac{1}{s_k}\sqrt{\frac{n}{k(n-k)}}\sum_{i=1}^{k}\left(Y_i-\overline{Y}_n\right)\right|\right\} \quad \text{(I.27)}$$

and

$$\max_{\lfloor\beta n\rfloor\le k\le\lfloor(1-\beta)n\rfloor}\left\{\left|\frac{1}{s_k}\sqrt{\frac{n}{k(n-k)}}\sum_{i=1}^{k}\left(Y_i-\overline{Y}_n\right)\right|\right\}, \quad \text{(I.28)}$$

where

$$s_k^2 = \frac{1}{n-2}\left(\sum_{i=1}^{k}\left(Y_i - \overline{Y}_k\right)^2 + \sum_{i=k+1}^{n}\left(Y_i - \overline{Y}_k^o\right)^2\right). \tag{I.29}$$

It is interesting to realize that

$$\sqrt{\frac{n}{k(n-k)}}\sum_{i=1}^{k}(Y_i - \overline{Y}_n) = \sqrt{\frac{k(n-k)}{n}}\left(\overline{Y}_k - \overline{Y}_k^o\right),$$

so that all variables, from which the maximum is taken, are the usual two-sample t-statistics that are distributed, under H, according to the t-distribution with $n-2$ degrees of freedom. This fact may be used when the critical values are approximated by the Bonferroni inequality.

Instead of statistic (I.27), we may also use the statistic

$$\max_{1\le k\le n-1}\left\{\left|\frac{1}{\widehat{\sigma}_n}\sqrt{\frac{n}{k(n-k)}}\sum_{i=1}^{k}\left(Y_i - \overline{Y}_n\right)\right|\right\}, \tag{I.30}$$

where

$$\widehat{\sigma}_n^2 = \min_{2\le k\le n-2}\left\{\frac{1}{n}\left(\sum_{i=1}^{k}\left(Y_i - \overline{Y}_k\right)^2 + \sum_{i=k+1}^{n}\left(Y_i - \overline{Y}_k^o\right)^2\right)\right\}. \tag{I.31}$$

The statistics (I.27) and (I.30) are asymptotically the same.

For n large, the approximations (I.20) and (I.21) hold also for the statistics (I.27) and (I.30). Aside that, for normally distributed random variables Y_i and for large values x, James et al. (1987) suggested to use the approximation

$$P\left(\max_{\lfloor\beta n\rfloor\le k\le\lfloor(1-\beta)n\rfloor}\left\{\left|\frac{1}{s_k}\sqrt{\frac{n}{k(n-k)}}\sum_{i=1}^{k}\left(Y_i - \overline{Y}_n\right)\right|\right\} > x\right) \approx$$

$$\sqrt{\frac{2n}{\pi}}\int_{\frac{x}{\sqrt{n}}}^{1}\left(1-y^2\right)^{\frac{n-4}{2}}dy + \frac{x\sqrt{2}}{\sqrt{\pi}}\left(1-\frac{x^2}{n}\right)^{\frac{n-4}{2}}\int_{x\sqrt{\frac{\beta}{(n-x^2)(1-\beta)}}}^{x\sqrt{\frac{1-\beta}{(n-x^2)\beta}}}\frac{1}{y}\nu\left(y+\frac{x^2}{(n-x^2)y}\right)dy.$$

Tables 12. – 15. contain critical values obtained by simulations for statistics (I.27) and (I.28) in the case that σ^2 was not known and was estimated using (I.29).

n	(I.27)	(I.28)		
		$\beta = 0.01$	$\beta = 0.05$	$\beta = 0.1$
50	2.857	2.856	2.775	2.683
100	2.891	2.864	2.778	2.694
200	2.934	2.893	2.801	2.715
300	2.961	2.914	2.805	2.718
500	2.993	2.931	2.820	2.728

Table 12. Simulated 10 % critical values for statistics (I.27) and (I.28).

n	(I.27)	(I.28)		
		$\beta = 0.01$	$\beta = 0.05$	$\beta = 0.1$
50	3.157	3.157	3.079	2.992
100	3.164	3.139	3.061	2.984
200	3.196	3.159	3.071	2.992
300	3.213	3.172	3.076	2.994
500	3.241	3.189	3.088	3.003

Table 13. Simulated 5 % critical values for statistics (I.27) and (I.28).

n	(I.27)	(I.28)		
		$\beta = 0.01$	$\beta = 0.05$	$\beta = 0.1$
50	3.421	3.421	3.359	3.279
100	3.402	3.383	3.311	3.248
200	3.428	3.391	3.314	3.245
300	3.452	3.409	3.324	3.246
500	3.462	3.420	3.324	3.251

Table 14. Simulated 2.5 % critical values for statistics (I.27) and (I.28).

n	(I.27)	(I.28)		
		$\beta = 0.01$	$\beta = 0.05$	$\beta = 0.1$
50	3.747	3.747	3.695	3.625
100	3.696	3.678	3.615	3.558
200	3.719	3.690	3.611	3.558
300	3.737	3.700	3.628	3.555
500	3.735	3.700	3.617	3.547

Table 15. Simulated 1% critical values for statistics (I.27) and (I.28).

3.2. Change in variance

We suppose that the observations $Y_1, \ldots, Y_n$ are independent normally distributed with a known mean μ and unknown variances. Supposing that the mean remains the same, the problem of the detection of a change in variance can be formulated as the following testing problem, i.e. we test the null hypothesis H against the alternative A:

$$\begin{aligned} &H : Y_1 \ldots, Y_n \sim N(\mu, \sigma^2) \\ &A : \exists \mathsf{m} \in \{1, \ldots, n-1\} \text{ such that} \\ &\quad Y_1, \ldots, Y_{\mathsf{m}} \quad \sim N(\mu, \sigma_1^2), \\ &\quad Y_{\mathsf{m}+1}, \ldots, Y_n \sim N(\mu, \sigma_2^2), \end{aligned} \tag{I.32}$$

where $\sigma_1^2 \neq \sigma_2^2$.

It is evident that the variables $\{(Y_i - \mu)^2, i = 1, \ldots, n\}$ follow a gamma distribution with the shape parameter $\alpha = 1/2$ and the scale parameter $\beta = 2\sigma^2$. It means that our problem leads to a special case of the detection of change in the scale parameter of random variables distributed according to the gamma law with a known and constant shape parameter.

The maximum-type test statistics obtained by the maximum likelihood approach have the form

$$\max_{1 \le k \le n-1} \{ | Z_k | \} \quad \text{and} \quad \max_{\lfloor \beta n \rfloor \le k \le \lfloor (1-\beta) n \rfloor} \{ | Z_k | \}, \tag{I.33}$$

where

$$\begin{aligned} Z_k^2 &= n \log \left(\frac{1}{n} \sum_{i=1}^{n} (Y_i - \mu)^2 \right) - k \log \left(\frac{1}{k} \sum_{i=1}^{k} (Y_i - \mu)^2 \right) - \\ &- (n-k) \log \left(\frac{1}{n-k} \sum_{i=k+1}^{n} (Y_i - \mu)^2 \right) = \\ &= k \log \frac{k}{n} \frac{\sum_{i=1}^{n} (Y_i - \mu)^2}{\sum_{i=1}^{k} (Y_i - \mu)^2} + (n-k) \log \frac{n-k}{n} \frac{\sum_{i=1}^{n} (Y_i - \mu)^2}{\sum_{i=k+1}^{n} (Y_i - \mu)^2}. \end{aligned} \tag{I.34}$$

For smaller n, the approximate critical values may be obtained by the Bonferroni inequality realizing that

$$P\left(Z_k^2 < t\right) = P\left(a_k(t) < \frac{\sum_{i=1}^k \left(Y_i - \mu\right)^2}{\sum_{i=1}^n \left(Y_i - \mu\right)^2} < b_k(t)\right),$$

where $a_k(t)$ and $b_k(t)$ are the solutions of the equation

$$-k \log\left(x\frac{n}{k}\right) - (n-k)\log\left((1-x)\frac{n}{n-k}\right) = t.$$

Clearly, the random variable $\sum_{i=1}^k \left(Y_i - \mu\right)^2 \Big/ \sum_{i=1}^n \left(Y_i - \mu\right)^2$ is distributed according to the beta distribution with parameters $k/2$ and $(n-k)/2$. This fact may be used if the approximate critical values are calculated by the Bonferroni inequality, for details see Worsley (1986).

For n large, the limit behavior of the studied probabilities is the same as before, i.e.

$$P\left(\max_{1\le k\le n-1} \{|\, Z_k \,|\} > \frac{x + b_n}{a_n}\right) \approx 1 - \exp\{-2e^{-x}\}, \quad x \in \mathcal{R}^1, \qquad \text{(I.35)}$$

and

$$P\left(\max_{\lfloor \beta n\rfloor \le k \le \lfloor (1-\beta)n\rfloor} \{|\, \dot{Z}_k \,|\} > x\right) \approx (1 - \Phi(x)) + 2x\phi(x)\log\frac{1-\beta}{\beta}. \qquad \text{(I.36)}$$

Notice that (I.35) was derived in Gombay and Horváth (1990).

Denote for a while $S_\mu^2 = n^{-1}\sum_{j=1}^n \left(Y_j - \mu\right)^2$. Then the sum-type test statistic has the form

$$\sum_{k=1}^n \frac{1}{n}\left(\frac{1}{\sqrt{n}}\sum_{i=1}^k \frac{\left(Y_i - \mu\right)^2 - S_\mu^2}{\sqrt{2}S_\mu^2}\right)^2$$

and for large n the critical values may be computed from the approximation

$$P\left(\sum_{k=1}^n \frac{1}{n}\left(\frac{1}{\sqrt{n}}\sum_{i=1}^k \frac{\left(Y_i - \mu\right)^2 - S_\mu^2}{\sqrt{2}S_\mu^2}\right)^2 > x\right) \approx P\left(\int_0^1 B^2(t)\,dt > x\right)$$

$$= P\left(\sum_{j=1}^\infty \frac{1}{j^2\pi^2}\,\chi_j^2(1) > x\right) \qquad \text{(I.37)}$$

For derivation of (I.37) see MacNeill (1974); see also (I.25) and Csörgő and Horváth (1997).

Let us remark that in case the mean μ is unknown, it can be replaced by its estimator $\overline{Y}_n$ and the approximations (I.35) – (I.37) are still valid.

	$\alpha = 0.10$		$\alpha = 0.05$		$\alpha = 0.01$	
n	Simul.	Bonf.	Simul.	Bonf.	Simul.	Bonf.
20	2.679	2.907	2.953	3.129	3.474	3.596
50	2.816	3.153	3.077	3.357	3.605	3.791
100	2.898	3.332	3.152	3.524	3.657	3.938

Table 16. Critical values for over-all (non-trimmed) maximum-type statistic (I.33) calculated using simulations and Bonferroni inequality.

3.3. Change in mean and/or variance

Sometimes it can happen that the change may occur either in one parameter or in both (simultaneously) . Then we test the null hypothesis H against the alternative A:

$$\begin{aligned} H &: Y_1 \ldots, Y_n \sim N(\mu, \sigma^2) \\ A &: \exists \mathrm{m} \in \{2, \ldots, n-2\} \text{ such that} \\ &\quad Y_1, \ldots, Y_{\mathrm{m}} \quad \sim N(\mu_1, \sigma_1^2), \\ &\quad Y_{\mathrm{m}+1}, \ldots, Y_n \sim N(\mu_2, \sigma_2^2), \end{aligned} \tag{I.38}$$

where $(\mu_1, \sigma_1^2) \neq (\mu_2, \sigma_2^2)$. The test statistic derived by likelihood ratio approach has the form

$$\max_{1 \le k \le n-1} \left\{ | \widetilde{Z}_k | \right\} \quad \text{and} \quad \max_{\lfloor \beta n \rfloor \le k \le \lfloor (1-\beta) n \rfloor} \left\{ | \widetilde{Z}_k | \right\}, \tag{I.39}$$

where

$$\begin{aligned} \widetilde{Z}_k^2 = n \log \left(\frac{1}{n} \sum_{i=1}^{n} (Y_i - \overline{Y}_n)^2 \right) - k \log \left(\frac{1}{k} \sum_{i=1}^{k} (Y_i - \overline{Y}_k)^2 \right) \\ - (n-k) \log \left(\frac{1}{n-k} \sum_{i=k+1}^{n} (Y_i - \overline{Y}_k^{\circ})^2 \right). \end{aligned} \tag{I.40}$$

For n large, the approximate critical values may be computed from the following approximations, i.e.,

$$P \left(\max_{1 \le k \le n-1} \left\{ | \widetilde{Z}_k | \right\} > \frac{x + b_{n,2}}{a_n} \right) \approx 1 - \exp\left\{ -2e^{-x} \right\}, \quad x \in \mathcal{R}^1, \tag{I.41}$$

where

$$a_n = \sqrt{2 \log \log n} \quad \text{and} \quad b_{n,2} = 2 \log \log n + \log \log \log n,$$

and

$$P \left(\max_{\lfloor \beta n \rfloor \le k \le \lfloor (1-\beta) n \rfloor} \left\{ | \widetilde{Z}_k | \right\} > x \right) \approx e^{-x^2/2} + e^{-x^2/2} x^2 \log \frac{1-\beta}{\beta}. \tag{I.42}$$

The formulas (I.41) and (I.42) follow from the approximation of the sequence of $\{\widetilde{Z}_k^2\}$ by the process $\{(B_1^2(t) + B_2^2(t))/(t(1-t))\}$, see Part III Sections 11 and 12. The approximation (I.41) is due to Horváth (1993), for (I.42) see Albin (1990). Unfortunately, we do not know how to derive the sum-type test statistic in this case.

4. Detection of changes in a location model (general type of errors)

This section deals with detection of changes in a location model when the distribution of errors is not necessarily normal. Three basic situations, i.e. change in location, change in scale and change in location and/or scale are considered. We concentrate on the maximum-type test statistics.

4.1. Changes in location

We consider the same problem as in 3.1., however we will not assume that the error terms have a normal distribution. Of course, the situation when e_i are normally distributed represents a special case.

More precisely, we are interested in testing the null hypothesis H against the alternative A:

$$\begin{aligned}
&H : Y_i = \mu + e_i, && i = 1, \ldots, n, && \text{(I.43)}\\
&A : \exists \mathsf{m} \in \{1, \ldots, n-1\} \quad \text{such that} && && \\
&\qquad Y_i = \mu + e_i, && i = 1, \ldots, \mathsf{m}, && \\
&\qquad Y_i = \mu + \delta + e_i, && i = \mathsf{m}+1, \ldots, n, \ \ \delta \neq 0, &&
\end{aligned}$$

where μ, $\delta \neq 0$ and m are parameters, $e_1, \ldots, e_n$ are iid random variables with distribution function F, zero mean, nonzero variance σ^2 and with $\mathsf{E}\,|e_i|^{2+\Delta} < \infty$ for some $\Delta > 0$.

If the distribution of the errors is known up to some parameters and only one parameter is subject to a change, we can apply the likelihood ratio method (see Section 2.1. for the description of the general principle), which means that we can derive $\max_{1 \le k < n} \Lambda_k$ in such a case. Under quite mild assumptions on the distribution of e_i one can use the same approximations to the distribution of the corresponding test statistic, see Csörgő and Horváth (1997) for details. Of course, the approximation is usually poorer than in the case of a normal distribution in Section 3.1..

If our knowledge about the distribution of the error terms e_i is poor or their distribution is too complicated, we can still use the test statistics described in Section 3 (see (I.17) – (I.19)) with the same approximation to the distribution of the test statistics. However, the quality of these approximations is (much) worse than in the case of a normal distribution.

Nevertheless, it can be shown that, under quite mild assumptions, these tests are consistent, i.e. they are sensitive to the considered alternatives. We are not going to prove it, but we demonstrate this on the expectations and variances of the partial sums

$$S_j = \sum_{i=1}^{j} \left(Y_i - \overline{Y}_n\right), \quad j = 1, \ldots, n, \tag{I.44}$$

under both the null hypothesis H and the alternative A. (Notice that most of the tests considered in Section 3 were based on S_j).

$$\begin{aligned} \mathsf{E}_H\, S_j &= 0, && 1 \le j \le n, \\ \mathsf{E}_A\, S_j &= \begin{cases} -j\delta\,\dfrac{n-\mathsf{m}}{n}, & \\ -\mathsf{m}\delta\,\dfrac{n-j}{n}, & \end{cases} && \begin{aligned} 1 \le j &\le \mathsf{m}, \\ \mathsf{m} < j &\le n, \end{aligned} \\ var_H\, S_j = var_A\, S_j &= \sigma^2\,\frac{j(n-j)}{n}, && 1 \le j \le n. \end{aligned}$$

It is seen that the expectation under the null hypothesis H is zero while the expectation under A is nonzero with the extreme reached for $j = \mathsf{m}$, i.e.

$$\max_{1 \le j < n} \left|\mathsf{E}_A S_j\right| = \left|\mathsf{E}_A S_\mathsf{m}\right| = |\delta|\,\frac{\mathsf{m}(n-\mathsf{m})}{n}.$$

We should also remark that by the central limit theorem S_k has approximately (for k large) normal distribution $N(E\,S_k,\, var\,S_k)$ both under H and A.

To test H against A, other procedures were also developed. Similarly, as in the two-sample problem, there are various robust and nonparametric test procedures. Let us mention M-tests, rank-based R-tests, Kolmogorov-Smirnov type tests among others.

Noticing that the two-sample problem coincides with our problem if k is known, we can develop M- and R- test procedures, as well as Kolmogorov-Smirnov type tests for our problem along the lines of the two-sample tests.

4.1.1. M-test procedures

We consider again the testing problem (I.43) assuming that e_i have a common distribution function F that is symmetric around zero, i.e. generally there is no need to have zero mean and a finite variance.

We remind the definition of M-estimators with the score function ψ, that is usually assumed to be monotone and skew symmetric, i.e. $\psi(x) = -\psi(x)\ \forall x \in \mathcal{R}^1$. The M-estimator $\widehat{\mu}_{n,M}(\psi)$ of μ is defined as any solution of the equation

$$\sum_{i=1}^{n} \psi\left(Y_i - t\right) = 0.$$

The basic theory of M-procedures was developed by Huber, for details see his book Huber (1981), and further studied in a number of papers and books.

The test procedures for the change point problem are based on the partial sums

$$S_{k,M}(\psi) = \sum_{i=1}^{k} \psi\Big(Y_i - \widehat{\mu}_{n,M}(\psi)\Big), \quad k = 1, \ldots, n-1. \tag{I.45}$$

The role of the scale plays $\sigma(\psi)$ defined by

$$\sigma^2(\psi) = \int_{-\infty}^{\infty} \psi^2(x)\, dF(x),$$

which can be estimated for example by the estimator

$$\widehat{\sigma}^2_{n,M}(\psi) = \tag{I.46}$$

$$= \min_{2 \le k \le n-2} \left\{ \frac{1}{n} \left(\sum_{i=1}^{k} \psi^2\Big(Y_i - \widehat{\mu}_{k,M}(\psi)\Big) + \sum_{i=k+1}^{n} \psi^2\Big(Y_i - \widehat{\mu}^o_{k,M}(\psi)\Big) \right) \right\},$$

where $\widehat{\mu}_{k,M}(\psi)$ and $\widehat{\mu}^o_{k,M}(\psi)$ are the M-estimators calculated from $Y_1, \ldots, Y_k$ and $Y_{k+1}, \ldots, Y_n$, respectively.

Then, the maximum-type M-test statistic is defined as

$$\max_{1 \le k < n} \left\{ \frac{1}{\widehat{\sigma}_{n,M}(\psi)} \left| \sqrt{\frac{n}{k(n-k)}} \sum_{i=1}^{k} \psi\Big(Y_i - \widehat{\mu}_{n,M}(\psi)\Big) \right| \right\}. \tag{I.47}$$

The quantities $\psi(Y_i - \widehat{\mu}_{n,M}(\psi))$, $i = 1, \ldots, n$, play the role of robust residuals. It can be easily checked that for $\psi(x) = x$, $x \in \mathcal{R}^1$, statistic (I.47) reduces to (I.27). Similarly, as in Section 3 we can approximate under certain assumptions the distribution of (I.47) under the null hypothesis using

$$P\left(\max_{1 \le k < n} \left\{ \frac{1}{\widehat{\sigma}_{n,M}(\psi)} \left| \sqrt{\frac{n}{k(n-k)}} \sum_{i=1}^{k} \psi\Big(Y_i - \widehat{\mu}_{n,M}(\psi)\Big) \right| \right\} > \frac{x + b_n}{a_n} \right) \approx$$
$$\approx 1 - \exp\{-2e^{-x}\}, \quad x \in \mathcal{R}^1, \tag{I.48}$$

where

$$a_n = \sqrt{2 \log \log n} \quad \text{and} \quad b_n = 2 \log \log n + \frac{1}{2} \log \log \log n - \frac{1}{2} \log \pi.$$

Typical ψ-functions are summarized in Table 17. For detailed discussion concerning the choice of parameters A, B, C and D see, e.g., Huber (1981), Hampel et al. (1986) or Antoch and Víšek (1992). Notice that, letting A, B, C and $D \to \infty$ gives in all four cases $\psi(x) = x$, i.e. the classical least squares estimator.

If we choose $\psi(x) = -f'(x)/f(x), x \in \mathcal{R}^1$, where f and f' are, respectively, density and its derivative, then under some assumptions the resulting maximum-type test statistic is asymptotically equivalent to that obtained by the likelihood ratio method.

	$\varrho(x)$	$\psi(x)$	
Fair	$A^2\left(\frac{\lvert x\rvert}{A} - \ln\left(1 + \frac{\lvert x\rvert}{A}\right)\right)$	$\frac{x}{1+\lvert x\rvert/A}$	$x \in \mathcal{R}_1$
Huber	$\frac{x^2}{2}$ $B\mid x\mid - \frac{B^2}{2}$	x $B\,\mathrm{sign}\,(x)$	$\mid x\mid \le B$ $\mid x\mid > B$
Tukey	$\frac{C^2}{6}\left(1-\left(1-(\frac{x}{C})^2\right)^3\right)$ $\frac{C^2}{6}$	$x\left(1-(\frac{x}{C})^2\right)^2$ 0	$\mid x\mid \le C$ $\mid x\mid > C$
Welsh	$\frac{D^2}{2}\left(1 - \exp\left\{-(\frac{x}{D})^2\right\}\right)$	$x\exp\{-(\frac{x}{D})^2\}$	$x \in \mathcal{R}_1$

Table 17. Typical ψ-functions generating M-estimators.

Concerning the choice of the score function ψ, statisticians often use

$$\psi(x) = \begin{cases} x, & |x| \le K, \\ K\,\mathrm{sign}(x), & |x| \ge K, \end{cases}$$

the so-called *Huber function*, where K is a suitably chosen constant. Letting $K \to \infty$ one has the classical least squares estimator while, for $K \to 0$ the so-called L_1-norm estimator is attained. Huber's ψ-function was proposed as a function leading to the estimators not influenced by outliers as, e.g., least squares estimator is. The use of Huber's ψ-function is wise because the procedures considered in Section 4.1. are sensitive with respect to outliers which could be wrongly detected as change points.

We mention here the important case with $\psi(x) = \mathrm{sign}(x)$, $x \in \mathcal{R}^1$, the so called L_1 test procedure. The test statistic is very easy to calculate in this case. We assume that the distribution of the error terms e_i is symmetric around zero and has the density positive in the neighborhood of zero.

Then the test statistic has the form

$$\max_{1\le k<n} \sqrt{\frac{n}{k(n-k)}}\left|\sum_{i=1}^{k} \mathrm{sign}\,(Y_i - \widetilde{\mu}_n)\right|,$$

where $\widetilde{\mu}_n$ is the sample median based on all observations. This means that the partial sums are just the differences of the number of the observations Y_i above and below the sample median $\widetilde{\mu}_n$.

4.1.2. R-test procedures

Now we turn to the rank-based test procedures. The test statistics are based on the simple linear rank statistics

$$S_{k,R} = \sum_{i=1}^{k} \left(a_n(R_i) - \overline{a}_n\right), \quad k = 1, \ldots, n, \tag{I.49}$$

where $R_1, \ldots, R_n$ is the vector of ranks corresponding to the observations $Y_1, \ldots, Y_n$, $a_n(1), \ldots, a_n(n)$ are scores the and $\overline{a}_n = \frac{1}{n}\sum_{i=1}^{n} a_n(i)$.

Here we need the continuity of the distribution function F of the observations, no other assumption on F is needed. Typical choices of the scores are, e.g.,

(1) Wilcoxon scores $\quad a_n(i) = i/(n+1),\ i = 1, \ldots, n;$

(2) van der Waerden scores $\quad a_n(i) = \Phi^{-1}(i/(n+1)),\ i = 1, \ldots, n$, etc.

The role of the scale is played by

$$\sigma_{n,R}^2 = \frac{1}{n-1}\sum_{i=1}^{n}\Big(a_n(i) - \overline{a}_n\Big)^2, \tag{I.50}$$

which needs not be estimated, while the role of residuals by $\{a_n(R_i) - \overline{a}_n,\ i = 1, \ldots, n\}$.

The maximum-type R-test statistic is defined as

$$\max_{1\le k<n}\left\{\frac{1}{\sigma_{n,R}}\left|\sqrt{\frac{n}{k(n-k)}}\sum_{i=1}^{k}\Big(a_n(R_i) - \overline{a}_n\Big)\right|\right\}. \tag{I.51}$$

Similarly, as in the previous section, we can approximate under certain assumptions the distribution of (I.51) under the null hypothesis using

$$P\left(\max_{1\le k<n}\left\{\frac{1}{\sigma_{n,R}(\psi)}\left|\sqrt{\frac{n}{k(n-k)}}\sum_{i=1}^{k}\Big(a_n(R_i) - \overline{a}_n\Big)\right|\right\} > \frac{x+b_n}{a_n}\right) \approx$$
$$\approx 1 - \exp\{-2e^{-x}\}, \quad x \in \mathcal{R}^1, \tag{I.52}$$

where

$$a_n = \sqrt{2\log\log n} \quad \text{and} \quad b_n = 2\log\log n + \frac{1}{2}\log\log\log n - \frac{1}{2}\log\pi.$$

However, this approximation is reasonable only when n is large. Nevertheless, under the null hypothesis H, the distribution of the R-test statistics does not depend on F. Therefore, we can get the approximation to the distribution of the R-test statistics via simulations, i.e., we can simulate the samples for example from the uniform distribution $R(0,1)$. The basic advantage is that these simulated critical values give good approximations even for n small. One can also use various modifications developed along the lines of Section 3.

EXAMPLE: One of the most often used rank-based tests is the test with the Wilcoxon scores. Corresponding (change point) R-test, which is easy to calculate, behaves reasonably well for a broad spectrum of the distributions F. Here

$$\sigma_{n,R}^2 = \frac{n}{12(n+1)} \approx \frac{1}{12},$$

hence the test statistic (I.51) becomes

$$\max_{1\le k<n}\left\{\sqrt{\frac{n}{k(n-k)}}\,\frac{\sqrt{12}}{n}\left|\sum_{i=1}^{k}\Big(R_i - \frac{n+1}{2}\Big)\right|\right\}.$$

REMARK: There were developed also U-test statistics and Kolmogorov-Smirnov type test statistics for our problem. For more information we refer to the book Csörgő and Horváth (1997).

4.1.3. MOSUM type test statistics

Now, we introduce two different classes of test statistics for our problem. They are based on the moving sums (MOSUM) of statistics S_k introduced in (I.44) and defined as

$$\max_{G<k\le n} \frac{1}{\sqrt{G}} \frac{1}{\widehat{\sigma}_n} \left|S_k - S_{k-G}\right| = \max_{G<k\le n} \frac{1}{\sqrt{G}} \frac{1}{\widehat{\sigma}_n} \left| \sum_{i=k-G+1}^{k} (Y_i - \overline{Y}_n)\right| \tag{I.53}$$

and

$$\max_{G<k\le n-G} \frac{1}{\sqrt{2G}} \frac{1}{\widehat{\sigma}_n} \left|S_{k+G} - 2S_k + S_{k-G}\right| = \tag{I.54}$$

$$\max_{G<k\le n-G} \frac{1}{\sqrt{2G}} \frac{1}{\widehat{\sigma}_n} \left| \sum_{i=k+1}^{k+G} Y_i - \sum_{i=k-G+1}^{k} Y_i \right|,$$

where $\widehat{\sigma}_n$ is any consistent estimator of σ, e.g. that defined by (I.31) and G/n small. Typically we choose $G/n \sim 0.05$ or 0.10.

Notice that (I.53) is the first order difference (the first order derivative) of S_k's while (I.54) corresponds to the second order difference (the second order derivative) of S_k's.

If n is large and G/n is small (e.g. $G/n \sim 0.1$ or 0.05) we can use the following approximation to the distributions of (I.53) and (I.54) under the null hypothesis H, i.e.

$$P\left(\max_{G<k\le n} \left\{ \frac{1}{\sqrt{G}} \frac{1}{\widehat{\sigma}_n} \left|S_k - S_{k-G}\right| \right\} > \frac{x + b_{n,G}}{a_{n,G}} \right) \approx 1 - \exp\left\{-2e^{-x}\right\}, \; x \in \mathcal{R}^1,$$

and

$$P\left(\max_{G<k\le n-G} \left\{ \frac{1}{\sqrt{2G}} \frac{1}{\widehat{\sigma}_n} \left|S_{k+G} - 2S_k + S_{k-G}\right| \right\} > \frac{x + b_{n,G} - \log\frac{2}{3}}{a_{n,G}} \right) \approx$$

$$\approx 1 - \exp\left\{-2e^{-x}\right\}, \quad x \in \mathcal{R}^1,$$

where

$$a_{n,G} = \sqrt{2\log\frac{n}{G}} \quad \text{and} \quad b_{n,G} = 2\log\frac{n}{G} + \frac{1}{2}\log\log\frac{n}{G} - \frac{1}{2}\log\pi.$$

Large values of these test statistics indicate that at least one change has occurred. Using the above approximations we get the approximation to the critical value with level α. Namely, we obtain the critical regions with approximate level α

$$\max_{G<k\le n}\left\{\frac{1}{\sqrt{G}}\frac{1}{\widehat{\sigma}_n}\left|S_k-S_{k-G}\right|\right\}>$$
$$>\frac{-\log\log\frac{1}{\sqrt{1-\alpha}}+2\log\frac{n}{G}+\frac{1}{2}\log\log\frac{n}{G}-\frac{1}{2}\log\pi}{\sqrt{2\log\frac{n}{G}}}$$

and

$$\max_{G<k\le n-G}\left\{\frac{1}{\sqrt{2G}}\frac{1}{\widehat{\sigma}_n}\left|S_{k+G}-2S_k+S_{k-G}\right|\right\}>$$
$$>\frac{-\log\log\frac{1}{\sqrt{1-\alpha}}+2\log\frac{n}{G}+\frac{1}{2}\log\log\frac{n}{G}-\frac{1}{2}\log\pi-\log\frac{2}{3}}{\sqrt{2\log\frac{n}{G}}}.$$

To get a simple picture on the sensitivity of the test statistics (I.53) and (I.54) with respect to alternatives, we present the expectations and variances of the moving sums both under the null hypothesis H and alternative A:

$$\mathsf{E}_H(S_j-S_{j-G})=0,\qquad G<j\le n,$$
$$\mathsf{E}_A(S_j-S_{j-G})=\begin{cases}-\delta G\dfrac{n-\mathsf{m}}{n}, & G<j\le \mathsf{m},\\ \delta\left(j-\mathsf{m}-G\dfrac{n-\mathsf{m}}{n}\right), & \mathsf{m}<j\le\mathsf{m}+G,\\ \delta G\dfrac{\mathsf{m}}{n}, & \mathsf{m}+G<j\le n.\end{cases}$$
$$\mathsf{E}_H\left(S_{j+G}-2S_j+S_{j-G}\right)=0,\qquad G<j\le n-G,$$
$$\mathsf{E}_A(S_{j+G}-2S_j+S_{j-G})=\begin{cases}0, & G<j\le\mathsf{m}-G,\\ \delta(j+G-\mathsf{m}), & \mathsf{m}-G<j\le\mathsf{m},\\ \delta(\mathsf{m}+G-j), & \mathsf{m}<j\le\mathsf{m}+G,\\ 0, & \mathsf{m}+G<j\le n-G,\end{cases}$$
$$var_H(S_j-S_{j-G})=var_A(S_j-S_{j-G})=\sigma^2\frac{G(n-G)}{n},\quad G<j\le n,$$
$$var_H(S_{j+G}-2S_j+S_{j-G})=var_A(S_{j+G}-2S_j+S_{j-G})=2\sigma^2G,\ G<j\le n-G.$$

We see that the expectations under H are zero, while under alternatives the expectations of S_j-S_{j-G} are nonzero. Concerning $S_{j+G}-2S_j+S_{j-G}$, their expectations (under A) are nonzero only for j close to the change point m ($|j-\mathsf{m}|\le G$). Hence, both test statistics are sensitive with respect to the considered alternatives. For this particular reason the test statistic (I.54) is suitable if we expect more changes and it is useful as a diagnostic tool.

♣ Similarly, the MOSUM type M-test statistics can be proposed, just replacing in (I.53) and (I.54) S_k and $\widehat{\sigma}_n$ by their M-type counterparts $S_{k,M}(\psi)$ and $\widehat{\sigma}_{n,M}(\psi)$ defined by (I.45) and (I.46), i.e., one can use

$$\max_{G<k\le n}\left\{\frac{1}{\sqrt{G}}\frac{1}{\widehat{\sigma}_{n,M}(\psi)}\left|S_{k,M}(\psi)-S_{k-G,M}(\psi)\right|\right\}\tag{I.55}$$

and

$$\max_{G<k\leq n-G}\left\{\frac{1}{\sqrt{2G}}\frac{1}{\widehat{\sigma}_{n,M}(\psi)}\left|S_{k+G,M}(\psi)-2S_{k,M}(\psi)+S_{k-G,M}(\psi)\right|\right\}. \tag{I.56}$$

♣ The MOSUM R-type statistics can be derived quite analogously. Namely we use, instead of (I.53) and (I.54), the test statistics

$$\max_{G<k\leq n}\left\{\frac{1}{\sqrt{G}}\frac{1}{\sigma_{n,R}}\left|S_{k,R}-S_{k-G,R}\right|\right\} \tag{I.57}$$

and

$$\max_{G<k\leq n-G}\left\{\frac{1}{\sqrt{2G}}\frac{1}{\sigma_{n,R}}\left|S_{k+G,R}-2S_{k,R}+S_{k-G,R}\right|\right\}, \tag{I.58}$$

where $S_{k,R}$ and $\sigma_{n,R}^2$ were introduced in (I.49) and (I.50).

4.2. Change in scale

We consider the same problem as in Section 3.2., however, we will not assume that the error terms have a normal distribution. Supposing that the mean remains the same, the problem of the detection of a change in variance can be formulated as the following testing problem, i.e. we test the null hypothesis H against the alternative A:

$$\begin{aligned} &H: Y_i=\mu+e_i, && i=1,\ldots,n, \\ &A: \exists\, \mathsf{m}\in\{1,\ldots,n-1\} \quad \text{such that} && \\ &\qquad Y_i=\mu+e_i, && i=1,\ldots,\mathsf{m}, \\ &\qquad Y_i=\mu+(1+h)e_i, && i=\mathsf{m}+1,\ldots,n, \end{aligned} \tag{I.59}$$

where $\mu\neq 0, h\neq 0,-1$ and m are parameters (realize that only the scale can change, not the mean), $e_1,\ldots,e_n$ are iid random variables with distribution function F, zero mean, nonzero variance σ^2 and $\mathsf{E}\,e_i^{4+\Delta}<\infty$ for some $\Delta>0$. Of course, the situation with e_i normally distributed as $N(0,\sigma^2)$ is a special case.

If the distribution of the error terms is known, we can apply the maximum likelihood principle in order to get a test procedure. Nevertheless, here we construct another test procedure. Testing H against A means that we are looking for a change in the expected values of $\{(Y_i-\mu)^2, i=1,\ldots,n\}$, and, since μ is assumed unknown, we replace it by its estimator $\overline{Y}_n$. In this light we propose to use tests for H against A based on the partial sums

$$S_{k,sc}=\sum_{i=1}^{k}\left(\left(Y_i-\overline{Y}_n\right)^2-\frac{1}{n}\sum\nolimits_{j=1}^{n}\left(Y_j-\overline{Y}_n\right)^2\right), \quad k=1,\ldots,n. \tag{I.60}$$

The differences

$$\left(Y_i-\overline{Y}_n\right)^2-\frac{1}{n}\sum\nolimits_{j=1}^{n}\left(Y_j-\overline{Y}_n\right)^2, \quad i=1,\ldots,n,$$

play the role of the residuals. The maximum-type test statistics are constructed in the same way as in the previous section, i.e. we use the test statistics

$$\max_{1\le k<n}\left\{\left|\sqrt{\frac{n}{k(n-k)}}\frac{1}{\widehat{\sigma}_{n,sc}}\sum_{i=1}^{k}\left((Y_i-\overline{Y}_n)^2-\frac{1}{n}\sum_{j=1}^{n}(Y_j-\overline{Y}_n)^2\right)\right|\right\}, \qquad \text{(I.61)}$$

where $\widehat{\sigma}^2_{n,sc}$ is an estimator of $var\,(Y_1-\mu)^2$, for example

$$\begin{aligned}\widehat{\sigma}^2_{n,sc} = \min_{1\le k<n}\Bigg\{\frac{1}{n}\Bigg(&\sum_{i=1}^{k}\Big[(Y_i-\overline{Y}_k)^2-\frac{1}{k}\sum\nolimits_{j=1}^{k}(Y_j-\overline{Y}_k)^2\Big]^2+\\ &+\sum_{i=k+1}^{n}\Big[(Y_i-\overline{Y}^{o}_k)^2-\frac{1}{n-k}\sum\nolimits_{j=k+1}^{n}(Y_j-\overline{Y}^{o}_k)^2\Big]^2\Bigg)\Bigg\}.\end{aligned} \qquad \text{(I.62)}$$

Under H the distribution of the test statistics (I.61) can be approximated similarly as in the analogous maximum-type test statistics, i.e. for large n the critical values may be computed from the approximation

$$\begin{aligned}P\Bigg(\max_{1\le k<n}&\left\{\left|\sqrt{\frac{n}{k(n-k)}}\frac{1}{\widehat{\sigma}_{n,sc}}\sum_{i=1}^{k}\Big[(Y_i-\overline{Y}_n)^2-\frac{1}{n}\sum_{j=1}^{n}(Y_j-\overline{Y}_n)^2\Big]\right|\right\}>\\ &>\frac{x+b_n}{a_n}\Bigg)\approx 1-\exp\left\{-e^{-x}\right\}, \quad x\in\mathcal{R}^1,\end{aligned}$$

where

$$a_n=\sqrt{2\log\log n} \quad\text{and}\quad b_n=2\log\log n+\frac{1}{2}\log\log\log n-\frac{1}{2}\log\pi.$$

♣ Concerning robust and nonparametric test procedures, they can be developed again along the lines in the previous section. We present here only the test based on an L_1-procedure that is really useful in practice. Concerning the error terms, we assume that they have symmetric distribution function F (no need of finiteness of any moment). The test statistic is of the form

$$\max_{1\le k<n}\sqrt{\frac{n}{k(n-k)}}\sum_{i=1}^{k}\operatorname{sign}\left(\left|Y_i-\widetilde{\mu}_n\right|-\widetilde{\nu}_n\right),$$

where $\widetilde{\mu}$ and $\widetilde{\nu}_n$ are sample median based on $Y_1,\dots,Y_n$ and the sample median of $\left|Y_i-\widetilde{\mu}_n\right|$, $i=1,\dots,n$. This is a very simple test statistics. Again the same approximation to its distribution under H can be used.

REMARK: We could also develop MOSUM type and Bayesian test statistic here. In fact, it suffices to replace the partial sums of residuals S_k and the scale estimator $\widehat{\sigma}$ by $S_{k,sc}$ and $\widehat{\sigma}_{k,sc}$, respectively. The approximations to the distribution of these new test statistics are the same as above.

4.3. Change in location and/or scale

Now we consider the same situation as in the previous section, however, aside the scale also the mean can change. This can be formulated as the following testing problem, i.e., we test the null hypothesis H against the alternative A:

$$\begin{aligned} H: Y_i &= \mu + e_i, & i &= 1,\dots,n, \\ A: \exists \mathsf{m} &\in \{1,\dots,n-1\} \quad \text{such that} & & \\ Y_i &= \mu + e_i, & i &= 1,\dots,\mathsf{m}, \\ Y_i &= \mu + \delta + (1+h)e_i, & i &= \mathsf{m}+1,\dots,n, \end{aligned} \tag{I.63}$$

where μ, $(\delta, h) \neq (0,0), h \neq -1$ and m are parameters, $e_1,\dots,e_n$ are iid random variables with distribution function F, zero mean, nonzero variance σ^2 and $\mathsf{E}\, e_i^{4+\Delta} < \infty$ for some $\Delta > 0$. Again, the situation with e_i having the normal distribution $N(0,\sigma^2)$ is a special case.

The basic test procedures are based either on (I.40) or on the quadratic form of the partial sums S_k and $S_{k,sc}$ introduced in (I.44) and (I.60). More precisely, the corresponding maximum-type statistic is defined as

$$\max_{1\le k<n} \left\{ \sqrt{\frac{n}{k(n-k)}} \sqrt{\frac{S_k^2}{\widehat{\sigma}_n^2} + \frac{S_{k,sc}^2}{\widehat{\sigma}_{n,sc}^2}} \right\}, \tag{I.64}$$

where $\widehat{\sigma}_n^2$ and $\widehat{\sigma}_{n,sc}^2$ were introduced in (I.31) and (I.62).

Comparing the test statistics with their counterparts for a change in location only and for a change in variance, we see that the test statistic (I.64) is the maximum over m of the roots of quadratic forms of the terms for testing a change in location and a change in scale separately. It can be shown, assuming $\mathsf{E}\, e_i^3 = 0$, that

$$cov\,(S_k, S_{k,sc}) = 0, \quad k = 1,\dots,n.$$

Further, under H and for k large it can be shown that

$$\frac{n}{k(n-k)} \left(\frac{S_k^2}{\widehat{\sigma}_n^2} + \frac{S_{k,sc}^2}{\widehat{\sigma}_{n,sc}^2} \right) \tag{I.65}$$

has approximately a χ^2-distribution with two degrees of freedom.

Moreover, it can be shown that under H

$$\begin{aligned} P\left(\max_{1\le k<n} \left\{ \sqrt{\frac{n}{k(n-k)}} \sqrt{\frac{S_k^2}{\widehat{\sigma}_n^2} + \frac{S_{k,sc}^2}{\widehat{\sigma}_{n,sc}^2}} \right\} > \frac{x + b_{n,2}}{a_n} \right) &\approx \\ \approx 1 - \exp\{-2e^{-x}\}, \quad x \in \mathcal{R}^1, \end{aligned} \tag{I.66}$$

where

$$a_n = \sqrt{2\log\log n} \quad \text{and} \quad b_{n,2} = 2\log\log n + \log\log\log n.$$

This approximation can be again used to get the approximation to the critical values. Concerning test statistic (I.40) under H_0 and under additional assumptions $\mathsf{E}\, e_i = 0, \mathsf{E}\, e_i^3 = 0$ and $0 < \mathsf{E}\, e_i^4 = 3\, var\ e_i < \infty$, the approximation (I.41) is still reasonable. For details see Horváth (1995).

Notice that introduction of the L_1 test procedures is very simple (we have to assume only that the error terms have a symmetric distribution around zero and a positive continuous density in a neighborhood of zero). The test procedure is based on the statistic

$$\max_{1\le k<n} \left\{ \sqrt{\frac{n}{k(n-k)}} \sqrt{\sum_{i=1}^{k} \Big(\operatorname{sign}(Y_i - \widetilde{\mu}_n)\Big)^2 + \sum_{i=1}^{k} \Big(\operatorname{sign}(|Y_i - \widetilde{\mu}_n| - \widetilde{\nu}_n)\Big)^2} \right\},$$

where $\widetilde{\mu}$ and $\widetilde{\nu}_n$ are the sample median based on $Y_1, \dots, Y_n$ and the sample median of $|Y_i - \widetilde{\mu}_n|$, $i = 1, \dots, n$. Under H, its distribution can be approximated in the same way as in (I.66).

However, as soon as we can assume that the error terms have the first four moments (almost) the same as the double exponential distribution, then it is more convenient to use the test statistic

$$\max_{1\le k<n} \left\{ \sqrt{-2\log \frac{\widetilde{\sigma}_k^k (\widetilde{\sigma}_k^o)^{n-k}}{\widetilde{\sigma}_n^n}} \right\},$$

where

$$\widetilde{\sigma}_k = \frac{1}{k}\sum_{i=1}^{k} \left|Y_i - \widetilde{\mu}_k\right| \quad \text{and} \quad \widetilde{\sigma}_k^o = \frac{1}{n-k}\sum_{i=k+1}^{n} \left|Y_i - \widetilde{\mu}_k^o\right|$$

with $\widetilde{\mu}_k$ and $\widetilde{\mu}_k^o$ being medians based on $X_1, \dots, X_k$ and $X_{k+1}, \dots, X_n$, respectively. Under H, the distribution can be approximated again as in (I.66).

4.4. Change in mean with unknown starting value – dependent observations

Consider the case of testing null hypothesis H against the alternative A:

$$\begin{aligned} H &: Y_i = \mu + e_i, && i = 1, \dots, n, \\ A &: \exists \mathsf{m} \in \{1, \dots, n-1\} \text{ such that} && \\ &\quad Y_i = \mu + e_i, && i = 1, \dots, \mathsf{m}, \\ &\quad Y_i = \mu + \delta + e_i, && i = \mathsf{m}+1, \dots, n,\ \delta \neq 0, \end{aligned} \tag{I.67}$$

where μ, $\delta \neq 0$ and m are parameters, and where the variables e_i are not anymore independent, but form a stationary sequence.

Here we can again apply the tests described in Section 3.1. when σ is replaced by a proper estimator or, in other words, a different standardization is needed. We will discuss it further. More precisely, we assume that in (I.67) the error terms e_i form a linear process satisfying

$$e_i = \sum_{t=0}^{\infty} w_t \epsilon_{i-t}, \quad i = 1, 2, \ldots,$$

where ϵ_i are iid random variables, $\mathsf{E}\,\epsilon_i = 0$, $var\,\epsilon_i = \sigma^2 > 0$, $\mathsf{E}\,|\epsilon_i|^{2+\triangle} < \infty$ for some $\triangle > 0$ and the weights $\{w_j\}_{j=0}^{\infty}$ satisfy

$$\sum_{j=0}^{\infty} j\,|w_j| < \infty, \quad \sum_{j=0}^{\infty} w_j \neq 0.$$

Notice that AR and ARMA processes fulfill these assumptions.

In procedures described in Section 3.1. σ^2 has to be replaced by

$$\sigma_0^2 = \sigma^2 \Big(\sum_{j=0}^{\infty} w_j \Big)^2.$$

For example, if the sequence e_n is an AR(1) sequence with the coefficient $\rho \in (-1, 1)$, then σ^2 in (II.9) has to be replaced by $\sigma^2/(1-\rho)$ or by its estimator.

Generally, σ_0^2 can be estimated, e.g., by

$$\widehat{\sigma}_{0,n}^2(L) = \widehat{R}(0) + 2\sum_{k=1}^{L} \left(1 - \frac{k}{L}\right) \widehat{R}(k), \quad L < n,$$

where for $k \geq 0$

$$\widehat{R}(k) = \frac{1}{n} \left\{ \sum_{t=1}^{\widehat{m}-k} (Y_t - \overline{Y}_{\widehat{m}})\,(Y_{t+k} - \overline{Y}_{\widehat{m}}) + \sum_{t=\widehat{m}+1}^{n-k} \left(Y_t - \overline{Y}_{\widehat{m}}^{o}\right) \left(Y_{t+k} - \overline{Y}_{\widehat{m}}^{o}\right) \right\}.$$

If we have more information about the type of dependency, it is advisable to use the estimator specific for the particular model since the above mentioned estimator $\widehat{\sigma}_{0,n}^2(L)$ behaves quite poorly for small and moderate sample sizes. Further discussions the on choice of L together with other theoretical results as well as results from a simulation study can be found in Antoch et al. (1997).

There exists an extensive literature about these procedures, mostly in econometrically oriented papers.

5. Change in simple linear regression

The basic change point detection problem in regression concerns the decision whether a relationship among some variables changed during the observation time. The simplest case that can be treated is that of the simple linear regression when only the relationship between two variables is studied with the explained variable depending linearly on the explanatory one. Even in such a simple setting we can consider many different situations, e.g.:

- either one or both parameters (intercept, slope) can change;
- either the starting parameters before the change point are known or they are unknown;

- either the continuity of the regression function at the change point is assumed or there can be discontinuity etc.

Analogously as above, we can consider for the testing problem both maximum-type statistics as well as sum-type statistics. Nevertheless, we will deal here with the maximum-type statistics only. Provided the number of observations is large, the decision whether to reject the null hypothesis may be based on the asymptotic distribution of the test statistic under H. However, we have to admit that, for many problems that appear in application, the limit behavior of test statistics is not known because the limit behavior depends on the values attained by the explanatory variable. In what follows, we treat several special cases for two basic situations. In the first one we suppose that the design points $\{x_i\}$ are randomly chosen, e.g. they represent a realization of a sequence of independent random variables, or more generally a realization of a stationary ARMA sequence. In the second case the design points are equally spaced, i.e. $x_i = i/n$, $i = 1, \ldots, n$.

5.1. Change in intercept – random design

We consider the problem of testing the null hypothesis H against the alternative A, i.e.

$$\begin{aligned} &H: Y_i = a + bx_i + e_i, && i = 1, \ldots, n, && \text{(I.68)} \\ &A: \exists \mathsf{m} \in \{2, \ldots, n-2\} \quad \text{such that} && && \\ &\qquad Y_i = a + bx_i + e_i, && i = 1, \ldots, \mathsf{m}, && \\ &\qquad Y_i = a^o + bx_i + e_i, && i = \mathsf{m}+1, \ldots, n, && \end{aligned}$$

where $a \neq a^o$. Here, and in the whole section about the changes in linear regression, the random errors e_i are supposed to be iid with $\mathsf{E}\, e_i = 0, \mathsf{E}\, e_i^2 = \sigma^2$ and $\mathsf{E}|e_i|^{2+\Delta} < \infty$ for some $\Delta > 0$ and independent of x_i. Moreover, the variance σ^2 is supposed to be known. If it is unknown, it can be replaced by its usual estimator

$$\widehat{\sigma}^2 = \frac{1}{n-2} \sum_{i=1}^{n} \left(Y_i - \widehat{a} - \widehat{b}x_i\right)^2,$$

where $\widehat{a}$ and $\widehat{b}$ are the least squares estimators of the parameters a and b calculated under H.

The maximum-type test statistics have the form

$$\max_{2 \le k \le n-2} \left\{ |U_k| \right\} \quad \text{and} \quad \max_{\lfloor \beta n \rfloor \le k \le \lfloor (1-\beta) n \rfloor} \left\{ |U_k| \right\}, \qquad \text{(I.69)}$$

where

$$U_k = \frac{1}{\sigma} \sqrt{\frac{k}{1 - k/n}} \frac{\overline{Y}_k - \overline{Y}_n - (\overline{x}_k - \overline{x}_n)\widehat{b}}{\sqrt{1 - k(\overline{x}_k - \overline{x}_n)^2 (1 - k/n) Q_{xx}^{-1}}}$$

Here $Q_{xx}^{-1} = \sum_{i=1}^{n} \left(x_i - \overline{x}_n\right)^2$.

Notice that the statistics $\{U_k\}$ are standardized partial sums of regression residuals, i.e.

$$U_k = \frac{n^{-1/2}\sum_{i=1}^{k}(Y_i - \widehat{Y}_i)}{\sqrt{var\left(n^{-1/2}\sum_{i=1}^{k}(Y_i - \widehat{Y}_i)\right)}},$$

where $\widehat{Y}_i = \widehat{a} + \widehat{b}x_i$. To find critical values of the test statistic (I.69), the Bonferroni inequality may be used if n is moderate whereas, for n large, the asymptotic behavior of the proposed test is applied.

Supposing that $\{x_i\}$ represent a realization of a stationary ARMA sequence, it follows that, as k and $n-k$ tend to ∞,

$$\frac{1}{k}\sum_{i=1}^{k} x_i \xrightarrow{\mathcal{P}} \zeta, \ \frac{1}{n-k}\sum_{i=k+1}^{n} x_i \xrightarrow{\mathcal{P}} \zeta, \ \frac{1}{k}\sum_{i=1}^{k} x_i^2 \xrightarrow{\mathcal{P}} \eta, \frac{1}{n-k}\sum_{i=k+1}^{n} x_i^2 \xrightarrow{\mathcal{P}} \eta \tag{I.70}$$

for some $\zeta \in \mathcal{R}_1$ and $\eta > 0$. Moreover, under additional assumptions about the rate of convergence of the above averages, for more information see Csörgő and Horváth (1997), we have

$$P\left(\max_{2\le k\le n-2}\left\{|U_k|\right\} > \frac{x+b_n}{a_n}\right) \approx 1 - \exp\left\{-2e^{-x}\right\}, \quad x \in \mathcal{R}^1, \tag{I.71}$$

where

$$a_n = \sqrt{2\log\log n} \quad \text{and} \quad b_n = 2\log\log n + \frac{1}{2}\log\log\log n - \frac{1}{2}\log\pi,$$

and

$$P\left(\max_{\lfloor\beta n\rfloor\le k\le\lfloor(1-\beta)n\rfloor}\left\{|U_k|\right\} > x\right) \approx 2\big(1-\Phi(x)\big) + 2x\phi(x)\log\frac{1-\beta}{\beta}\cdot \tag{I.72}$$

5.2. Change in intercept – equally spaced design

We consider again the problem (I.68) with $x_i = i/n$, $i = 1,\dots,n$, and the maximum-type statistics (I.69). Under H, the limit behavior of (I.69) is again given by (I.71), while

$$P\left(\max_{\lfloor\beta n\rfloor\le k\le\lfloor(1-\beta)n\rfloor}\left\{|U_k|\right\} > x\right) \approx \tag{I.73}$$

$$\approx 2\big(1-\Phi(x)\big) + 2x\phi(x)\int_{\beta}^{1-\beta}\frac{1}{t(1-t)\big(1-3t(1-t)\big)}\,dt.$$

For deatils see Kim and Siegmund (1989).

5.3. Change in both intercept and slope – random design

We test the null hypothesis H against the alternative A in the form

$$\begin{aligned} H &: Y_i = a + bx_i + e_i, && i = 1,\dots,n, \\ A &: \exists \mathsf{m} \in \{2,\dots,n-2\} \ \text{such that} && \\ &\quad Y_i = a + bx_i + e_i, && i = 1,\dots,\mathsf{m}, \\ &\quad Y_i = a^o + b^o x_i + e_i, && i = \mathsf{m}+1,\dots,n, \end{aligned} \tag{I.74}$$

where $a \neq a^o$ and/or $b \neq b^o$.

Let us denote

$$\boldsymbol{X}_k = \begin{pmatrix} 1 & x_1 \\ & \dots \\ 1 & x_k \end{pmatrix}, \qquad \boldsymbol{X}_k^o = \begin{pmatrix} 1 & x_{k+1} \\ & \dots \\ 1 & x_n \end{pmatrix}$$

and

$$\begin{aligned} \chi^2(k) &= \frac{1}{\sigma^2}\big(\widehat{a} - \widehat{a}^o, \widehat{b} - \widehat{b}^o\big)\Big(\big(\boldsymbol{X}_k'\boldsymbol{X}_k\big)^{-1} + \big(\boldsymbol{X}_k^{o\prime}\boldsymbol{X}_k^o\big)^{-1}\Big)^{-1}\big(\widehat{a} - \widehat{a}^o, \widehat{b} - \widehat{b}^o\big)' \\ &= \frac{1}{\sigma^2}\left(\frac{nk\big(\overline{Y}_k - \overline{Y}_n\big)^2}{n-k} + \frac{Q_{xy}^2(k)}{Q_{xx}(k)} + \frac{Q_{xy}^{o2}(k)}{Q_{xx}^o(k)} - \frac{Q_{xy}^2(n)}{Q_{xx}(n)}\right), \end{aligned}$$

where $\widehat{a}$, $\widehat{a}^o, \widehat{b}$ and $\widehat{b}^o$ are the least squares estimators of corresponding quantities under A and

$$Q_{xx}(k) = \sum_{i=1}^{k}\big(x_i - \overline{x}_k\big)\big(x_i - \overline{x}_k\big), \qquad Q_{xy}(k) = \sum_{i=1}^{k}\big(x_i - \overline{x}_k\big)\big(Y_i - \overline{Y}_k\big),$$

$$Q_{xx}^o(k) = \sum_{i=k+1}^{n}\big(x_i - \overline{x}_k^o\big)\big(x_i - \overline{x}_k^o\big), \quad Q_{xy}^o(k) = \sum_{i=k+1}^{n}\big(x_i - \overline{x}_k^o\big)\big(Y_i - \overline{Y}_k^o\big).$$

The maximum-type test statistics are of the form

$$\max_{2 \le k \le n-2}\big\{\chi_k^2\big\} \quad \text{and} \quad \max_{\lfloor \beta n \rfloor \le k \le \lfloor (1-\beta)n \rfloor}\big\{\chi_k^2\big\}. \tag{I.75}$$

It holds that

$$P\left(\max_{2 \le k \le n-2}\big\{\chi_k^2\big\} > \Big(\frac{x + b_{n,2}}{a_n}\Big)^2\right) \approx 1 - \exp\big\{-2e^{-x}\big\}, \quad x \in \mathcal{R}^1, \tag{I.76}$$

where

$$a_n = \sqrt{2\log\log n} \quad \text{and} \quad b_{n,2} = 2\log\log n + \frac{1}{2}\log\log\log n,$$

and

$$P\left(\max_{\lfloor \beta n \rfloor \le k \le \lfloor (1-\beta)n \rfloor}\big\{\chi_k^2\big\} > x^2\right) \approx e^{-x^2/2} + x^2 e^{-x^2/2}\log\frac{1-\beta}{\beta}. \tag{I.77}$$

The approximation (I.76) can be found in Csörgő and Horváth (1997), see also Albin (1990).

5.4. Change in both intercept and slope – equally spaced design

We consider the problem (I.74) with $x_i = i/n$, $i = 1, \dots, n$. Then for the maximum-type test statistics (I.75) hold

$$P\left(\max_{2\le k\le n-2}\left\{\chi_k^2\right\} > \left(\frac{x+b_{n,2}}{a_n}\right)^2\right) \approx 1-\exp\{-2e^{-x}\}, \quad x\in\mathcal{R}^1,$$

where

$$a_n = \sqrt{2\log\log n} \quad \text{and} \quad b_{n,2} = 2\log\log n + \frac{1}{2}\log\log\log n,$$

and

$$P\left(\max_{\lfloor\beta n\rfloor\le k\le\lfloor(1-\beta)n\rfloor}\left\{\chi_k^2\right\} > x^2\right) \approx e^{-x^2/2} + 2x^2e^{-x^2/2}\log\frac{1-\beta}{\beta}.$$

For more details see Albin et al. (2000).

5.5. Appearance of gradual trend – equally spaced design

Sometimes it is important to decide whether all observations have the same mean and whether at some unknown time point a gradual (continuous) trend appeared. When the appearing trend is supposed to be linear, we are to test the null hypothesis H against the alternative A:

$$\begin{aligned} &H: Y_i = a + e_i, && i = 1,\dots,n, \\ &A: \exists\, \mathsf{m} \in \{1,\dots,n-1\} \text{ such that} && \\ &\qquad Y_i = a + e_i, && i = 1,\dots,\mathsf{m}, \\ &\qquad Y_i = a + b\cdot\frac{i-\mathsf{m}}{n} + e_i, && i = \mathsf{m}+1,\dots,n, \end{aligned} \tag{I.78}$$

where $b \neq 0$.

The maximum-type statistics have the form

$$\max_{1\le k<n}\left\{\frac{|\widehat{b}_k|}{\sqrt{var\,\widehat{b}_k}}\right\} \quad \text{and} \quad \max_{1\le k\le\lfloor(1-\beta)n\rfloor}\left\{\frac{|\widehat{b}_k|}{\sqrt{var\,\widehat{b}_k}}\right\}, \tag{I.79}$$

where $\widehat{b}_k$ is the least squares estimator of b under A supposing that the change occurred at the moment k. Notice that

$$\frac{|\widehat{b}_k|}{\sqrt{var\,\widehat{b}_k}} = \frac{\left|\frac{1}{\sigma}\frac{1}{\sqrt{n}}\sum_{i=k+1}^{n}(Y_i-\overline{Y}_n)\frac{i-k}{n}\right|}{\sqrt{\frac{(n-k)(n-k+1)(n-k+1/2)}{3n^3}-\frac{(n-k)^2(n-k+1)^2}{4n^4}}}.$$

For n large, critical values can be attained using the approximation

$$P\left(\max_{1\le k\le n-1}\left\{\frac{|\widehat{b}_k|}{\sqrt{var\,\widehat{b}_k}}\right\} > \frac{x+b_{n,3}}{a_n}\right) \approx 1-\exp\{-2e^{-x}\}, \quad x\in\mathcal{R}^1, \tag{I.80}$$

where

$$a_n = \sqrt{2 \log \log n} \quad \text{and} \quad b_{n,3} = 2 \log \log n + \log \frac{\sqrt{3}}{4\pi},$$

and

$$P\left(\max_{1 \le k \le \lfloor (1-\beta)n \rfloor} \left\{ \frac{|\widehat{b}_k|}{\sqrt{var\, \widehat{b}_k}} \right\} > x \right) \approx \tag{I.81}$$

$$\approx 2\big(1 - \Phi(x)\big) + 2 \frac{1}{\sqrt{\pi}}\, \phi(x) \int_0^{1-\beta} \frac{\sqrt{6t}}{(1-t)(1+3t)}\, dt$$

$$= 2\big(1 - \Phi(x)\big) + \frac{1}{\sqrt{\pi}}\, \phi(x) \left(\sqrt{\frac{3}{2}} \log \frac{1 + \sqrt{1-\beta}}{1 - \sqrt{1-\beta}} - \sqrt{2} \arctan \sqrt{3(1-\beta)} \right).$$

The approximation (I.80) can be found in Jarušková (1998a), for (I.81) see Jarušková (1997).

5.6. Continuous change in linear regression

In this section we suppose that the regression slope may continuously change and thus test the hypothesis H against the alternative A:

$$H : Y_i = a + b \cdot \frac{i}{n} + e_i, \qquad i = 1, \ldots, n, \tag{I.82}$$

$$A : \exists\, \mathsf{m} \in \{2, \ldots, n-2\} \quad \text{such that}$$

$$Y_i = a + b \cdot \frac{i}{n} + e_i, \qquad i = 1, \ldots, \mathsf{m},$$

$$Y_i = a + b \cdot \frac{i}{n} + c \cdot \frac{i - \mathsf{m}}{n} + e_i, \qquad i = \mathsf{m}+1, \ldots, n,$$

where $c \neq 0$.

The maximum-type test statistics are of the form

$$\max_{2 \le k \le n-2} \left\{ \frac{|\widehat{c}_k|}{\sqrt{var\, \widehat{c}_k}} \right\} \quad \text{and} \quad \max_{\lfloor \beta n \rfloor \le k \le \lfloor (1-\beta)n \rfloor} \left\{ \frac{|\widehat{c}_k|}{\sqrt{var\, \widehat{c}_k}} \right\}, \tag{I.83}$$

where $\widehat{c}_k$ is the least squares estimator of c under A supposing that the change occurred at the moment k. It holds that

$$P\left(\max_{2 \le k \le n-2} \left\{ \frac{|\widehat{c}_k|}{\sqrt{var\, \widehat{c}_k}} \right\} > \frac{x + b_{n,3}}{a_n} \right) \approx 1 - \exp\left\{ -4e^{-x} \right\}, \quad x \in \mathcal{R}^1, \tag{I.84}$$

where

$$a_n = \sqrt{2 \log \log n} \quad \text{and} \quad b_{n,3} = 2 \log \log n + \log \frac{\sqrt{3}}{4\pi},$$

and

$$P\left(\max_{\lfloor\beta n\rfloor\le k\le\lfloor(1-\beta)n\rfloor}\left\{\frac{|\widehat{c}_k|}{\sqrt{var\,\widehat{c}_k}}\right\}>x\right)\approx$$

$$\approx 2\big(1-\Phi(x)\big)+2\frac{1}{\sqrt{\pi}}\,\phi(x)\int_{\beta}^{1-\beta}\sqrt{\frac{3}{8}}\,\frac{1}{t(1-t)}\,dt \qquad \text{(I.85)}$$

$$=2\big(1-\Phi(x)\big)+\sqrt{\frac{6}{\pi}}\,\phi(x)\,\log\frac{1-\beta}{\beta}.$$

6. Nonparametric and robust procedures in simple regression

The M- and R-procedures can be constructed along the lines of the procedures described in Sections 5.1.–5.6.. One has only to replace the least squares estimators and corresponding residuals by their M- and R-counterparts.

6.1. M-test procedures

We start with the M-test procedures. Consider the problem of testing the null hypothesis H against the alternative A:

$$\begin{aligned}
&H: Y_i = a + bx_i + e_i, && i=1,\dots,n, \qquad \text{(I.86)}\\
&A: \exists\,\mathsf{m}\in\{2,\dots,n-2\}\ \text{ such that}\\
&\qquad Y_i = a + bx_i + e_i, && i=1,\dots,\mathsf{m},\\
&\qquad Y_i = \mu^o + b^o x_i + e_i, && i=\mathsf{m}+1,\dots,n,
\end{aligned}$$

where $(a,b)\ne(a^o,b^o)$. The random errors e_i are supposed to be iid with symmetric distribution function, the moment assumptions need not to be satisfied and the errors be independent of $\{x_i\}$.

We assume that ψ is a monotone and skew symmetric $\big(\psi(x)=-\psi(x)\ \forall x\in\mathcal{R}^1\big)$ score function. For information on the choice of ψ see Section 4.1.1. Then the M-estimators $\widehat{a}_{n,M}(\psi)$ and $\widehat{b}_{n,M}(\psi)$ based on all n observations can be defined as the solution of the equations

$$\sum_{i=1}^{n}\psi\big(Y_i-a-bx_i\big)=0 \quad\text{and}\quad \sum_{i=1}^{n}x_i\psi\big(Y_i-a-bx_i\big)=0,$$

the corresponding M-residuals as

$$\widehat{e}_{i,M}(\psi)=\psi\Big(Y_i-\widehat{a}_{n,M}(\psi)-x_i\widehat{b}_{n,M}(\psi)\Big),\quad i=1,\dots,n,$$

and the scale estimator can be introduced, e.g., as

$$\widehat{\sigma}^2_{n,M}(\psi) = \min_{2\le k<n-2} \frac{1}{n}\Big\{ \sum_{i=1}^{k} \psi^2\Big(Y_i - \widehat{a}_{n,M}(\psi) - x_i\widehat{b}_{n,M}(\psi)\Big) + \sum_{i=k+1}^{n} \psi^2\Big(Y_i - \widehat{a}^o_{n,M}(\psi) - x_i\widehat{b}^o_{n,M}(\psi)\Big)\Big\},$$

where $\widehat{a}^o_{n,M}(\psi)$ and $\widehat{b}^o_{n,M}(\psi)$ are the M-estimators of a and b based on $Y_{m+1},\dots,Y_n$.

The maximum-type test statistic for the case $b = b^o$ (change in the intercept a only) has the form

$$\max_{1\le k<n}\left\{ \frac{1}{\widehat{\sigma}_{n,M}(\psi)} \left| \sqrt{\frac{n}{k(n-k)}} \sum_{i=1}^{k} \widehat{e}_{i,M}(\psi) \right| \right\}. \tag{I.87}$$

Under mild assumptions on ψ, on the error distribution F, on design points (satisfying (I.70)) and if H holds true, we can use the same approximation to the distribution of (I.87) as in previous case, i.e.

$$P\left(\max_{1\le k<n}\left\{ \frac{1}{\widehat{\sigma}_{n,M}(\psi)} \left| \sqrt{\frac{n}{k(n-k)}} \sum_{i=1}^{k} \widehat{e}_{i,M}(\psi) \right| \right\} > \frac{x+b_n}{a_n} \right) \approx \approx 1 - \exp\big\{-2e^{-x}\big\}, \quad x\in\mathcal{R}^1, \tag{I.88}$$

where

$$a_n = \sqrt{2\log\log n} \quad\text{and}\quad b_n = 2\log\log n + \frac{1}{2}\log\log\log n - \frac{1}{2}\log\pi.$$

Next, we consider the testing problem (I.86) when both parameters can change. Let the error terms e_i have a symmetric distribution function. Then the M-test statistic is of the form

$$\max_{2\le k\le n-2}\big\{\widetilde{\chi}^2(k)\big\}, \tag{I.89}$$

where

$$\widetilde{\chi}^2(k) = \frac{1}{\widehat{\sigma}^2_{n,M}(\psi)}\left\{ \frac{n}{k(n-k)}\Big(\sum_{i=1}^{k}\widehat{e}_{i,M}(\psi)\Big)^2 + \frac{\Big[\sum_{i=1}^{n}(x_i-\overline{x}_n)^2\Big]\cdot\Big[\sum_{i=1}^{k}(x_i-\overline{x}_n)\widehat{e}_{i,M}(\psi)\Big]^2}{\Big[\sum_{i=1}^{k}(x_i-\overline{x}_n)^2\Big]\cdot\Big[\sum_{i=k+1}^{n}(x_i-\overline{x}_n)^2\Big]}\right\}.$$

The approximation of the distribution of the statistic (I.89) analogous as in (I.76) applies here too, i.e., we can use

$$P\left(\max_{2\le k\le n-2}\big\{\widetilde{\chi}^2(k)\big\} > \Big(\frac{x+b_{n,2}}{a_n}\Big)^2\right) \approx 1-\exp\big\{-e^{-x}\big\}, \quad x\in\mathcal{R}^1, \tag{I.90}$$

where

$$a_n = \sqrt{2 \log \log n} \quad \text{and} \quad b_{n,2} = 2 \log \log n + \frac{1}{2} \log \log \log n.$$

6.2. R-test procedures

We consider the testing problem (I.86) with the error terms having continuous distribution function F. Here we need an estimator of the slope parameter b. We can use either some R-estimator if we want to have purely rank procedure or, which is more often the case, we can use any consistent estimator (least squares, L_1-norm etc.). We denote this estimator by $\widetilde{b}_n$. Then we calculate the ranks $R_1(\widetilde{b}_n), \ldots, R_n(\widetilde{b}_n)$ of $Y_1 - x_1\widetilde{b}_n, \ldots, Y_n - x_n\widetilde{b}_n$. Of course, in practice we estimate both parameters a and b. However, it is clear that the ranks of $(Y_i - x_i\widetilde{b}_n)$'s do not depend on the estimate of the intercept.

The maximum-type R-statistic for testing problem (I.86) is defined as

$$\max_{1 \le k < n} \left\{ \widetilde{\widetilde{\chi}}^2(k) \right\}, \tag{I.91}$$

where

$$\widetilde{\widetilde{\chi}}^2(k) = \frac{1}{\widehat{\sigma}^2_{n,R}(\psi)} \left\{ \frac{n}{k(n-k)} \Big(\sum_{i=1}^{k} \left[a(R_i(\widetilde{b}_n)) - \overline{a}_n \right] \Big)^2 + \right.$$
$$\left. + \frac{\left[\sum_{i=1}^{n} (x_i - \overline{x}_n)^2 \right] \cdot \left[\sum_{i=1}^{k} (x_i - \overline{x}_n) a\big(R_i(\widehat{b}_n)\big) \right]^2}{\left[\sum_{i=1}^{k} (x_i - \overline{x}_n)^2 \right] \cdot \left[\sum_{i=k+1}^{n} (x_i - \overline{x}_n)^2 \right]} \right\}$$

with

$$\widehat{\sigma}^2_{n,R} = \frac{1}{n-1} \sum_{i=1}^{n} \Big(a_n(i) - \overline{a}_n \Big)^2.$$

Unfortunately, see Section 4.1.2., this rank test is not distribution free under H as it was the case in the location model. Under the same assumptions on the design matrix as in Section 5.1., the distribution of (I.91) can be under H approximated as in (I.90), where the statistic (I.89) is replaced by statistic (I.91).

The R-procedures for problems in Sections 5.1.–5.6. can be constructed along the same line. However, they are not very attractive. Nevertheless, the R-test statistic for the testing problem (I.78), where the error terms e_i have a continuous distribution function F, is quite appealing because under H it is distribution free. More precisely, the R-test statistic has the form

$$\max_{1 \le k < n} \left\{ \frac{1}{\widehat{\sigma}_{n,R}} \left| \sum_{i=k+1}^{n} (i-k)\big(a(R_i) - \overline{a}_n\big) \right| \left(\frac{(n-k)^3}{3} - \frac{(n-k)^4}{4n} \right)^{-1/2} \right\}, \tag{I.92}$$

where $R_1, \ldots, R_n$ are the ranks of $Y_1, \ldots, Y_n$.

Under H, the distribution of (I.92) does not depend on the error distribution F. Therefore, similarly as in the case of the rank-based procedures for changes in location described earlier, the approximation as in (I.90) (where (I.89) is replaced by (I.92)) remains valid for our test statistic, too. Aside that, we can obtain the approximation also through simulations.

7. Change in general linear regression

We briefly survey procedures for detection changes in general linear models. We consider the testing problem:

$$\begin{aligned} H : Y_i &= \boldsymbol{x}_i'\boldsymbol{\beta} + e_i, && i = 1, \ldots, n, \\ A : \exists\, \mathsf{m} &\in \{0, \ldots, n-1\} \quad \text{such that} && \\ Y_i &= \boldsymbol{x}_i{}'\boldsymbol{\beta} + e_i, && i = 1, \ldots, \mathsf{m}, \\ &= \boldsymbol{x}_i'\boldsymbol{\beta} + \boldsymbol{x}_i'\boldsymbol{\delta} + e_i, && i = \mathsf{m}+1, \ldots, n, \end{aligned} \tag{I.93}$$

where $\boldsymbol{\beta} = (\beta_1, \ldots, \beta_p)'$ and $\boldsymbol{\delta} = (\delta_1, \ldots, \delta_p)' \neq \mathbf{0}$ are parameters, $\boldsymbol{x}_1, \ldots, \boldsymbol{x}_n$ are $p-$dimensional vectors; $e_1, \ldots, e_n$ are random errors that are iid with zero mean, nonzero variance σ^2 and $E|e_i|^{2+\Delta} < \infty$ for some $\Delta > 0$. Concerning the sets of assumptions on the design points $\boldsymbol{x}_i$, $i = 1 \ldots, n$ slightly stronger then for classical tests on the parameters $\boldsymbol{\beta}$ are needed. Here are some possibilities:

- $\boldsymbol{x}_i = \boldsymbol{g}(i/n) = (g_1(i/n), \ldots, g_p(i/n))'$, $i = 1, \ldots, n$, where $g_1, \ldots, g_p$ are continuous functions on $[0, 1]$ such that $\int_a^b \boldsymbol{g}(x)\,dx$ is a positive definite matrix for each $0 \leq a < b \leq 1$.
- There are fixed p linearly independent p-dimensional vectors $\boldsymbol{x}_1{}^0, \ldots, \boldsymbol{x}_p{}^0$ and the design points are chosen according to some rule from this p vectors.
- The design points can be a realization of AR or ARMA sequences fulfilling certain moment assumptions.

Notice that as a special case one gets the testing problems considered in Section 5.

Similarly as in the previous sections the test procedures are developed using either likelihood ratio approach or Bayesian one (or their modifications). We present likelihood ratio type test statistics only.

If one assumes that e_i's have $N(0, \sigma^2)$ distribution the likelihood ratio type test statistic is

$$V_{n0} = \max_{p<k<n-p} \left\{ -2 \log \frac{\frac{k}{n}\widehat{\sigma}^2_{1,k} + \frac{n-k}{n}\widehat{\sigma}^2_{k+1,n}}{\widehat{\sigma}^2_{k+1,n}} \right\},$$

where

$$\widehat{\sigma}^2_{1,k} = \frac{1}{k}\sum_{i=1}^{k} \left(Y_i - \boldsymbol{x}_i'\widehat{\boldsymbol{\beta}}_{1,k}\right)^2, \quad \widehat{\sigma}^2_{k+1,n} = \frac{1}{n-k}\sum_{i=k+1}^{n} \left(Y_i - \boldsymbol{x}_i'\widehat{\boldsymbol{\beta}}_{k+1,n}\right)^2,$$

$\widehat{\boldsymbol{\beta}}_{1,k}$ and $\widehat{\boldsymbol{\beta}}_{k+1,n}$ are the least squares estimators of $\boldsymbol{\beta}$ based on $Y_1, \ldots, Y_k$ and $Y_{k+1}, \ldots, Y_n$, respectively.

A slight modification leads to the test statistics

$$V_n = \max_{p \le k < n} \left\{ \frac{1}{\widetilde{\sigma}_n^2} \frac{n}{(n-k)k} \boldsymbol{S}_k' (\boldsymbol{X}_n' \boldsymbol{X}_n)^{-1} \boldsymbol{S}_k \right\},$$

where $\boldsymbol{X}_n = (\boldsymbol{x}_1, \ldots, \boldsymbol{x}_n)'$, $\boldsymbol{S}_k = \sum_{i=1}^k x_i (Y_i - \boldsymbol{x}_i' \widehat{\boldsymbol{\beta}}_{1,n})$, $k = 1, \ldots, n$ and, $\widetilde{\sigma}_n^2$ is an estimator of σ^2; for instance

$$\widetilde{\sigma}_n^2 = \frac{1}{n-p} \min_{p<k<n-p} \left\{ k \widehat{\sigma}_{1,k}^2 + (n-k) \widehat{\sigma}_{k+1,n}^2 \right\}$$

can be used.

Under the null hypothesis H and mild assumptions we have the following approximations

$$P\Big(V_{n0} > \Big(\frac{x + b_{n,p}}{a_n}\Big)^2\Big) \approx 1 - \exp\{-2e^{-x}\}, \quad x \in \mathcal{R}^1$$

and

$$P\Big(V_n > \Big(\frac{x + b_{n,p}}{a_n}\Big)^2\Big) \approx 1 - \exp\{-2e^{-x}\}, \quad x \in \mathcal{R}^1,$$

where

$$b_{n,p} = 2 \log\log n + \frac{p}{2} \log\log\log n - \log \Gamma(p/2) \quad \text{and} \quad a_n = \sqrt{2 \log\log n}.$$

Similarly as in the previous sections these approximations can be used to get an approximation to the critical values.

In econometrically oriented literature one often meets the test procedures based on partial sums of residuals, namely the following test statistic is used

$$\max_{1 \le k < n} \left\{ \frac{1}{\sqrt{n} \widetilde{\sigma}_n} \Big| \sum_{i=1}^k (Y_i - \boldsymbol{x}_i' \widehat{\boldsymbol{\beta}}_{1,n}) \Big| \right\}$$

This test statistic is much simpler than V_n and V_{n0} but it is sensitive only to some $\delta \neq 0$.

Under the null hypothesis H and mild assumptions (e_i can be even dependent)

$$P\Big(\max_{1 \le k < n} \frac{1}{\sqrt{n} \widetilde{\sigma}_n} \Big| \sum_{i=1}^k (Y_i - \boldsymbol{x}_i' \widehat{\boldsymbol{\beta}}_{1,n}) \Big| \le x\Big) \approx P\Big(\sup_{0<t<1} |B(t)| \le x\Big) =$$

$$= 1 - 2 \sum_{j=1}^{\infty} (-1)^{j+1} e^{-2j^2 x^2}, \quad x \in \mathcal{R}^1,$$

where $\{B(t);\ t \in (0,1)\}$ is a Brownian bridge and $\widetilde{\sigma}_n$ is a proper standardization. More information can be found e.g. in Jandhyala and MacNeill (1991), Jandhyala and MacNeill (1992), MacNeill (1978), Horváth (1995), Kim and Siegmund (1989) or Ploberger (1992).

Part II. Estimators of changes

Part I deals with tests on the stability of statistical models. The problem was formulated in terms of testing the null hypothesis H against the alternative hypothesis A. The null hypothesis H claims that the model remains the same during the whole observational period, usually it means that the parameters of the model do not change. The alternative hypothesis A claims that, at an unknown time point, the model changes, which means that some of the parameters of the model are subject to a change. In case we reject the null hypothesis H, i.e. we decide that there is a change in the model, a number of questions arise:

- when has the model changed;
- is there just one change or are there more changes;
- what is the total number of changes.

The time moment when the model has changed is usually called *change point.* In this part our primary interest is to *estimate the change point(s)* in various models. Of course, we also estimate other parameters of the models. We will show approximations to the distributions of the change point estimators and finally construct the interval estimators for the change points. As we will see below, the estimators of the change points are usually closely related to some of the test statistics treated in Part I.

8. Change in location

We assume that the observations $Y_1, \ldots, Y_n$ follow the model

$$Y_i = \begin{cases} \mu + \sigma e_i, & i = 1, 2, \ldots, \mathsf{m}, \\ \mu + \delta + \sigma e_i, & i = \mathsf{m}+1, \ldots, n, \end{cases} \tag{II.1}$$

where μ, $\sigma^2 > 0$, $\delta \neq 0$ and $\mathsf{m}(< n)$ are parameters, and $e_1, \ldots, e_n$ are iid random variables with zero mean, unit variance and with $\mathsf{E}\,|e_i|^{2+\Delta} < \infty$ with some $\Delta > 0$. The distribution of e_i will be denoted by F.

It is supposed that the observations $Y_1, \ldots, Y_n$ are obtained at the ordered time moments $t_1 < \cdots < t_n$. Hence, this model says that there is just one change after the mth observation, and the corresponding time moment is called the *change point.* More exactly, the change occurs in the time period $(t_{\mathsf{m}}, t_{\mathsf{m}+1}]$. However, we will work here only with integer valued t_{m}.

We assume the change point m satisfies

$$\mathsf{m} = \lfloor n\gamma \rfloor, \quad \gamma \in (0, 1), \tag{II.2}$$

where $\lfloor a \rfloor$ denotes the integer part of a. This means that the change can occur neither at the very beginning nor at the very end of the observational period. To estimate the unknown parameters we can apply some of the general methods, e.g. the maximum likelihood method, however the distribution function F of the error terms e_i has to be known.

- Another possibility is to estimate the unknown parameters m, μ and δ by the least squares (LS) method, that gives a simple solution. The least squares estimators $\widehat{m}_{LS}$, $\widehat{\mu}_{LS}$ and $\widehat{\delta}_{LS}$ of the parameters m, μ and δ are defined as solutions of the minimization problem

$$\min\left\{\sum_{i=1}^{k}(Y_i-\mu)^2+\sum_{i=k+1}^{n}(Y_i-\mu-\delta)^2;\right. \tag{II.3}$$

$$\left. k\in\{1,\dots,n-1\},\ \mu\in\mathcal{R}^1,\ \delta\in\mathcal{R}^1\right\},$$

In other words, the parameters m, μ and δ are estimated in such a way that the sum of squares of residuals is minimal.

Direct calculation gives the explicit expressions for the estimators $\widehat{\mu}_{LS}$ and $\widehat{\delta}_{LS}$:

$$\widehat{\mu}_{LS}=\overline{Y}_{\widehat{m}_{LS}}, \tag{II.4}$$

$$\widehat{\delta}_{LS}=\overline{Y}^{o}_{\widehat{m}_{LS}}-\overline{Y}_{\widehat{m}_{LS}}, \tag{II.5}$$

while $\widehat{m}_{LS}$ is a solution of the maximization problem

$$\max\left\{\sqrt{\frac{n}{k(n-k)}}\,\big|S_k\big|;\ k\in\{1,\dots,n-1\}\right\},$$

where $\overline{Y}_k$ and $\overline{Y}_k^{o}$ are defined in (I.2). We recall the definition of the partial sums S_k:

$$S_k=\sum_{i=1}^{k}(Y_i-\overline{Y}_n),\quad k=1,\dots,n. \tag{II.6}$$

Notice that if the error terms e_i have $N(0,1)$ distribution, the least squares estimators coincide with the maximum likelihood estimators.

The solution of the maximization problem need not to be unique, therefore we make the following convention.

CONVENTION. We will typically treat a random variable V defined as a solution of the maximization problem

$$\max\left\{W(t);\ t\in T\right\},$$

where $\{W(t);\ t\in T\}$ is a random process or a sequence of random variables. If the solution is not uniquely determined, we always take the smallest solution and write shortly

$$V=\arg\max\left\{W(t);\ t\in T\right\}.$$

Using this convention, the estimator $\widehat{m}_{LS}$ is defined as

$$\widehat{m}_{LS}=\arg\max\left\{\sqrt{\frac{n}{k(n-k)}}\,\big|S_k\big|;\ k\in\{1,\dots,n-1\}\right\}. \tag{II.7}$$

Moreover, the estimator $\widehat{m}_{LS}$ can be equivalently defined as

$$\widehat{\mathsf{m}}_{LS} = \arg\max \left\{ \frac{k(n-k)}{n} \left(\overline{Y}_k - \overline{Y}_k^{o} \right)^2 ;\ k \in \{1, \ldots, n-1\} \right\}.$$

• Another type of estimators of m is based on moving sums (MOSUM). These estimators will be denoted by $\widehat{\mathsf{m}}_{LS}(G)$ and defined as

$$\widehat{\mathsf{m}}_{LS}(G) = \arg\max \left\{ \left| S_{k+G} - 2S_k + S_{k-G} \right| ;\ k \in \{G+1, \ldots, n-G\} \right\}. \qquad \text{(II.8)}$$

The bandwidth G is usually chosen much smaller than n, typically $g = 0.1n$ or $0.05n$. Some tuning has to be done for the real data.

The connection of the estimators $\widehat{\mathsf{m}}_{LS}$ and $\widehat{\mathsf{m}}_{LS}(G)$ with the test statistics (I.17) and (I.54) is clearly visible. Calculating the tests statistics we usually calculate the corresponding estimators as a "free" by-product.

To get an idea about the behavior of the estimators, we present the expectations and variances of the partial sums of residuals and the moving sums:

$$\mathsf{E}\, S_k = \begin{cases} -\delta k \dfrac{n-\mathsf{m}}{n}, & 1 \le k \le \mathsf{m}, \\ -\delta (n-k) \dfrac{\mathsf{m}}{n}, & \mathsf{m} < k \le n, \end{cases}$$

$$var\, S_k = \sigma^2 \frac{k(n-k)}{n}, \qquad 1 \le k \le n,$$

$$\mathsf{E}\left(S_{k+G} - 2S_k + S_{k-G} \right) = \begin{cases} 0, & G < k \le \mathsf{m} - G, \\ \delta\left(k + G - \mathsf{m}\right), & \mathsf{m} - G < k \le \mathsf{m}, \\ \delta\left(\mathsf{m} + G - k\right), & \mathsf{m} < k \le \mathsf{m} + G, \\ 0, & \mathsf{m} + G < k \le n - G, \end{cases}$$

$$var \left(S_{k+G} - 2S_k + S_{k-G} \right) = 2\sigma^2 G.$$

In both cases the maximum of the absolute values of the expectations is reached for $k = \mathsf{m}$. Moreover, if the random parts of $(S_k - \mathsf{E}S_k)\sqrt{\frac{n}{k(n-k)}}$ and $\left(S_{k+G} - 2S_k + S_{k-G}\right) - \mathsf{E}\left(S_{k+G} - 2S_k + S_{k-G}\right)$ are stochastically smaller than the expectations, then the estimators $\widehat{\mathsf{m}}_{LS}$ and $\widehat{\mathsf{m}}_{LS}(G)$ are "close" to the change point. More exactly, under certain assumptions, as $n \to \infty$,

$$\frac{\widehat{\mathsf{m}}_{LS} - \mathsf{m}}{n} \xrightarrow{\mathcal{P}} 0 \quad \text{and} \quad \frac{\widehat{\mathsf{m}}_{LS}(G) - \mathsf{m}}{n} \xrightarrow{\mathcal{P}} 0,$$

i.e., the estimators are consistent.For details see Antoch and Hušková (1998).

Finding the exact distribution of either type of estimators is a very complex problem (similarly as in the testing problem); up to now no exact solution has been found. The limit distribution of these estimators has been treated and this provides a reasonable approximation to the exact distribution.

If $\delta \neq 0$ can be assumed small with respect to the number of observations n, mathematically expressed as $\delta \equiv \delta_n \to 0$ as $n \to \infty$, we speak about a *local change*. In such a case we have the following approximation, i.e.:

$$P\left(\frac{\delta^2}{\sigma^2}\left(\widehat{\mathsf{m}}_{LS} - \mathsf{m}\right) \leq x\right) \approx P(V \leq x), \quad x \in \mathcal{R}^1, \tag{II.9}$$

where the random variable V is defined by

$$V = \arg\max \left\{W(s) - |s|/2\,;\ s \in \mathcal{R}^1\right\}. \tag{II.10}$$

Here $\{W(s); s \in \mathcal{R}^1\}$ is a two-sided standard Wiener process, i.e.,

$$W(s) = \begin{cases} W_1(-s), & s < 0, \\ W_2(s), & s \geq 0, \end{cases}$$

where $\{W_1(t);\ t \in [0,\infty)\}$ and $\{W_2(t);\ t \in [0,\infty)\}$ are independent standard Wiener processes. This approximation is "distribution free", which means that it does not depend on the distribution of the error terms e_i.

Concerning the MOSUM type estimators, it holds

$$P\left(\frac{2\delta^2}{3\sigma^2}\left(\widehat{\mathsf{m}}_{LS}(G) - \mathsf{m}\right) \leq x\right) \approx P(V \leq x), \quad x \in \mathcal{R}^1, \tag{II.11}$$

with V given by (II.10).

The parameter σ^2 can be estimated, e.g., by

$$\widehat{\sigma}_n^2 = \frac{1}{n-2}\left\{\sum_{i=1}^{\widehat{\mathsf{m}}} \left(Y_i - \overline{Y}_{\widehat{\mathsf{m}}}\right)^2 + \sum_{i=\widehat{\mathsf{m}}+1}^{n} \left(Y_i - \overline{Y}^{o}_{\widehat{\mathsf{m}}}\right)^2\right\}, \tag{II.12}$$

where $\widehat{\mathsf{m}}$ is any of the above estimators of m. The approximations described in (II.9) and (II.11) remain valid if δ and σ are replaced by their estimators (II.5) and (II.12), respectively.

The distribution of the random variable V is known. It was independently derived by several authors. Stryhn (1996) showed that random variable V has the distribution function

$$P(V \leq x) = \begin{cases} 1 + \sqrt{\frac{x}{2\pi}}e^{-x/8} - \frac{1}{2}(x+5)\Phi(-\frac{1}{2}\sqrt{x}) + \frac{3}{2}e^{x}\Phi(-\frac{3}{2}\sqrt{x}), & x \geq 0, \\ 1 - P(V \leq -x), & x < 0. \end{cases} \tag{II.13}$$

This result can be used for construction of a confidence interval for the change point m. The tables needed for this task can be found in the book of Csörgő and Horváth (1997); here we give only a few useful quantiles.

β	0.9	0.95	0.975	0.99	0.995
v_β	4.696	7.687	11.033	15.868	19.767

Table 18. Selected quantiles of V, $P(V \leq v_\beta) = \beta$.

8.1. Dependent observations

Now, we shortly discuss the problem of estimation of m when in the model (II.1) the error terms are dependent. More precisely, we consider the model (II.1) with m fulfilling (II.2), where the error terms e_i form a linear process satisfying

$$e_i = \sum_{t=0}^{\infty} w_t \epsilon_{i-t}, \quad i = 1, 2, \ldots,$$

where ϵ_i are iid random variables, $\mathsf{E}\,\epsilon_i = 0$, $var\,\epsilon_i = \sigma^2 > 0$, $\mathsf{E}\,|\epsilon_i|^{2+\Delta} < \infty$ for some $\Delta > 0$ and the weights $\{w_j\}_{j=0}^{\infty}$ satisfy

$$\sum_{j=0}^{\infty} j\,|w_j| < \infty, \quad \sum_{j=0}^{\infty} w_j \neq 0.$$

Notice that AR and ARMA processes fulfill these assumptions.

The maximum likelihood method can be applied here, however, we need to know the distribution of the error terms e_i. It appears that we can apply the estimators $\widehat{\mathsf{m}}_{LS}$ and $\widehat{\mathsf{m}}_{LS}(G)$ defined by (II.7) and (II.8), respectively. Their distributions can be approximated as in the case of independent errors, i.e. (II.9) and (II.11) apply, however, *different standardization is needed.* Namely, σ^2 has to be replaced by

$$\sigma_0^2 = \sigma^2 \Big(\sum_{j=0}^{\infty} w_j \Big)^2.$$

For example, if the sequence e_n is an AR(1) sequence with the coefficient $\rho \in (-1, 1)$, then σ^2 in (II.9) has to be replaced by $\sigma^2/(1-\rho)$ or by its estimator.

Generally, σ_0^2 can be estimated, e.g., by

$$\widehat{\sigma}_{0,n}^2(L) = \widehat{R}(0) + 2 \sum_{k=1}^{L} \left(1 - \frac{k}{L}\right) \widehat{R}(k), \quad L < n,$$

where for $k \geq 0$

$$\widehat{R}(k) = \frac{1}{n} \left\{ \sum_{t=1}^{\widehat{\mathsf{m}}-k} (Y_t - \overline{Y}_{\widehat{\mathsf{m}}})\,(Y_{t+k} - \overline{Y}_{\widehat{\mathsf{m}}}) + \sum_{t=\widehat{\mathsf{m}}+1}^{n-k} \left(Y_t - \overline{Y}^{o}_{\widehat{\mathsf{m}}}\right) \left(Y_{t+k} - \overline{Y}^{o}_{\widehat{\mathsf{m}}}\right) \right\}.$$

If we have more information about the type of dependency, it is advisable to use the estimator specific for the particular model since the above mentioned estimator $\widehat{\sigma}_{0,n}^2(L)$ behaves quite poorly for small and moderate sample sizes. Further discussions the on choice of L together with other theoretical results as well as results from a simulation study can be found in Antoch et al. (1997).

♣ Now, we turn to the *estimators based on M-statistics* and *R-statistics.* Analogously as in the testing problem H against A, we can construct the estimators of the change point m replacing in the least squares type estimators

(II.7) and (II.8) the partial sums S_k by their M-counterparts $S_{k,M}(\psi)$ or their R-counterparts $S_{k,R}$, cf Part I.

8.2. M-estimators of the change point

We assume that the observations $Y_1, \ldots, Y_n$ follow the model (II.1) with m fulfilling (II.2) and with the error term e_i having a common distribution F that is symmetric about zero (no need of zero mean and a finite variance). The M- estimator $\widehat{\mu}_n(\psi)$ of μ generated by the score function ψ is defined as the solution of the equation

$$\sum_{i=1}^{n} \psi(Y_i - t) = 0.$$

The score function ψ is assumed to be monotone and skew symmetric, i.e. $\psi(x) = -\psi(-x)$, $\forall x \in \mathcal{R}^1$.

The M-estimator $\widehat{\mathsf{m}}_M(\psi)$ of m is defined as

$$\widehat{\mathsf{m}}_M(\psi) = \arg\max \left\{ \sqrt{\frac{n}{k(n-k)}}\, \left|S_{k,M}(\psi)\right| ;\ k \in \{1, \ldots, n-1\} \right\}, \quad \text{(II.14)}$$

where

$$S_{k,M}(\psi) = \sum_{i=1}^{k} \psi\Big(Y_i - \widehat{\mu}_n(\psi)\Big), \quad k = 1, \ldots, n. \quad \text{(II.15)}$$

Under some mild assumptions on ψ and F we have the following approximation to the distribution of these errors:

$$P\left(\frac{\delta^2 \lambda^2(\psi, F)}{\int \psi^2(x)\, dF(x)} \Big(\widehat{\mathsf{m}}_M(\psi) - \mathsf{m}\Big) \le x \right) \approx P(V \le x), \quad x \in \mathcal{R}^1, \quad \text{(II.16)}$$

where $\lambda(\psi, F)$ is the value of the derivative, in t, of the function $\int -\psi(x - t)dF(x)$ at $t = 0$. This means that the approximation to the distribution of $\widehat{\mathsf{m}}_M(\psi)$ is the same as that to the distribution of $\widehat{\mathsf{m}}_{LS}$ or $\widehat{\mathsf{m}}_{LS}(G)$ up to the multiplicative constant that depends on the choice of ψ and on the distribution of the error terms. The unknown constants can be replaced by suitable estimators, e.g., by:

$$\widehat{\delta}_{n,M}(\psi) = \widehat{\mu}^{o}_{\widehat{\mathsf{m}}_M}(\psi) - \widehat{\mu}_{\widehat{\mathsf{m}}_M}(\psi),$$

$$\widehat{\sigma}^2_{n,M}(\psi) = \frac{1}{n-2} \left\{ \sum_{i=1}^{\widehat{\mathsf{m}}_M} \psi^2(Y_i - \widehat{\mu}_{\widehat{\mathsf{m}}_M}(\psi)) + \sum_{i=\widehat{\mathsf{m}}_M+1}^{n} \psi^2(Y_i - \widehat{\mu}^{o}_{\widehat{\mathsf{m}}_M}(\psi)) \right\},$$

$$\begin{aligned} \widehat{\lambda}_{n,M}(\psi) =& \frac{1}{2nc_n} \left\{ \sum_{i=1}^{\widehat{\mathsf{m}}_M} \Big(\psi(Y_i - \widehat{\mu}_{\widehat{\mathsf{m}}_M}(\psi) + c_n) - \psi(Y_i - \widehat{\mu}_{\widehat{\mathsf{m}}_M}(\psi) - c_n) \Big) \right. \\ & \left. + \sum_{i=\widehat{\mathsf{m}}_M+1}^{n} \Big(\psi(Y_i - \widehat{\mu}^{o}_{\widehat{\mathsf{m}}_M}(\psi) + c_n) - \psi(Y_i - \widehat{\mu}^{o}_{\widehat{\mathsf{m}}_M}(\psi) - c_n) \Big) \right\}, \end{aligned}$$

where $\widehat{\mu}_{\widehat{\mathsf{m}}_M}(\psi)$ and $\widehat{\mu}^o_{\widehat{\mathsf{m}}_M}(\psi)$ are M-estimators based on $Y_1, \ldots, Y_{\widehat{\mathsf{m}}_M}$ and $Y_{\widehat{\mathsf{m}}_M+1}, \ldots, Y_n$ and $\{c_n\}_n$ is a sequence of positive numbers tending to 0 not faster than $n^{-1/2}$.

Motivated by the definition of the least squares estimators in (II.3), we can also define other M-estimators. Let ρ be a non-negative convex function on $\mathcal{R}^1$ and let ψ be its derivative. Then the M-estimators $\widetilde{\mathsf{m}}_M(\rho), \widetilde{\delta}_M(\rho)$ and $\widetilde{\mu}_M(\rho)$ are defined as a solution of the minimization problem

$$\min\left\{\sum_{i=1}^{k} \rho(Y_i - \mu) + \sum_{i=k+1}^{n} \rho(Y_i - \mu - \delta); \mu \in \mathcal{R}^1, \delta \in \mathcal{R}^1, k \in \{1, \ldots, n-1\}\right\}$$

The distribution of the estimator $\widetilde{\mathsf{m}}_M(\rho)$ can be approximated in the same way as that of $\widehat{\mathsf{m}}_M(\psi)$, i.e. (II.16) remains true if $\widehat{\mathsf{m}}_M(\psi)$ is replaced by $\widetilde{\mathsf{m}}_M(\rho)$.

Typical choices of the score functions are discussed in subsection 4.1.1. of the present paper or in some books or papers devoted to the $M-$ estimators.

Next, we shortly discuss the particular choice $\psi(x) = \operatorname{sign} x$, $x \in \mathcal{R}^1$. This is usually called L_1-procedure and the corresponding estimators are called L_1-estimators. This is due to the fact that the sample median based on $Y_1, \ldots, Y_n$ is a solution of the minimization problem

$$\min\left\{\sum_{i=1}^{n} |Y_i - t| \,;\ t \in \mathcal{R}^1\right\}.$$

We should recall that the sample median minimizes the sum of the L_1-distances, while the least squares estimator minimizes the sum of the L_2-distances. We denote this L_1-estimator of the change point m by $\widehat{\mathsf{m}}_{L_1}$ and we receive the approximation to its distribution in the following form:

$$P\left(\frac{\widehat{\mathsf{m}}_{L_1} - \mathsf{m}}{4f^2(0)} \leq x\right) \approx P(V \leq x), \quad x \in \mathcal{R}^1,$$

where $f(0)$ is the density of the error term at the point 0. If the density is unknown, we have to replace it by an estimator.

There is also another possibility to define the L_1 estimators that is more transparent, however, more difficult to calculate. Namely, we define $\widetilde{\mathsf{m}}_{L_1}, \widetilde{\delta}_{L_1}$ and $\widetilde{\mu}_{L_1}$ as a solution of the minimization problem

$$\min\left\{\sum_{i=1}^{k} |Y_i - \mu| + \sum_{i=k+1}^{n} |Y_i - \mu + \delta| \,;\ \mu \in \mathcal{R}^1, \delta \in \mathcal{R}^1, k \in \{1, \ldots, n-1\}\right\}.$$

The M-estimators of a change in other models can be introduced along the lines explained above.

8.3. *R*-estimators of the change point

We assume the model (II.1) with m fulfilling (II.2) and with the error terms having absolutely continuous density f with the Fisher information

$$0 < I(f) = \int_{-\infty}^{\infty} \frac{(f'(x))^2}{f(x)} dx < \infty,$$

where f' is the derivative of f.

The rank-based estimator of m is based on the partial sums

$$S_{k,R} = \sum_{i=1}^{k} \left(a(R_i) - \overline{a}_n \right), \quad k = 1, \dots, n, \tag{II.17}$$

where $R_1, \dots, R_n$ are the ranks corresponding to $Y_1, \dots, Y_n$ and

$$\overline{a}_n = n^{-1} \sum_{i=1}^{n} a(i).$$

The scores $a(1), \dots, a(n)$ are usually related to a function ξ defined on the interval $(0,1)$ such that

$$0 < \int_0^1 |\xi(u)|^{2+\Delta} du < \infty$$

with some $\Delta > 0$; typically,

$$a(i) = \xi(i/(n+1)), \quad i = 1, \dots, n. \tag{II.18}$$

In this section we will work with these scores; for other possibilities consult books on rank-based procedures.

The R-estimator of m is defined as

$$\widehat{\mathsf{m}}_R = \arg\max \left\{ \sqrt{\frac{n}{k(n-k)}} \, |S_{k,R}|; \; k \in \{1, \dots, n-1\} \right\}. \tag{II.19}$$

We have the following approximation to the distribution of $\widehat{\mathsf{m}}_R$:

$$P\left(\frac{\delta^2 b_R^2(\xi)}{\sigma_R^2} \left(\widehat{\mathsf{m}}_R - \mathsf{m} \right) \le x \right) \approx P(V \le x), \quad x \in \mathcal{R}^1,$$

where

$$\sigma_R^2 = \frac{1}{n-1} \sum_{i=1}^{n} \left(a(R_i) - \overline{a}_n \right)^2 \approx \int_0^1 \xi^2(u) du - \left(\int_0^1 \xi(u)\, du \right)^2$$

and

$$b_R^2(\xi) = \left(\int_{-\infty}^{\infty} \xi(F(x)) f'(x)\, dx \right)^2.$$

If F is unknown, we replace $b_R^2(\xi)$ by an estimator, however, as this is going beyond the scope of this paper, the interested reader should consult some advanced material on the rank-based procedures.

Comparing the rank-based tests discussed in Part I with rank-based estimators just introduced, we see that the estimators are not anymore distribution-free even asymptotically and, moreover, we have stronger assumptions on the scores than in case of tests.

The MOSUM type M- and R-estimators can be introduced analogously and the same holds for the approximations to their distributions.

8.4. Multiple changes

Consider the location model with multiple changes

$$Y_i = \mu_j + e_i, \quad \lfloor n\gamma_{j-1} \rfloor < i \le \lfloor n\gamma_j \rfloor, \; j = 1, \dots, q+1, \tag{II.20}$$

where $\mu_j \in \mathcal{R}^1$, $\mu_j \neq \mu_{j+1}$, $j = 1, \dots, q$, $0 = \gamma_0 < \gamma_1 < \cdots < \gamma_{q+1} = 1$ and $e_1, \dots, e_n$ are iid random variables with zero mean, nonzero variance $var\ e_i = \sigma^2$ and $\mathsf{E}\,|e_i|^{2+\Delta} < \infty$ for some $\Delta > 0$. The change points are $\lfloor n\gamma_j \rfloor$, $j = 1, \dots, q$. Their number q can be known or unknown, however an upper bound q_o for q is supposed to be known.

We should point out, that if we test the null hypothesis

$$H : Y_1, \dots, Y_n \quad \text{are iid with mean } \mu$$

against the alternative A corresponding to the multiple changes described in (II.20), where only the upper bound q_0 for the number of possible changes is known, we can apply any of the test procedures developed for the alternative with one change only. These tests are consistent. We could construct, e.g., the likelihood ratio test, however this brings big problems with finding an approximation to the critical values. A Bayesian approach is studied, e.g., in Chib (1999). We describe three types of estimators for the number of changes. They also provide the estimators for the change points.

The first one was proposed by Yao (1988). This is a modification of the Schwarz' criterion for determining the dimension of regression. The estimator $\widehat{q}$ of the number of change q is defined as a solution of the minimization problem

$$\min \left\{ \frac{n}{2} \log \widetilde{\sigma}_q^2 + q \log n \;\middle|\; q = 1, \dots, q_0 \right\}, \tag{II.21}$$

where q_0 is the possible largest number of changes and

$$\widetilde{\sigma}_q^2 = \min \Bigg\{ \sum_{j=1}^{q+1} \sum_{i=n_{j-1}+1}^{n_j} (Y_i - \mu_j)^2 \;\Bigg|\; \mu_j \in \mathcal{R}^1,\, j = 1, \dots, q,\, 0 = n_0 < n_1 < \cdots < n_q < n_{q+1} = n \Bigg\}.$$

The estimators of the parameters $n_j = \lfloor n\gamma_j \rfloor$ and μ_j are then obtained as a solution of the minimization problem

$$\min \Bigg\{ \sum_{j=1}^{\widehat{q}+1} \sum_{i=n_{j-1}+1}^{n_j} (Y_i - \mu_j)^2 \;\Bigg|\; \mu_j \in \mathcal{R}^1,\, j = 1, \dots, \widehat{q}+1,\, 0 = n_0 < n_1 < \cdots < n_{\widehat{q}} < n_{\widehat{q}+1} = n \Bigg\}.$$

Computational difficulties are evident and therefore some modifications were proposed. Notice, however, that if the observations follow a normal distribution, then these estimators are consistent and coincide with the maximum likelihood estimators.

The second and the third method are related to the methods for one change only developed, and treated, in previous sections. Thus the estimators $\widehat{m}_{LS}$ and $\widehat{m}_{LS}(G)$ introduced in (II.7) and (II.8) can be used to estimate multiple changes, too.

Direct calculation gives

$$\mathsf{E}S_k = \sum_{j=1}^{v} \Big(\lfloor n\gamma_j \rfloor - \lfloor n\gamma_{j-1} \rfloor\Big)(\mu_j - \overline{\mu}) + \Big(k - \lfloor n\gamma_v \rfloor\Big)(\mu_v - \overline{\mu}),$$
$$\lfloor n\gamma_{v-1} \rfloor < k \le \lfloor n\gamma_v \rfloor, \; v = 1, \ldots, q+1,$$

where

$$\overline{\mu} = \frac{1}{n} \sum_{j=1}^{q+1} \Big(\lfloor n\gamma_j \rfloor - \lfloor n\gamma_{j-1} \rfloor\Big)\mu_j$$

and

$$\mathsf{E}\Big(S_{k+G} - 2S_k + S_{k-G}\Big) = \tag{II.22}$$

$$= \begin{cases} 0, & G < |k - \lfloor n\gamma_j \rfloor|, \\ (\mu_j - \mu_{j-1})(\lfloor n\gamma_j \rfloor - k + G), & \lfloor n\gamma_j \rfloor - G < k \le \lfloor n\gamma_j \rfloor, \\ (\mu_j - \mu_{j-1})(k - \lfloor n\gamma_j \rfloor + G), & \lfloor n\gamma_j \rfloor < k \le \lfloor n\gamma_j \rfloor + G, \end{cases}$$

for $j = 1, \ldots, q$. The extremes in both expectations can be reached only for $k = \lfloor n\gamma_j \rfloor$, $j = 1, \ldots, q$. This leads to the following estimation procedures.

At first, find

$$\widetilde{m}^{(1)} = \arg\max \left\{ \sqrt{\frac{n}{k(n-k)}} \, |S_k|; \; k \in \{1, \ldots, n-1\} \right\}, \tag{II.23}$$

where S_k is given by (II.6). At second divide observations into two groups $Y_1, \ldots, Y_{\widetilde{m}^{(1)}}$ and $Y_{\widetilde{m}^{(1)}+1}, \ldots, Y_n$ and find the estimator in each group using (II.7). The whole procedure is repeated for each subset until "a (constant) mean is obtained", i.e. until there is a statistically significant difference in the means of groups. We apply the test based on

$$\max_{k \in D} \left\{ \sqrt{\frac{n}{k(n-k)}} \, \frac{1}{\widehat{\sigma}_n} \left| \sum_{i \in D} \Big(Y_i - \frac{1}{\#D} \sum_{j \in D} Y_j\Big) \right| \right\},$$

where D denotes the indexes corresponding to the particular group, $\#D$ its cardinality and $\widehat{\sigma}_n^2$ is a suitable estimator of σ^2. Critical values can be determined from the approximation (I.20) and the level α_n has to be chosen such that, as $n \to \infty$, $\alpha_n \to 0$. Then, this procedure consistently estimates all change points and also the number of changes. This type of procedure was developed by Vostrikova (1981).

Motivated by (II.22) and by the behavior of $\widehat{m}_{LS}(G)$, we propose the following estimation procedure. We find all pairs of indices v_j, w_j, $j = 1, \ldots, \widehat{q}$, such that $w_j - v_j \ge G/2$, $j = 1, \ldots, \widehat{q}$ and such that for $k = v_j, \ldots, w_j$

$$\begin{aligned}
\left|S_{k+G} - 2S_k + S_{k-G}\right| &\le D(n; G, \alpha_n),\\
\left|S_{v_j-1+G} - 2S_{v_j-1} + S_{v_j-1-G}\right| &\ge D(n; G, \alpha_n),\\
\left|S_{w_j+1+G} - 2S_{w_j+1} + S_{w_j+1-G}\right| &\ge D(n; G, \alpha_n),
\end{aligned}$$

where

$$D(n; G, \alpha_n) = \sigma_n \sqrt{2G}\ \frac{2\log \frac{n}{G} + \frac{1}{2}\log\log\frac{n}{G} - \frac{1}{2}\log\frac{4\pi}{9} - \log\log\frac{1}{1-\alpha_n}}{\sqrt{2\log(n/G)}}$$

and $\alpha_n \to 0$ as $n \to \infty$. This means that we find all segments where a certain critical value is exceeded. This critical value corresponds to test whether a change occurred or not based on

$$\max_{G<k<n_G}\left\{\left|S_{k+G} - 2S_k + S_{k-G}\right|\right\}$$

with asymptotic level α_n. The estimator of the number of change points is $\widehat{q}$.

Then $\widehat{q}$, estimate of the number of change points, and the index $k \in [v_j, w_j]$ for which the maximum over the set $\{v_j, \dots, w_j\}$ is reached, can serve as the estimator of one of the change points.

These two procedures, particularly the second one, are quite simple and can certainly serve as a simple diagnostic tool. There is a problem to find a reasonable estimator of the variance σ^2. For that, we need an upper bound, say q_o, of the number q of change points. Then we can estimate σ^2 as follows:

$$\begin{aligned}
\sigma_n^2 = &\frac{1}{n - q_o - 1} \min_{1\le k_1<k_2<\dots<k_{q_o}\le n}\left\{\sum_{i=1}^{k_1}\left(Y_i - \overline{Y}_{0,k_1}\right)^2 + \right.\\
&\left. + \sum_{i=k_1+1}^{k_2}\left(Y_i - \overline{Y}_{k_1,k_2}\right)^2 + \dots + \sum_{i=k_{q_o}+1}^{n}\left(Y_i - \overline{Y}_{k_{q_o},n}\right)^2\right\},
\end{aligned}$$

where

$$\overline{Y}_{k,s} = \frac{1}{s-k}\sum_{i=k+1}^{s} Y_i.$$

There exist more sophisticated procedures for detection and identification of multiple changes, however, they need stronger assumptions, for details see, e.g., Yao (1988).

Particular attention has been paid to the *epidemic alternatives* corresponding to the model (II.20) with $q = 2$ and $\mu_1 = \mu_3 \ne \mu_2$. For more information see, among others, Bhattacharya and Brockwell (1976), Yao (1993), Hušková (1995), Antoch and Hušková (1996).

8.5. Change in location or/and scale

The present section deals with changes in location or/and scale in the location model:

$$Y_i = \begin{cases} \mu + \sigma e_i, & i = 1, \dots, \mathsf{m}, \\ \mu + \delta + (\sigma + h) e_i, & i = \mathsf{m} + 1, \dots, n, \end{cases} \tag{II.24}$$

where m, μ, $(\delta, h) \neq (0,0)$ and $\sigma^2 > 0$ are parameters, and $e_1, \dots, e_n$ are iid random variables with symmetric distribution around zero, unit variance and $\mathsf{E}\,|e_i|^{4+\Delta} < \infty$ with some $\Delta > 0$. The change point m fulfills (II.2).

If the distribution function F of e_i is known, we can try to derive the maximum likelihood estimators (MLE). For example, if $e_1, \dots, e_n$ are iid with $N(0,1)$ distribution, direct calculation gives the MLE $\widehat{\mathsf{m}}_{MLE}$ of the change point m in the form

$$\widehat{\mathsf{m}}_{MLE} = \arg\max \left\{ \frac{\widetilde{\sigma}_n^n}{\widetilde{\sigma}_k^k (\widetilde{\sigma}_k^o)^{n-k}};\ k \in \{1, \dots, n\} \right\},$$

where for $k = 1, \dots, n-1$,

$$\widetilde{\sigma}_k^2 = \frac{1}{k} \sum_{i=1}^{k} (Y_i - \overline{Y}_k)^2 \quad \text{and} \quad \widetilde{\sigma}_k^{o2} = \frac{1}{n-k} \sum_{i=k+1}^{n} (Y_i - \overline{Y}_k^o)^2,$$

i.e. $\widehat{\sigma}_k^2$ and $\widehat{\sigma}_k^{o2}$ are estimators of σ^2 based on $Y_1, \dots, Y_k$ and $Y_{k+1}, \dots, Y_n$, respectively.

The distribution of $\widehat{\mathsf{m}}_{MLE}$ can be approximated using, $x \in \mathcal{R}^1$,

$$P\left(\frac{\left(\frac{\delta^2}{\sigma^2} + \frac{1}{2}\left(\frac{h^2}{\sigma^2}\right)\right)^2}{\frac{\delta^2}{\sigma^2} + \frac{1}{2}\left(\frac{h^2}{\sigma^2}\right)^2 var\, e_1^2} \left(\widehat{\mathsf{m}}_{MLE} - \mathsf{m}\right) \le x \right) \approx P(V \le x), \tag{II.25}$$

where the random variable V is defined by (II.10).

Despite the fact, that the estimator $\widehat{\mathsf{m}}_{MLE}$ was developed for observations with normal distribution, it can be used for other underlying distributions, too. The same holds for the approximation (II.25). Moreover, the approximation (II.25) can be used if we replace δ^2, σ^2 and $var\, e_1^2$ by suitable estimators.

Notice, that the estimator $\widehat{\mathsf{m}}_{LS}$ can be used also when we a priori know that the change occurs only in the location (i.e. $h = 0$) or only in the scale σ (i.e. $\delta = 0$); in the approximation, we replace h and δ, respectively, by zero.

Next, we propose an estimator of m that works for a quite general situation described in the model (II.24). The estimator is based on the pairs of the partial sums S_k and $S_{k,sc}$, $k = 1, \dots, n$, where S_k is defined by (II.6) and

$$S_{k,sc} = \sum_{i=1}^{k} \left((Y_i - \overline{Y}_n)^2 - \frac{1}{n} \sum_{j=1}^{n} (Y_j - \overline{Y}_n)^2 \right), \quad k = 1, \dots, n. \tag{II.26}$$

The estimator itself has the form

$$\widetilde{\mathsf{m}}_{SC} = \arg\max \left\{ \frac{n}{k(n-k)} \Big(\frac{S_k^2}{\widehat{\sigma}_k^2} + \frac{S_{k,sc}^2}{\widehat{\sigma}_{k,sc}^2} \Big);\ k \in \{1, \dots, n-1\} \right\}, \qquad \text{(II.27)}$$

where $\widehat{\sigma}_k^2$ and $\widehat{\sigma}_{k,sc}^2$ are, respectively, estimators of σ and $var\,(Y_1 - \mu)^2$. We can use, e.g., the estimators

$$\widehat{\sigma}_k^2 = \frac{1}{n-2} \left\{ \sum_{i=1}^{k} (Y_i - \overline{Y}_k)^2 + \sum_{i=k+1}^{n} (Y_i - \overline{Y}_k^o)^2 \right\}$$

and

$$\widehat{\sigma}_{k,sc}^2 = \frac{1}{n-2} \Bigg(\sum_{i=1}^{k} \Big((Y_i - \overline{Y}_k)^2 - \frac{1}{k} \sum_{j=1}^{k} (Y_j - \overline{Y}_k)^2 \Big) + $$
$$+ \sum_{i=k+1}^{n} \Big((Y_i - \overline{Y}_k^o)^2 - \frac{1}{n-k} \sum_{j=k+1}^{k} (Y_j - \overline{Y}_k^o)^2 \Big) \Bigg).$$

The approximation to the distribution is of a similar type as above, i.e.:

$$P\left(\frac{\frac{\delta^2}{\sigma^2} + \frac{1}{2}\left(\frac{h^2}{\sigma^2}\right)^2}{\left(\frac{\delta^2}{\sigma^2} + \frac{1}{2}\left(\frac{h^2}{\sigma^2}\right)^2\right)^2 var\, e_1^2} \left(\widetilde{\mathsf{m}}_{SC} - \mathsf{m} \right) \le x \right) \approx P(V \le x), \quad x \in \mathcal{R}^1, \qquad \text{(II.28)}$$

where the random variable V is defined by (II.10).

Finally, we present an L_1-estimator that is asymptotically equivalent with (II.27). This estimator has the form

$$\arg\max \left\{ \frac{n}{k(n-k)} \sum_{i=1}^{k} \Big(\mathrm{sign}\,(Y_i - \widetilde{\mu}_n) \Big)^2 + \sum_{i=1}^{k} \Big(\mathrm{sign}\,\big(|Y_i - \widetilde{\mu}_n| - \widetilde{\nu}_n \big) \Big)^2 ; \right.$$
$$\left. k \in \{1, .., n-1\} \right\}, \qquad \text{(II.29)}$$

where $\widetilde{\mu}_n$ is the sample median of $Y_1, \dots, Y_n$ and $\widetilde{\nu}_n$ is the sample median of $|Y_1 - \widetilde{\mu}_n|, \dots, |Y_n - \widetilde{\nu}_n|$. This estimator is very simple to calculate. The approximation to the distribution is the same as in (II.28).

8.6. Gradual changes

We assume the observations $Y_1, \dots, Y_n$ follow the model:

$$Y_i = \begin{cases} \mu + \sigma e_i, & i = 1, \dots, \mathsf{m}, \\ \mu + \delta \frac{i-\mathsf{m}}{n} + \sigma e_i, & i = \mathsf{m}+1, \dots, n, \end{cases} \qquad \text{(II.30)}$$

where μ, $\sigma^2 > 0$, $\delta \neq 0$ and $\mathsf{m}\,(< n)$ are unknown parameters, $e_1, \dots, e_n$ are iid random errors with zero mean, nonzero variance σ^2 and $\mathsf{E}\,|e_i|^{2+\Delta} < \infty$ with some $\Delta > 0$. Again, m is the change point which is assumed to fulfill

(II.2). This type of change is called *gradual.* The least squares method leads to the following estimator of m, i.e.:

$$\widetilde{\mathsf{m}}_{gr} = \arg\max \left\{ \frac{\left(\sum_{i=k+1}^{n} (Y_i - \overline{Y}_n) \frac{i-k}{n} \right)^2}{\sum_{i=k+1}^{n} \frac{(i-k)^2}{n^2} - \frac{1}{n} \left(\sum_{i=k+1}^{n} \frac{i-k}{n} \right)^2}; \; k \in \{1, \dots, n-1\} \right\},$$

and the parameters μ and δ are estimated by

$$\widehat{\mu}_n = \overline{Y}_n + \frac{\widehat{\delta}_n}{n} \sum_{i=\widehat{\mathsf{m}}_{gr}+1}^{n} \frac{i - \widehat{\mathsf{m}}_{gr}}{n}$$

and

$$\widehat{\delta}_n = \frac{\sum_{i=\widehat{\mathsf{m}}_{gr}+1}^{n} (Y_i - \overline{Y}_n) \frac{i - \widehat{\mathsf{m}}_{gr} + 1}{n}}{\sum_{i=\widehat{\mathsf{m}}_{gr}+1}^{n} \left(\frac{i - \widehat{\mathsf{m}}_{gr}}{n} \right)^2 - \left(\sum_{i=\widehat{\mathsf{m}}_{gr}+1}^{n} \frac{i - \widehat{\mathsf{m}}_{gr}}{n} \right)^2}.$$

The distribution of this estimator is approximately normal, more exactly,

$$P\left(\sqrt{\frac{\theta(1-\theta)}{1+3\theta}} \frac{\delta}{\sqrt{n}\sigma} \left(\widetilde{\mathsf{m}}_{gr} - \mathsf{m} \right) \le x \right) \approx \Phi(x), \quad x \in \mathcal{R}^1.$$

Notice that the approximation to the distribution of the estimator $\widetilde{\mathsf{m}}_{gr}$, in this particular case, completely differs from the approximations we had so far. This is due to the fact that we have a gradual (continuous) change while all other considered changes are abrupt (jump).

9. Change in regression

9.1. Change point estimators in regression models

We assume that the observations $Y_1, \dots, Y_n$ follow the regression model:

$$Y_i = \begin{cases} \boldsymbol{x}_i'\boldsymbol{\beta} + \sigma e_i, & i = 1, \dots, \mathsf{m}, \\ \boldsymbol{x}_i'\boldsymbol{\beta} + \boldsymbol{x}_i'\boldsymbol{\delta} + \sigma e_i, & i = \mathsf{m}+1, \dots, n, \end{cases}$$

where $\boldsymbol{\beta}$, $\sigma^2 > 0$, $\boldsymbol{\delta} \neq \mathbf{0}$, $\mathsf{m}(< n)$, are parameters, $\boldsymbol{\beta}$ and $\boldsymbol{\delta}$ are p-dimensional vectors and $e_1, \dots, e_n$ are iid random variables with zero mean, nonzero variance σ^2, and with $\mathsf{E}\,|e_i|^{4+\Delta} < \infty$ with some $\Delta > 0$. The change point m fulfills (II.2).

Similarly as in the testing problem, there are many variants of the assumptions on the design points $\boldsymbol{x}_i$, $i = 1, \dots, n$. Generally, in the usual linear regression setup the assumptions are stronger than in the estimation problem. We give here only three possible sets of assumptions:

- $\boldsymbol{x}_i = g(i/n) = \big(g_1(i/n), \ldots, g_p(i/n)\big)'$, $i = 1, \ldots, n$, where $g_1, \ldots, g_p$ are continuous functions on $[0, 1]$ such that $\int_a^b g(\boldsymbol{x})\, dx$ is a positive definite matrix for each $0 \le a < b \le 1$.
- There are fixed p linearly independent p-dimensional vectors $\boldsymbol{x}_1^0, \ldots,$ $\boldsymbol{x}_p^0$ and the design points are chosen according to some rule from this p vectors.
- The design points can be a realization of AR or ARMA sequences fulfilling certain moment assumptions.

The estimator is defined as follows:

$$\widehat{\mathsf{m}}_{regr} = \arg\min \left\{ \sum_{i=1}^{k} (Y_i - \boldsymbol{x}'\widehat{\boldsymbol{\beta}}_k)^2 + \sum_{i=k+1}^{n} (Y_i - \boldsymbol{x}'\widehat{\boldsymbol{\beta}}_k^o)^2;\ k = 1, \ldots, n \right\},$$

where $\widehat{\boldsymbol{\beta}}_k$ and $\widehat{\boldsymbol{\beta}}_k^o$ are least squares estimators of $\boldsymbol{\beta}$ based on $Y_1, \ldots, Y_k$ and $Y_{k+1}, \ldots, Y_n$, respectively. Equivalently, the estimators can be also defined as

$$\widehat{\mathsf{m}}_{regr} = \arg\max \left\{ \sum_{i=1}^{k} \boldsymbol{x}_i'(Y_i - \boldsymbol{x}_i'\widehat{\boldsymbol{\beta}}_n) \cdot \boldsymbol{C}_k^{-1}\boldsymbol{C}_n(\boldsymbol{C}_n - \boldsymbol{C}_k) \cdot \right.$$

$$\left. \cdot \sum_{i=1}^{k} \boldsymbol{x}_i'(Y_i - \boldsymbol{x}_i'\widehat{\boldsymbol{\beta}}_n);\ k \in \{1, \ldots, n\} \right\},$$

where $\boldsymbol{C}_n = \sum_{i=1}^{n} \boldsymbol{x}_i \boldsymbol{x}_i'$. Here $\sum_{i=1}^{k} \boldsymbol{x}_i'(Y_i - \boldsymbol{x}_i^T \widehat{\boldsymbol{\beta}}_n), i = 1, \ldots, n$, are vectors of sums of partial residuals.

The approximation to the distribution of the estimators

$$P\left(\frac{1}{\sigma^2} \frac{1}{2k_n} \sum_{i=\widehat{\mathsf{m}}-k_n+1}^{\widehat{\mathsf{m}}+k_n} (\boldsymbol{x}_i'\boldsymbol{\delta})^2 \left(\widehat{\mathsf{m}}_{regr} - \mathsf{m}\right) \le x \right) \approx P(V \le x), \quad x \in \mathcal{R}^1,$$

with $k_n \to \infty$, $k_n/n \to 0$ and the random variable V is defined by (II.10).

The readers interested in this problem should consult some more advanced text, e.g., Csörgő and Horváth (1997), Horváth and Kokozska (1997) and Bai (1994).

9.2. Change in regression parameters and/or scale

We assume that the observations $Y_1, \ldots, Y_n$ follow the regression model:

$$Y_i = \begin{cases} \boldsymbol{x}_i'\boldsymbol{\beta} + \sigma e_i, & i = 1, \ldots, \mathsf{m}, \\ \boldsymbol{x}_i'\boldsymbol{\beta} + \boldsymbol{x}_i'\boldsymbol{\delta} + (\sigma + h)e_i, & i = \mathsf{m} + 1, \ldots, n, \end{cases}$$

where $\boldsymbol{\beta}$, $\sigma^2 > 0$, $(\boldsymbol{\delta}, h) \ne \boldsymbol{0}$, $\mathsf{m}(< n)$, are parameters, and $e_1, \ldots, e_n$ are iid random variables with unit variance, and with $\mathsf{E}\,|e_i|^{4+\Delta} < \infty$ with some $\Delta > 0$, the distribution of e_i is supposed to be symmetric around 0. The assumptions on the design points are the same as in the previous section and the change point m fulfills (II.2).

The estimator is defined as

$$\widetilde{\mathsf{m}}_{regr} = \arg\max \left\{ \frac{\widetilde{\sigma}_n^n}{\widetilde{\sigma}_k^k \widetilde{\sigma}_k^{o(n-k)}};\ k \in \{1, \ldots, n-1\} \right\},$$

where

$$\widetilde{\sigma}_k^2 = \frac{1}{k} \sum_{i=1}^{k} \left(Y_i - \boldsymbol{x}'\widehat{\boldsymbol{\beta}}_k\right)^2, \quad k = 1, \ldots, n-1,$$

$$\widetilde{\sigma}_k^{o2} = \frac{1}{n-k} \sum_{i=k+1}^{n} \left(Y_i - \boldsymbol{x}'\widehat{\boldsymbol{\beta}}_k^0\right)^2, \quad k = 1, \ldots, n-1,$$

and $\widehat{\boldsymbol{\beta}}_k$ and $\widehat{\boldsymbol{\beta}}_k^o$ are the least squares estimators of $\boldsymbol{\beta}$ based on $Y_1, \ldots, Y_k$ and $Y_{k+1}, \ldots, Y_n$. The approximation to the distribution of $\widetilde{\mathsf{m}}_{regr}$ is $\forall x \in \mathcal{R}^1$

$$P\left(\left(\frac{1}{\sigma^2} \frac{1}{2k_n} \sum_{i=\widehat{\mathsf{m}}-k_n+1}^{\widehat{\mathsf{m}}+k_n} (\boldsymbol{x}_i'\boldsymbol{\delta})^2 + \frac{1}{2}\left(\frac{h}{\sigma^2}\right)^2 \right)^2 \left(\widetilde{\mathsf{m}}_{regr} - \mathsf{m}\right) \le x \right) \approx P(V \le x),$$

where the random variable V is defined by (II.10).

10. Confidence intervals

We focus on the confidence intervals based on the estimator $\widehat{\mathsf{m}}_{LS}$, defined by (II.7), of the change points in the location model studied in Section 8. However, we attempt to keep in mind generalization to other estimators and models as well.

Three types of confidence intervals will be developed, one based on the limit distribution of the (point) estimator(s) of m and two based on the bootstrap methods. All three methods are suitable for local changes while only the bootstrap constructions apply also to fixed changes.

Asymptotic approach

Using the approximation to the distribution of $\widehat{\mathsf{m}}_{LS}$ described in (II.9), we get the $100(1-\alpha)\%$ approximate confidence interval

$$\left(\widehat{\mathsf{m}}_{LS} - v_{1-\alpha/2}\frac{\widehat{\sigma}_n^2}{\widehat{\delta}_n^2},\ \widehat{\mathsf{m}}_{LS} + v_{1-\alpha/2}\frac{\widehat{\sigma}_n^2}{\widehat{\delta}_n^2} \right), \tag{II.31}$$

where $\widehat{\mathsf{m}}_{LS}$, $\widehat{\delta}_n$ and $\widehat{\sigma}_n^2$ are defined in (II.7), (II.5) and (II.12), respectively, and $v_{1-\alpha/2}$ is the quantile corresponding to the random variable V defined in (II.10). (Selected quantiles are in Table 18.)

Now, we turn to the bootstrap approximations to the confidence intervals based on the estimator $\widehat{\mathsf{m}}_{LS}$ of m.

Bootstrap sampling scheme I.

Take two independent samples $Y_1^*, \ldots, Y_{\widehat{\mathsf{m}}}^*$ and $Y_{\widehat{\mathsf{m}}+1}^*, \ldots, Y_n^*$ from the empirical cumulative distribution function of $Y_1, \ldots, Y_{\widehat{\mathsf{m}}}$ and $Y_{\widehat{\mathsf{m}}+1}, \ldots, Y_n$, respectively. Then the bootstrap estimator $\widehat{\mathsf{m}}_{LS}^*$ (corresponding to the estimator $\widehat{\mathsf{m}}_{LS}$) is defined as

$$\widehat{\mathsf{m}}^*_{LS} = \arg\max \left\{ \sqrt{\frac{n}{k(n-k)}} \left| \sum_{i=1}^{k} (Y_i^* - \overline{Y}_n^*) \right|;\ k \in \{1, \ldots, n-1\} \right\}. \tag{II.32}$$

This means, that the bootstrap estimator is calculated exactly as $\widehat{\mathsf{m}}$, however, Y_i, $i = 1, \ldots, n$ are replaced by their bootstrap counterparts Y_i^*, $i = 1, \ldots, n$.

We can also modify the procedure taking the maximum only over the set $\{k : |k - \widehat{\mathsf{m}}| \le D_n\}$, with $\{D_n\}$ fulfilling, as $n \to \infty$,

$$D_n \le \min(\mathsf{m}, n - \mathsf{m}) \quad \text{and} \quad D_n \delta_n^2 \longrightarrow \infty. \tag{II.33}$$

If the amount of the change $\delta_n \equiv \delta \neq 0$ is fixed, $\{D_n\}$ can be chosen to tend to infinity arbitrary slowly, while in the case of local changes $\{D_n\}$ has to tend to infinity faster than δ_n^{-2}.

The modified bootstrap estimator $\widehat{\widehat{\mathsf{m}}}{}^*_{LS}$ related to $\widehat{\mathsf{m}}_{LS}$ is defined as

$$\widehat{\widehat{\mathsf{m}}}{}^*_{LS} = \arg\max \left\{ \sqrt{\frac{n}{k(n-k)}} \left| \sum_{i=1}^{k} (Y_i^* - \overline{Y}_n^*) \right|;\ |k - \widehat{\mathsf{m}}_{LS}| \le D_n \right\}, \tag{II.34}$$

where $\{D_n\}$ fulfills (II.33).

The bootstrap estimators $\widehat{\mathsf{m}}^*_{LS}(G)$ and $\widehat{\widehat{\mathsf{m}}}{}^*_{LS}(G)$ related to $\widehat{\mathsf{m}}_{LS}(G)$ are defined accordingly.

Bootstrap sampling scheme II.

Define the estimated residuals

$$\widetilde{e}_i = \begin{cases} Y_i - \overline{Y}_{\widehat{\mathsf{m}}}, & i = 1, \ldots, \widehat{\mathsf{m}}, \\ Y_i - \overline{Y}^o_{\widehat{\mathsf{m}}}, & i = \widehat{\mathsf{m}} + 1, \ldots, n, \end{cases}$$

and the centered residuals

$$\widehat{e}_i = \widetilde{e}_i - \frac{1}{n} \sum_{j=1}^{n} \widetilde{e}_j, \quad i = 1, \ldots, n,$$

where $\overline{Y}_{\widehat{\mathsf{m}}}$ and $\overline{Y}^o_{\widehat{\mathsf{m}}}$ are defined by (I.2), respectively, with $\mathsf{m} = \widehat{\mathsf{m}}$.

Take $e_1^{**}, \ldots, e_n^{**}$ iid from the empirical cdf of $\widehat{e}_1, \ldots, \widehat{e}_n$ and consider the bootstrap observations

$$Y_i^{**} = \begin{cases} \widehat{Y}_{\widehat{\mathsf{m}}} + e_i^{**}, & i = 1, \ldots, \widehat{\mathsf{m}}, \\ \widehat{Y}^o_{\widehat{\mathsf{m}}} + e_i^{**}, & i = \widehat{\mathsf{m}} + 1, \ldots, n. \end{cases}$$

Then we proceed as in the bootstrap sampling scheme I, i.e. we apply (II.32) and (II.34) with Y_i^* replaced by Y_i^{**}.

Both these bootstrap schemes provide a reasonable approximation to the confidence intervals for the change point m, typically they provide a better approximation than the one based on the approximation (II.31).

Part III. Selected limit properties

11. Selected limit theorems for test statistics

The test statistics which appeared in the text are mostly functionals of partial sums of iid variables e_i. Their distributions are very complex. However, as we suppose that the number of observations n is large, their asymptotic distributions are of interest. The theory used for obtaining the limit distributions is based on the Donsker invariance principle and the theory of strong approximations. Therefore, the limit distributions do not depend on the assumption of the normality of e_i. Usually it is sufficient to suppose that the random errors e_i are iid satisfying $\mathsf{E}\, e_i = 0, \mathsf{E}\, e_i^2 = 1$ and $\mathsf{E}\,|e_i|^{2+\delta} < \infty$ for some $\delta > 0$.

The over-all maximum-type statistics go to infinity as $n \to \infty$ almost surely. The speed of the convergence of their critical values can be traced by approximating these statistics by the maximum of certain processes over an increasing interval, e.g.,

$$\left| \max_{1\le k\le n} \frac{|\sum_{i=1}^k e_i|}{\sqrt{k}} - \sup_{1/n\le t\le 1} \frac{|W(t)|}{\sqrt{t}} \right| = o_P\left(\frac{1}{\sqrt{2\log\log n}}\right),$$

$$\left| \max_{1\le k\le n-1} \left| \left(\tfrac{k}{n}\left(1-\tfrac{k}{n}\right)\right)^{-\eta} \frac{1}{\sqrt{n}} \sum_{i=1}^{k} \left(e_i - \bar{e}_n\right)\right| - \sup_{1/n\le t\le 1-1/n} \frac{|B(t)|}{(t(1-t))^{\eta}} \right| =$$

$$= o_P\left(\frac{1}{\sqrt{2\log\log n}}\right)$$

$\forall \eta \in [0, 1/2]$, and

$$\left| \max_{1\le k\le n} \frac{\left|\frac{1}{\sqrt{n}}\sum_{i=1}^k \frac{k-i}{n} e_i\right|}{\sqrt{\frac{1}{n}\sum_{i=1}^k \left(\frac{k-i}{n}\right)^2}} - \sup_{1/n\le t\le 1} \frac{\left|\int_0^t (t-x)\, dW(x)\right|}{\sqrt{t^3/3}} \right| = o_P\left(\frac{1}{\sqrt{2\log\log n}}\right)$$

Similar approximations may be derived for a multidimensional case, for details see the book Csörgő and Horváth (1997).

Next we present the limit behavior of some of over-all maximum type statistics. The type of limits are known as extreme value theorems. Under the assumptions formulated at the beginning of the section and with $a_n = \sqrt{2\log\log n}$ we have $\forall y \in \mathcal{R}^1$, as $n \to \infty$,

$$P\left(a_n \max_{1\le k<n} \frac{1}{\sqrt{k}} \sum_{i=1}^{k} e_i \le y + 2\log\log n + \frac{\log\log\log n}{2} - \frac{\log \pi}{2}\right) \to \exp\left\{\frac{e^{-y}}{2}\right\},$$

$$P\left(a_n \max_{1\le k<n} \frac{1}{\sqrt{k}} \left|\sum_{i=1}^{k} e_i\right| \le y + 2\log\log n + \frac{\log\log\log n}{2} - \frac{\log \pi}{2}\right) \to \exp\left\{e^{-y}\right\},$$

$$P\left(a_n \max_{1\le k<n}\sqrt{\frac{n}{k(n-k)}}\left|\sum_{i=1}^{k}(e_i-\bar{e}_n)\right|\le\right.$$
$$\left.\le y+2\log\log n+\frac{1}{2}\log\log\log n-\frac{1}{2}\log\pi\right)\to\exp\left\{2e^{-y}\right\},$$

$$P\left(a_n \max_{1\le k<n}\frac{\left|\sum_{i=1}^{k}(k-i)e_i\right|}{\sqrt{\sum_{i=1}^{k}(k-i)^2}}\le y+2\log\log n-\frac{1}{2}\log(4\pi/3)\right)\to\exp\{2e^{-y}\},$$

$$P\left(a_n \max_{1\le k<n}\frac{\left|\sum_{i=1}^{k}(k-i)(e_i-\bar{e}_n)\right|}{\sqrt{\sum_{i=1}^{k}(k-i)^2-(\sum_{i=1}^{k}(k-i))^2/n}}\le\right.$$
$$\left.\le y+2\log\log n-\frac{1}{2}\log(4\pi/3)\right)\to\exp\left\{2e^{-y}\right\}.$$

Some of the above results can be extended to the multivariate case. Particularly, if $\boldsymbol{Y}_1,\dots,\boldsymbol{Y}_n$ are iid p-dimensional random vectors with zero mean, positive variance matrix $\boldsymbol{\Sigma}$ and $\mathsf{E}|Y_{ij}|^{2+\Delta}<\infty$ with some $\Delta>0$, $i=1,\dots,n$, $j=1,\dots,p$, then, for any $y\in\mathcal{R}^1$,

$$P\left(\sqrt{2\log\log n}\max_{1\le k<n}\left\{\sqrt{\frac{n}{k(n-k)}}\sqrt{\boldsymbol{S}_k'\boldsymbol{\Sigma}^{-1}\boldsymbol{S}_k'}\right\}\le\right.$$
$$\left.\le y+2\log\log n+\frac{p}{2}\log\log\log n-\frac{1}{2}\log\Gamma(p)\right)\to\exp\left\{-2\exp\{-y\}\right\}$$

as $n\to\infty$, where

$$\boldsymbol{S}_k=\left(S_{1,k},\dots,S_{p,k}\right)',\quad \boldsymbol{S}_{j,k}=\sum_{i=1}^{k}\left(Y_{ij}-\overline{Y}_j\right),\quad k=1,\dots,n,\ j=1,\dots,p.$$

For maximum-type statistic of moving sums under the assumptions (denoting $G=G_n$)

$$\frac{n^{2/(2+\Delta)}\log n}{G}\to 0\quad\text{and}\quad G/n\to 0\quad\text{as } n\to\infty,$$

we have for any $y\in\mathcal{R}^1$, as $n\to\infty$,

$$P\left(\sqrt{2\log\frac{n}{G}}\max_{G<k<n-G}\left\{\frac{1}{\sqrt{G}}\frac{1}{\sigma}\left|S_{k+G}-2S_k+S_{k-G}\right|\right\}\le\right.$$
$$\left.\le y+2\log\frac{n}{G}+\frac{1}{2}\log\log\frac{n}{G}-\frac{1}{2}\log\frac{4\pi}{9}\right)\to\exp\left\{-2\exp\{-y\}\right\}.$$

For a trimmed maximum-type statistic as well as for a sum-type statistic we obtain the limit distributions by approximating them, respectively, by the maximum of some processes over a fixed interval, e.g.,

$$\sum_{k=1}^{n} \frac{1}{n} \left(\frac{1}{\sqrt{n}} \sum_{i=1}^{k} e_i \right)^2 \xrightarrow{\mathcal{D}} \int_0^1 W^2(t)\, dt,$$

$$\max_{\lfloor \beta n \rfloor \le k \le n} \frac{|\sum_{i=1}^{k} e_i|}{\sqrt{k}} \xrightarrow{\mathcal{D}} \max_{\beta \le t \le 1} \frac{|W(t)|}{\sqrt{t}}, \forall \beta \in (0,1),$$

$$\max_{\lfloor \beta n \rfloor \le k \le n} \frac{\left| \frac{1}{\sqrt{n}} \sum_{i=1}^{k} \frac{k-i}{n} e_i \right|}{\sqrt{\frac{1}{n} \sum_{i=1}^{k} \left(\frac{k-i}{n}\right)^2}} \xrightarrow{\mathcal{D}} \max_{\beta \le t \le 1} \frac{|\int_0^t (t-x)\, dW(x)|}{\sqrt{t^3/3}}, \forall \beta \in (0,1),$$

$$\max_{\lfloor \beta n \rfloor \le k \le \lfloor (1-\beta) n \rfloor} \left| \sqrt{\frac{n}{k(n-k)}} \sum_{i=1}^{k} (e_i - \bar{e}_n) \right| \xrightarrow{\mathcal{D}}$$

$$\max_{\beta \le t \le (1-\beta)} \frac{|B(t)|}{\sqrt{t(1-t)}}, \forall \beta \in (0, 1/2),$$

and

$$\max_{1 \le k \le n-1} \left| \left(\tfrac{k}{n} (1 - \tfrac{k}{n}) \right)^{-\eta} \frac{1}{\sqrt{n}} \sum_{i=1}^{k} \left(e_i - \bar{e}_n \right) \right| \xrightarrow{\mathcal{D}}$$

$$\sup_{1/n \le t \le 1-1/n} \frac{|B(t)|}{\left(t(1-t)\right)^{\eta}}, \forall \eta \in [0, 1/2).$$

Here $\xrightarrow{\mathcal{D}}$ denotes the convergence in distibution, $\{W(t),\ t \in (0,1)$ and $\{B(t),\ t \in (0,1)$ are Wiener process and Brownian bridge, respectively. The random variables on the rhs are functionals of Wiener process and Brownian bridge. The exact form of their distribution functions are usually explicitely unknown. Some approximations were developed, selected details are described bellow. For more information see Billingsley (1968) and Csörgő and Horváth (1993).

12. Properties of maximum of one-dimensional Gaussian processes

The limit distributions of the maximum-type statistics are given by distributions of a maximum of a Gaussian process either over a fixed or an increasing interval. As we wish to approximate the upper quantiles of our test statistics, we are interested in the high level exceedence probabilities of the considered limit processes. More specifically, approximate critical values can be found applying theorems for the exceedence probabilities of stationary Gaussian processes, see Theorem 12.2.9 and 12.3.5 of Leadbetter et al. (1983).

Let $\{\xi(t), t \geq 0\}$ be a zero-mean standardised stationary Gaussian process with the covariance function $\rho_\xi(\tau)$ that satisfies

$$\lim_{\tau \to \infty} \rho_\xi(\tau) \log \tau = 0 \tag{III.1}$$

and has the following expansion at zero, i.e.,

$$\rho_\xi(\tau) = 1 - C|\tau|^\alpha + o(|\tau|^\alpha) \quad \text{as} \quad \tau \to 0,$$

where $\alpha = 1$ or $\alpha = 2$. Then it holds:

1) for T being fixed and the exceedence level $u \to \infty$

$$\lim_{u \to \infty} \frac{P(\max_{0 \leq t \leq T} \xi(t) > u)}{u^{2/\alpha - 1} \phi(u)} = TC^{1/\alpha} H_\alpha; \tag{III.2}$$

2) for T tending to infinity and

$$x_T = \sqrt{2 \log T} + \frac{1}{\sqrt{2 \log T}} \left(x + \frac{2 - \alpha}{2\alpha} \log \log T + \log \left(C^{1/\alpha} H_\alpha \frac{2^{(2-\alpha)/2\alpha}}{\sqrt{2\pi}} \right) \right),$$

$$\lim_{T \to \infty} P\left(\max_{0 \leq t \leq T} \xi(t) > x_T \right) = 1 - \exp\left\{ -e^{-x} \right\},$$

$x \in \mathcal{R}^1, H_1 = 1, H_2 = 1/\sqrt{\pi}$.

If we are interested in the maximum of absolute values of studied processes then, under the same conditions as above, the following properties hold:

1) for T being fixed and the exceedence level $u \to \infty$

$$\lim_{u \to \infty} \frac{P(\max_{0 \leq t \leq T} |\xi(t)| > u)}{u^{2/\alpha - 1} \phi(u)} = 2TC^{1/\alpha} H_\alpha, \tag{III.3}$$

2) for T tending to infinity and x_T as above

$$\lim_{T \to \infty} P\left(\max_{0 \leq t \leq T} |\xi(t)| > x_T \right) = 1 - \exp\left\{ -2e^{-x} \right\}. \tag{III.4}$$

REMARK: For T fixed and u large, the following approximations seem to be more accurate than (III.2) and (III.3), i.e.

$$P\left(\max_{0 \leq t \leq T} \xi(t) > u \right) \approx (1 - \Phi(u)) + u^{2/\alpha - 1} \phi(u) TC^{1/\alpha} H_\alpha$$

and

$$P\left(\max_{0 \leq t \leq T} |\xi(t)| > u \right) \approx 2(1 - \Phi(u)) + 2u^{2/\alpha - 1} \phi(u) TC^{1/\alpha} H_\alpha.$$

EXAMPLE 1: Consider the process

$$\left\{ Y(t) = \frac{\int_0^t x^p \, dW(x)}{\sqrt{t^{2p+1}/(2p+1)}}, \; t \in (0, 1] \right\}, \quad p = 0, 1, \ldots,$$

and the process $\{U(t) = Y(e^{-t}), \; t \in [0, \infty)\}$. The covariance function of the process $\{Y(t), \; t \in (0, 1]\}$ is $R_Y(t, s) = (t/s)^{p+1}, 0 < t \leq s \leq 1$. It follows that $\{U(t), t \in [0, \infty)\}$ is a zero-mean standardized Gaussian process with the covariance function

$$R_U(\tau) = \mathsf{E}\, U(t)U(t+\tau) = \exp\left\{-\frac{2p+1}{2}\tau\right\}, \qquad \tau \geq 0,$$

that satisfies (III.1) and

$$R_u(\tau) = 1 - \frac{2p+1}{2}|\tau| + o(\tau) \quad \text{as} \quad \tau \to 0.$$

It follows that

$$P\left(\max_{\beta\leq t\leq 1} |Y(t)| > u\right) = P\left(\max_{0\leq t\leq -\log\beta} |U(t)| > u\right)$$
$$\approx 2(-\log\beta)\,\frac{2p+1}{2}\,u\phi(u).$$

for small β and large u. Particularly, for $p = 0$ we get

$$P\left(\max_{\beta\leq t\leq 1} \frac{|\,W(t)\,|}{\sqrt{t}} > u\right) \approx u\phi(u)\log\frac{1}{\beta}$$

A better approximation can be found in the subsection 13.

EXAMPLE 2: Consider the process

$$\left\{YC(t) = \frac{\int_0^t (t-x)^p\, dW(x)}{\sqrt{t^{2p+1}/(2p+1)}}, t \in (0,1]\right\}, \qquad p = 1, 2, \ldots,$$

with the covariance function

$$R_{YC}(t,s) = (2p+1)\sum_{i=0}^{p}\binom{p}{i}\frac{1}{p+i+1}\left(\frac{t}{s}\right)^{i+\frac{1}{2}}\left(1-\frac{t}{s}\right)^{p-i}, \quad 0 < t \leq s \leq 1.$$

The process $\{UC(t) = YC(e^{-t}),\ t \in [0,\infty)\}$ is a zero-mean standardized stationary Gaussian process with the covariance function

$$R_{UC}(t) = (2p+1)\sum_{i=0}^{p}\binom{p}{i}\frac{1}{p+i+1}e^{-(i+1/2)\tau}\left(1-e^{-\tau}\right)^{p-i}, \qquad \tau \geq 0$$

satisfying (III.1) and

$$R_{UC}(\tau) = 1 - \frac{2p+1}{8(2p-1)}\tau^2 + o(\tau^2) \quad \text{as} \quad \tau \to 0.$$

It follows that

$$P\left(\max_{\beta\leq t\leq 1} |YC(t)| > u\right) = P\left(\max_{0\leq t\leq -\log\beta} |UC(t)| > u\right) \approx$$
$$\approx 2(-\log\beta)\sqrt{\frac{2p+1}{8(2p-1)}}\,\frac{1}{\pi}\,\phi(u).$$

The former theory for stationary Gaussian processes were generalized for locally stationary Gaussian processes by Hüsler (1990), Hüsler (1993) and

Bräker (1993). Let $\{\xi(t), t \geq 0\}$ be a zero-mean locally stationary Gaussian process with the covariance function $\rho_\xi(t,s)$ that satisfies

$$\lim_{\tau\to\infty}\Big(\sup\,\{\rho_\xi(t,s), |\,t-s\,| > \tau\}\Big)\log\tau = 0$$

and has the following expansion at zero, i.e.

$$\rho_\xi(t,t+\tau) = 1 - C(t)|\,\tau\,|^\alpha + o(|\,\tau\,|^\alpha) \quad \text{as} \quad \tau \to 0,$$

where $\alpha = 1$ or $\alpha = 2$ and $0 < \inf C(t) \leq \sup C(t) < \infty$. Then

1) for T being fixed and the exceedence level $u \to \infty$

$$\lim_{u\to\infty}\frac{P(\max_{0\leq t\leq T}\xi(t) > u)}{u^{2/\alpha-1}\phi(u)} = H_\alpha\int_0^T\big(C(t)\big)^{1/\alpha}\,dt;$$

2) for $T \to \infty$ and

$$x_T = \sqrt{2\log T} + \frac{1}{\sqrt{2\log T}}\left(x + \frac{2-\alpha}{2\alpha}\log\log T + \log\left(C_T^\star H_\alpha\frac{2^{(2-\alpha)/2\alpha}}{\sqrt{2\pi}}\right)\right),$$

where $C_T^\star = T^{-1}\int_0^T\big(C(t)\big)^{1/\alpha}\,dt$,

$$\lim_{T\to\infty}P\left(\max_{0\leq t\leq T}\xi(t) > x_T\right) = 1 - \exp\{-e^{-x}\}, \quad x \in \mathcal{R}^1.$$

13. Exceedence level properties of Wiener process and Brownian bridge

The standardized Wiener process $\{W(t)/\sqrt{t}, \beta \leq t \leq 1\}$ and standardized Brownian bridge process

$$\left\{\frac{B(t)}{\sqrt{t(1-t)}} = \frac{W(t)-tW(1)}{\sqrt{t(1-t)}},\ \beta \leq t \leq 1-\beta\right\}$$

belong, for every $0 < \beta < 1/2$, to locally stationary random processes with $\alpha = 1$, so that the described exceedence theory for locally stationary random processes may be applied to get high level exceedence probabilities that we applied in our text. However, for these two, a better approximation is known, i.e., for small β and large x

$$P\left(\sup_{\beta\leq t\leq 1}\frac{|\,W(t)\,|}{\sqrt{t}} > x\right) \approx x\phi(x)\left((1-\frac{1}{x^2})\log\frac{1}{\beta} + \frac{4}{x^2}\right)$$

and

$$P\left(\sup_{\beta\leq t\leq 1}\frac{|\,B(t)\,|}{\sqrt{t(1-t)}} > x\right) \approx 2x\phi(x)\left((1-\frac{1}{x^2})\log\frac{1-\beta}{\beta} + \frac{2}{x^2}\right).$$

For the multivariate Wiener process $\boldsymbol{W}(t) = \{W_1(t), \ldots, W_d(t)\}$ with independent components and the multivariate Brownian bridge process $\boldsymbol{B}(t) = \{B_1(t), \ldots, B_d(t)\}$, the following approximations hold, i.e., for small β and large x

$$P\left(\sup_{\beta\le t\le 1}\frac{W_1^2(t)+\cdots+W_d^2(t)}{t}>x^2\right)\approx\frac{x^d e^{-x^2/2}}{2^{d/2}\Gamma(d/2)}\left(\left(1-\frac{d}{x^2}\right)\log\frac{1}{\beta}+\frac{4}{x^2}\right)$$

and

$$P\left(\sup_{\beta\le t\le (1-\beta)}\frac{B_1^2(t)+\cdots+B_d^2(t)}{t(1-t)}>x^2\right)\approx\frac{2x^d e^{-x^2/2}}{2^{d/2}\Gamma(d/2)}\left(\left(1-\frac{d}{x^2}\right)\log\frac{1-\beta}{\beta}+\frac{2}{x^2}\right)$$

14. Selected limit properties of change point estimators

In this section we formulate some important theoretical results on behavior of change point estimators described in Section 8.

If random variables $Y_1,\ldots,Y_n$ follow the model (II.1) with $\delta_n\equiv\delta\neq 0$ fixed, $\mathsf{m}=\lfloor n\gamma\rfloor, \gamma\in(0,1)$ and $e_1,\ldots,e_n$ be iid random variables with zero mean, nonzero variance σ^2 and $\mathsf{E}\,|e_i|^{2+\Delta}<\infty$ with some $\Delta>0$. We deal with $\widehat{\mathsf{m}}_{LS}$ and $\widehat{\mathsf{m}}_{LS}(G)$ defined in (II.7) and (II.8).

Then $\widehat{\mathsf{m}}_{LS}-\mathsf{m}$, as $n\to\infty$, has the limit distribution as

$$\arg\max\ \left\{\delta W_I(j)-\delta^2|j|/2;\ j=0,\pm1,\pm2,\ldots\right\},$$

where

$$W_I(j)=\begin{cases}0, & j=0,\\ \sum_{i=j}^{0}e_i, & j=-1,-2,\ldots,\\ -\sum_{i=1}^{j}e_i, & j=1,2,\ldots\end{cases}$$

This assertion says that the approximation depends on the distribution of the error terms e_i, which sometimes can create problems.

However, this is not the case with local changes i.e. when $\delta\equiv\delta_n\to 0$ as $n\to\infty$. Particularly, if

$$\delta_n\to 0\quad\text{and}\qquad |\delta_n|\sqrt{n}/\sqrt{\log\log n}\ \to\infty. \tag{III.5}$$

then, as $n\to\infty$, $\delta_n^2\sigma^{-2}\left(\widehat{\mathsf{m}}_{LS}-\mathsf{m}\right)$ has the same distribution as V, where V is a random variable defined by (II.10). The distribution function of V is known and it has the form (II.13).

If, moreover,

$$G/n\to 0\quad\text{and}\quad G^{-1}n^{2/(2+\Delta)}\log n\to 0\quad\text{as } n\to\infty, \tag{III.6}$$

then, as $n\to\infty$, $2\delta_n^2\left(3\sigma^2\right)^{-1}\left(\widehat{\mathsf{m}}_{LS}(G)-\mathsf{m}\right)$ has the same limit distribution as V, where V is a random variable defined by (II.10).

In case $\delta=0$ (no change) $\widehat{\mathsf{m}}_{LS}$ and $\widehat{\mathsf{m}}_{LS}(G)$ have different limit distributions. Particularly, we have $\forall\varepsilon\in(0,1/2]$

$$P\left(\widehat{\mathsf{m}}_{LS}<n\varepsilon\right)\approx 1/2,\qquad P\left(\widehat{\mathsf{m}}_{LS}>n(1-\varepsilon)\right)\approx 1/2$$

and $\widehat{\mathsf{m}}_{LS}(G)$ has approximately uniform distribution on $\left\{G+1,\ldots,n-G\right)$.

Part IV. Selected Matlab codes

We decided to offer the reader the possibility to use the methods described in details in this paper. Therefore, we prepared Matlab codes covering them. As an example, several of these macros including detailed description and links to the previous sections are included here. They cover the following models, i.e. detection of:

(1) change in mean with unknown starting value, maximum – type test statistic;
(2) change in mean with unknown starting value, sum – type test statistic;
(3) change in variance;
(4) change in mean and/or variance.

More codes are available from the authors.

Respective macros were written and tested in Matlab version 6.0.0.88 (R12 of September 22, 2000. However, we believe that they can be run under much older versions of the Matlab as well.

Each user should take into account that the critical values are not implemented into the codes. The reason is that, for each method, there exist several possibilities how to calculate them using either asymptotic results, Bonferroni inequality or simulations. Despite the fact that the authors personally prefer, for most situations, the simulated critical values, we leave the choice on the reader. For each of the methods, this paper reports all the possibilities.

As an example we used the following three data sets:

Nile data

This data set is probably one of the most frequently used in the change point setup, for details see, e.g., Cobb (1978) or Hinkley and Schechtman (1987). The data correspond to the annual flows (in billions of cubic meters) in the Nile river at Aswan (Egypt) during the years 1871 – 1970.

Simulated data I.

The data used as the testing example for the subroutine `m3_2.m` were generated as follows:

1) $X_1, \ldots, X_{20}$ follow standard normal distribution $N(0,1)$;
2) $X_{21}, \ldots, X_{50}$ follow a normal distribution $N(0,3^2)$;

Simulated data II.

The data used as the testing example for the subroutine `m3_3.m` were generated as follows:

1) $X_1, \ldots, X_{20}$ follow standard normal distribution $N(0,1)$;
2) $X_{21}, \ldots, X_{50}$ follow a normal distribution $N(4,3^2)$;

year	flow	year	flow	year	flow	year	flow
1871	1120	1896	1220	1921	768	1946	1040
1872	1160	1897	1030	1922	845	1947	860
1873	963	1898	1100	1923	864	1948	874
1874	1210	1899	774	1924	862	1949	848
1875	1160	1900	840	1925	698	1950	890
1876	1160	1901	874	1926	845	1951	744
1877	813	1902	694	1927	744	1952	749
1878	1230	1903	940	1928	796	1953	838
1879	1370	1904	833	1929	1040	1954	1050
1880	1140	1905	701	1930	759	1955	918
1881	995	1906	916	1931	781	1956	986
1882	935	1907	692	1932	865	1957	797
1883	1110	1908	1020	1933	845	1958	923
1884	994	1909	1050	1934	944	1959	975
1885	1020	1910	969	1935	984	1960	815
1886	960	1911	831	1936	897	1961	1020
1887	1180	1912	729	1937	822	1962	906
1888	799	1913	456	1938	1010	1963	901
1889	958	1914	824	1939	771	1964	1170
1890	1140	1915	702	1940	676	1965	912
1891	1100	1916	1120	1941	649	1966	746
1892	1210	1917	1100	1942	846	1967	919
1893	1150	1918	832	1943	812	1968	718
1894	1250	1919	764	1944	742	1969	714
1895	1260	1920	821	1945	801	1970	740

The annual flows (in billions of cubic meters) in river Nile at Aswan (Egypt) during the years 1871 – 1970. Data are used as the example in subroutines `m3_1_be.m.` and `m3_1_le.m.`

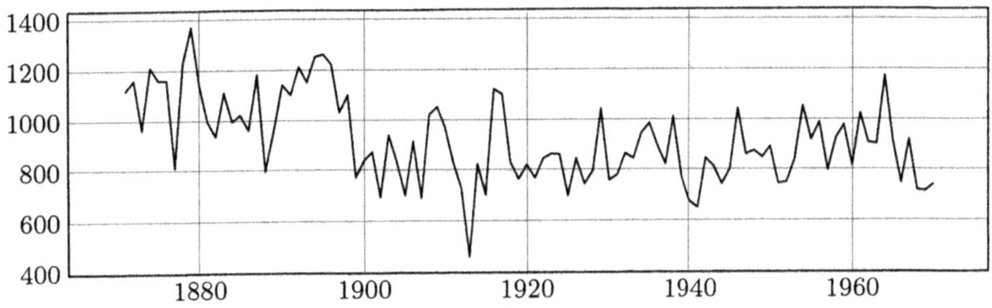

Nile data.

i	value	i	value	i	value	i	value	i	value
1	0.979	11	-0.438	21	-3.870	31	-1.116	41	0.962
2	-0.265	12	1.366	22	2.891	32	0.770	42	-0.578
3	-0.548	13	-0.872	23	1.957	33	2.432	43	4.673
4	-0.096	14	0.089	24	-1.232	34	-1.020	44	0.169
5	-1.380	15	0.247	25	-5.308	35	-3.955	45	-3.673
6	-0.728	16	0.407	26	1.318	36	-6.327	46	0.195
7	1.886	17	0.648	27	1.485	37	0.919	47	-3.284
8	-2.941	18	-0.164	28	2.053	38	-2.366	48	-3.208
9	0.980	19	0.811	29	-1.187	39	-1.511	49	0.649
10	-1.191	20	0.408	30	0.147	40	4.659	50	-2.317

Simulated data used as the example in subroutine `m3_2.m`.

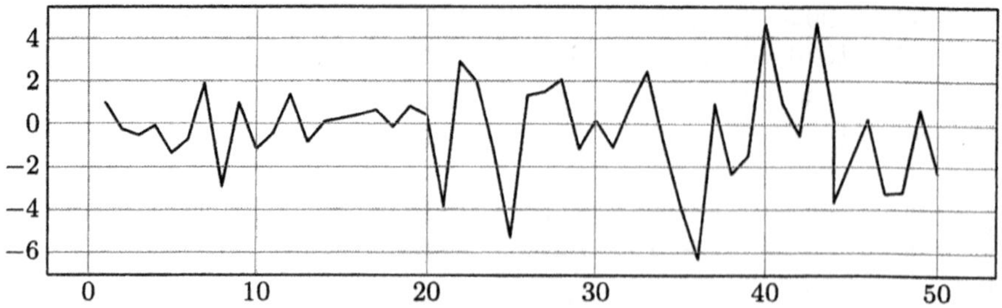

Simulated data I.

i	value	i	value	i	value	i	value	i	value
1	0.979	11	-0.438	21	0.129	31	2.883	41	4.962
2	-0.265	12	1.366	22	6.891	32	4.770	42	3.422
3	-0.548	13	-0.872	23	5.957	33	6.432	43	8.673
4	-0.096	14	0.089	24	2.767	34	2.979	44	4.169
5	-1.380	15	0.247	25	-1.308	35	0.044	45	0.326
6	-0.728	16	0.407	26	5.318	36	-2.327	46	4.195
7	1.886	17	0.648	27	5.485	37	4.919	47	0.715
8	-2.941	18	-0.164	28	6.053	38	1.633	48	0.791
9	0.980	19	0.811	29	2.812	39	2.488	49	4.649
10	-1.191	20	0.408	30	4.147	40	8.659	50	1.682

Simulated data used as the example in subroutine `m3_3.m`.

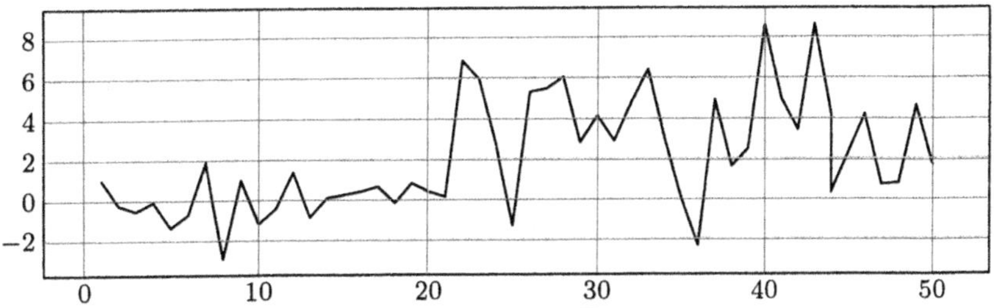

Simulated data II.

15. Change in mean with unknown starting value, max-type test statistics

```
function Results = m3_1_le(data, sigma2, trimming)

%%% SUBROUTINE m3_1.m
%%%
%%% PURPOSE To test the null hypothesis H against the alternative
%%%         A in the model:
```

$$
\begin{aligned}
H &: Y_i = \mu + e_i, && i = 1,\dots,n,\\
A &: \exists \mathrm{m} \in \{1,\dots,n-1\} \text{ such that} && \\
&\quad Y_i = \mu + e_i, && i = 1,\dots,\mathrm{m},\\
&\quad Y_i = \mu + \delta + e_i, && i = \mathrm{m}+1,\dots,n,\ \delta \neq 0.
\end{aligned} \tag{I.16}
$$

We suppose that the variables e_i are iid with the normal distribution $N(0,\sigma^2)$, σ^2 either known or unknown. If σ^2 is known, we apply statistics (I.17) or (I.18). On the other hand, if σ^2 is unknown, we apply statistics (I.27) or (I.28) with σ^2 estimated using (I.29).

```
%%%
%%% CALLING SEQUENCE
%%%
%%% function   Results = m3_1(data, sigma2, trimming)
%%%
%%% INPUT
%%%
%%% data              ...  data series to be tested for the presence
%%%                        of change point in model (I.16)
%%%
%%% sigma2            ...  variance of the error term
%%%
%%%    > 0            ...  user-supplied value for 'sigma^2' is used
%%%    <=0            ...  'sigma^2' is estimated using (I.29)
%%%
%%% trimming          ...  trimming proportion (default: beta = 0.05)
%%%
%%%   in (0,1/2)      ...  trimmed statistic (I.18) or (I.28) is used
```

```
%%%   otherwise     ... non-trimmed statistic (I.17) or (I.27) is used
%%%
%%% OUTPUT
%%%
%%% Results           :
%%%
%%%    chp_place    ... place of maximum of test statistics
%%%    ts_value     ... value of maximum of test statistics
%%%    sigma2_used ... value of 's_k^2 (sigma^2)' used for calculation
%%%    test_stats   ... values of test statistics (I.17) or (I.27),
%%%                     respectively (I.18) or (I.28)
%%%    sigma2_stats .. values of statistics (I.29)
%%%    trimming_used . value of 'trimming' used for calculation
%%%
%%% EXTERNAL PROCEDURES CALLED    none
%%%
%%% EXAMPLE
%%%
%%%    nile                              <- data are read in
%%%    Results = m3_1_le(nil, 0, 0.05) <- calculation
%%%
%%%    Results.chp_place     = 28       -> change point place
%%%    Results.ts_value      =  8.7143 -> value of test statistic (I.17)
%%%                                        or (I.18)
%%%    Results.sigma2_used   = 16293    -> 'sigma^2' used
%%%    Results.test_stats    = 1.1943, 1.8867, ...
%%%                                     -> values of test statistics
%%%    Results.sigma2_stats = 28503, 27904, ...
%%%                              -> values of estimators of 'sigma^2'
%%%    Results.trimming_used = 0.05
%%%
%%% CONTROL of PARAMETERS

if nargin    == 2, trimming   = 0;    end
if nargin    == 1, sigma2     = 0;    trimming = 0;  end
if trimming   < 0| trimming >= 0.5, trimming = 0;  end
if sigma2     < 0, sigma2     = 0;    end

%%% COMPUTATION

%%% CALCULATION OF NECESSARY CONSTANTS

data = data(:);
n    = length(data);
nm1  = n - 1;
nm2  = n - 2;
sk2  = zeros(nm1,1);
n1   = 1;
```

```
n2   = nm1;
if   trimming > 0
     n1 = round(n*trimming);
     n2 = n - n1;
     if  n1 < 1,    n1 = 1;   end
     if  n2 > nm1, n2 = nm1; end
end
konst1 = sqrt(n ./ ([1:nm1].*[nm1:-1:1]))';

%%% CALCULATION OF TEST STATISTICS

junk  = cumsum(data - mean(data));
stats = abs(konst1 .* junk(1:nm1));

if sigma2 == 0
   for i = 1:nm1
           data1  = data(1:i);
           data2  = data(i+1:n);
           sk2(i) = (sum((data1 - mean(data1)).^2)  + ...
                    sum((data2  - mean(data2)).^2)) / nm2;
   end
  else
   sk2 = sigma2;
end
test_stats = abs(stats./sqrt(sk2));

[ts_value, chp_place] = max(test_stats(n1:n2));
chp_place = chp_place + n1 - 1;

%%% STORING THE RESULTS

Results.chp_place = chp_place;
Results.ts_value  = ts_value;
if sigma2 > 0
     Results.sigma2_used = sigma2;
   else
     Results.sigma2_used = sk2(chp_place);
end
Results.test_stats     = test_stats;
Results.sigma2_stats   = sk2;
Results.trimming_used  = trimming;

%%% PLOTTING (RELEVANT FOR MATLAB USERS ONLY)

plot([n1:n2], test_stats(n1:n2))
axis([0 n floor(min(test_stats(n1:n2))) ceil(max(test_stats(n1:n2)))])
xlabel('Index')
if sigma2 > 0  & trimming == 0
```

```
   ylabel('Test statistics (I.17)')
end
if sigma2 > 0 & trimming > 0
   ylabel('Test statistics (I.18)')
end
if sigma2 == 0  & trimming == 0
   ylabel('Test statistics (I.27)')
end
if sigma2 == 0  & trimming > 0
   ylabel('Test statistics (I.28)')
end
title('MODEL (I.16)')
grid on

if sigma2 == 0
   figure
   plot([n1:n2], sk2(n1:n2))
   xlabel('Index')
   ylabel('Values s_k^2 calculated using (I.29)')
   title('MODEL (I.16)')
   grid on
   axis([0 n floor(min(sk2)) ceil(max(sk2))])
end
```

16. Change in mean with unknown starting value, sum-type test statistics

```
function Results = m3_1_be(data, sigma2)

%%% SUBROUTINE m3_1_be.m
%%%
%%% PURPOSE To test the null hypothesis H against the alternative
%%%         A in the model:
```

$$
\begin{aligned}
H &: Y_i = \mu + e_i, && i = 1, \dots, n,\\
A &: \exists \mathsf{m} \in \{1, \dots, n-1\} \text{ such that} &&\\
&\quad Y_i = \mu + e_i, && i = 1, \dots, \mathsf{m},\\
&\quad Y_i = \mu + \delta + e_i, && i = \mathsf{m}+1, \dots, n,\ \delta \neq 0.
\end{aligned}
$$

We suppose that the variables e_i are iid with the normal distribution $N(0, \sigma^2)$, σ^2 either known or unknown. If σ^2 is known, we apply Bayesian-type statistic (I.19). On the other hand, if σ^2 is unknown or supplied value of σ^2 is ≤ 0, we apply Bayesian-type statistic (I.19) with σ^2 estimated using (I.26).

```
%%% CALLING SEQUENCE   Results = m3_1_be(data, sigma2)
%%%
%%% INPUT
```

```
%%%
%%% data              ...  data series to be tested for the presence
%%%                        of change point
%%%
%%% sigma2            ...  variance of the error term
%%%     > 0           ...  statistic (I.19) is used with user-supplied
%%%                        value for 'sigma^2'
%%%     <=0           ...  statistic (I.19) is used and 'sigma^2'
%%%     or missing         estimated using (I.26)
%%%
%%% OUTPUT
%%%
%%% Results            :
%%%     test_stat     ...  value of the test statistic (I.19)
%%%     sigma2        ...  user-supplied value for 'sigma^2' or value of
%%%                        the statistic (I.26)
%%%
%%% EXTERNAL PROCEDURES CALLED  none
%%%
%%% EXAMPLE
%%%     nile                           <- data about Nile are read in
%%%     Results = m3_1_be(nil, 0)      <- calculation
%%%
%%%     Results.test_stat = 2.5276     -> value of test statistic (I.19)
%%%     Results.sigma2    = 2.8340e+04 -> value of statistic (I.26)
%%%
%%% CONTROL OF PARAMETERS

if nargin == 1, sigma2 = 0; end

%%% CALCULATION OF NECESSARY CONSTANTS

data = data(:);
n    = length(data);
nm1  = n - 1;

%%% CALCULATION OF THE TEST STATISTIC

junk = cumsum(data - mean(data));
stat = sum(junk.^2)/(n^2);

if sigma2 <= 0, sigma2 = sum((data - mean(data)).^2)/n; end
test_stat  = stat/sigma2;

%%% STORING THE RESULTS

Results.test_stat = test_stat;
Results.sigma2    = sigma2;
```

17. Change in variance

```
function   Results = m3_2(data, trimming, a)

%%% SUBROUTINE m3_2.m
%%%
%%% PURPOSE
```

We suppose that the observations $Y_1, \dots, Y_n$ are independent normally distributed with a known mean μ and unknown variances. Supposing that the mean remains the same, the problem of the detection of a change in variance can be formulated as the following testing problem, i.e. we test the null hypothesis H against the alternative A:

$$\begin{aligned} &H : Y_1 \dots, Y_n \sim N(\mu, \sigma^2) \\ &A : \exists \mathrm{m} \in \{1, \dots, n-1\} \text{ such that} \\ &\quad Y_1, \dots, Y_{\mathrm{m}} \quad \sim N(\mu, \sigma_1^2), \\ &\quad Y_{\mathrm{m}+1}, \dots, Y_n \sim N(\mu, \sigma_2^2), \end{aligned} \tag{I.32}$$

where $\sigma_1^2 \neq \sigma_2^2$.

```
%%% CALLING SEQUENCE   Results = m3_2(data, trimming, a)
%%%
%%% INPUT
%%%
%%% data               ...  data series to be tested for the presence of
%%%                         change point in variance in model (I.32)
%%%
%%% trimming           ...  trimming proportion (default: beta = 0.05)
%%%
%%%   if in (0,1/2) ..      trimmed statistic (I.33) is used
%%%   otherwise     ...     non-trimmed statistic (I.33) is used
%%%
%%% a                  ...  (expected) mean of observations
%%%
%%%  specified        ...   user-supplied value of 'a'
%%%  not specified ...      'a' will be estimated by the mean of the data
%%%
%%% OUTPUT
%%%
%%% Results                :
%%%
%%%     chp_place          ...  place of the maximum of test statistics
%%%     ts_value           ...  value of the maximum of test statistics
%%%     stats_zk           ...  values of statistics $Z_k$ from (I.34)
%%%     trimming_used      ...  value of 'trimming' used for calculation
%%%     a_used             ...  value of 'a' used for calculation
%%%
%%% EXTERNAL PROCEDURES CALLED     none
```

```
%%% EXAMPLE
%%%     randn('seed',1);                         <- seeding RNG
%%%     data = [randn(20,1); 3*randn(30,1)]      <- data are generated
%%%     Results = m3_2(data, 0.05, 0)            <- calculation
%%%
%%%     Results.chp_place      = 20     -> change point place
%%%     Results.ts_value       = 4.1267 -> value of test statistic (I.17)
%%%                                        or (I.18)
%%%     Results.stats_zk       = 0.9174, 1.6637, ...
%%%                              -> values of statistics Z_k
%%%     Results.a_used         = 0
%%%     Results.trimming_used  = 0.9174, 1.6637, ...
%%%
%%% CONTROL OF PARAMETERS

if nargin == 1,   trimming = 0; a = mean(data);  end
if nargin == 2,   a = mean(data);   end
if trimming < 0 | trimming >= 0.5, trimming = 0; end

%%% CALCULATION OF NECESSARY CONSTANTS

data = data(:);
n    = length(data);
nm1  = n - 1;
zk2  = zeros(n,1);
n1   = 1;
n2   = nm1;
if   trimming > 0
     n1 = round(n*trimming);
     n2 = n - n1;
     if  n1 < 1,    n1 = 1;   end
     if  n2 > nm1, n2 = nm1; end
end

%%% CALCULATION OF TEST STATISTICS

junk  = n*log(sum((data - a).^2)/n);
for k = n1:n2
        data1  = data(1:k);
        data2  = data(k+1:n);
        junk1  = sum((data1 - a).^2)/k;
        junk2  = sum((data2 - a).^2)/(n-k);
        zk2(k) = junk - k*log(junk1) - (n-k)*log(junk2);
end
stats_zk = sqrt(zk2);

[ts_value, chp_place] = max(stats_zk(n1:n2));
chp_place = chp_place + n1 - 1;
```

```
%%% STORING THE RESULTS

Results.chp_place = chp_place;
Results.ts_value  = ts_value;
Results.stats_zk  = stats_zk;
Results.a_used    = a;
Results.trimming_used = trimming;

%%% PLOTTING (RELEVANT FOR MATLAB USERS ONLY)

plot([n1:n2],stats_zk(n1:n2));
axis([0 n floor(min(stats_zk(n1:n2))) ceil(max(stats_zk))])
xlabel('Index')
ylabel('Test statistics (I.33)')
title('MODEL (I.32)')
grid on
```

18. Change in mean and/or variance

```
function   Results = m3_3(data, trimming)

%%% SUBROUTINE m3_3.m
%%%
%%% PURPOSE
```

Sometimes it can happen that the change may occur either in one parameter or in both (simultaneously) . Then we test the null hypothesis H against the alternative A:

$$\begin{aligned} &H : Y_1 \dots, Y_n \sim N(\mu, \sigma^2) \\ &A : \exists \mathrm{m} \in \{2, \dots, n-2\} \text{ such that} \\ &\qquad Y_1, \dots, Y_{\mathrm{m}} \quad \sim N(\mu_1, \sigma_1^2), \\ &\qquad Y_{\mathrm{m}+1}, \dots, Y_n \sim N(\mu_2, \sigma_2^2), \end{aligned} \tag{I.38}$$

where $(\mu_1, \sigma_1^2) \neq (\mu_2, \sigma_2^2)$.

```
%%% CALLING SEQUENCE   Results = m3_3(data, trimming)
%%%
%%% INPUT
%%%
%%% data            ...  data series to be tested for the presence of
%%%                      change point in variance in model (I.38)
%%%
%%% trimming        ...  trimming proportion (default: beta = 0.05)
%%%
%%%   if in (0,1/2) ..   trimmed statistic (I.39) is used
%%%   otherwise     ...  non-trimmed statistic (I.39) is used
```

```
%%%
%%% OUTPUT
%%%
%%% Results                 :
%%%
%%%     chp_place           ...  place of the maximum of test statistics
%%%     ts_value            ...  value of the maximum of test statistics
%%%     stats_zk            ...  values of statistics sqrt of (I.41)
%%%     trimming_used       ...  value of 'trimming' used for calculation
%%%     a1_used             ...  value of 'a1' used for calculation
%%%     a2_used             ...  value of 'a2' used for calculation
%%%
%%% EXTERNAL PROCEDURES CALLED    none
%%%
%%% EXAMPLE
%%%     randn('seed',1);                          <- seeding RNG
%%%     data = [randn(20,1); 4+3*randn(30,1)]     <- data are generated
%%%
%%%     Results = m3_3(data, 0.05)                <- calculation
%%%
%%%     Results.chp_place     = 21      -> change point place
%%%     Results.ts_value      =  6.639 -> value of test statistic (I.39)
%%%     Results.stats_zk      =  0, 2.212, ...
%%%                                     -> values of statistics Z_k
%%%     Results.trimming_used =  0.05
%%%     Results.a1_used       = -0.0321
%%%     Results.a2_used       =  3.5931
%%%
%%% CONTROL OF PARAMETERS

if nargin == 1,    trimming = 0; end
if trimming < 0 | trimming >= 0.5, trimming = 0; end

%%% CALCULATION OF NECESSARY CONSTANTS

data = data(:);
n    = length(data);
zk2  = zeros(n,1);
n1   = 2;
n2   = n - 2;
nm2  = n - 2;
if   trimming > 0
     n1   = round(n*trimming);
     n2   = n - n1;
     if  n1 < 2,    n1 = 2;   end
     if  n2 > nm2, n2 = nm2; end
end
```

```
%%% CALCULATION OF TEST STATISTICS

junk = n*log(sum((data - mean(data)).^2)/n);
for k = n1:n2
        data1  = data(1:k);
        data2  = data(k+1:n);
        junk1  = sum((data1 - mean(data1)).^2)/k;
        junk2  = sum((data2 - mean(data2)).^2)/(n-k);
        zk2(k) = junk - k*log(junk1) - (n-k)*log(junk2);
end
stats_zk = sqrt(zk2);

[ts_value, chp_place] = max(stats_zk(n1:n2));
chp_place = chp_place + n1 - 1;

%%% STORING THE RESULTS

Results.chp_place = chp_place;
Results.ts_value  = ts_value;
Results.stats_zk  = stats_zk;
Results.trimming_used = trimming;
Results.a1_used     = mean(data(1:chp_place));
Results.a2_used     = mean(data(chp_place+1:n));

%%% PLOTTING (RELEVANT FOR MATLAB USERS ONLY)

plot([n1:n2],stats_zk(n1:n2));
axis([0 n floor(min(stats_zk(n1:n2))) ceil(max(stats_zk))])
xlabel('Index')
ylabel('Test statistics (I.39)')
title('MODEL (I.38)')
grid on
```

References

Selected books and proceedings related to the problem

Basseville M. and Benveniste A. (eds.) (1986). *Detection of Abrupt Changes in Signals and Dynamic Systems.* Springer Verlag, Heidelberg.

Basseville M. and Nikiforov I. V. (1993). *Detection of Abrupt Changes. Theory and Applications.* Prentice Hall, London.

Broemling L. and Tsurumi H. (1987). *Econometrics and Structural Change over Time.* M. Dekker, New York.

Brodsky B. S. and Darkhovsky B. E. (1993). *Nonparametric Methods in Change Point Problem.* Kluwer, Dordrecht.

Carlstein E., Müller H. G. and Siegmund D. (eds.) (1994). *Change Point Problem.* IMS Lecture Notes – Monograph Series **23**.

Csörgő M. and Horváth L. (1997). *Limit Theorems in Change Point Analysis.* J. Wiley, New York.

Hackl P. (1980). *Testing the Constancy of Regression Relationship over Time.* Vandenhoeck and Ruprecht, Göttingen.

Hackl P. (ed.) (1989). *Statistical Analysis and Forecasting of Economic Structural Change.* Springer Verlag, New York.

Hackl P. and Westlund A. H. (eds.) (1991). *Economic Structural Changes. Analysis and Forecastings.* IIASA & Springer Verlag, Heidelberg.

Selected articles in journals and proceedings; additional monographs

Albin J. M. P. (1990). *On extremal theory for stationary processes.* Ann. Probab. **18**, 92 – 108.

Albin J. M. P., Carlsson B. and Jarušková D. (2000). *On the asymptotic distribution of a statistic for test of appearance of linear trend.* Extremes (submitted).

Anderson T. W. (1971). *The Statistical Analysis of Time Series.* J. Wiley, New York.

Anderson T. W. and Darling D. A. (1952). *Asymptotic theory of certain goodness of fit criteria based on stochastic processes.* Ann. Math. Statist. **23**, 193 – 212.

Anderson T. W. and Darling D. A. (1954). *A test of goodness of fit.* J. American Statistical Association **49**, 765 – 769.

Antoch J. and Ekblom H. (1993). *Recursive robust regression – computational aspects and comparison.* Computational Statistics and Data Analysis **19**, 115 – 128.

Antoch J. and Hušková M. (1993). *Change point problem.* In: Computational Aspects of Model Choice, Antoch J. (ed.), Physica-Verlag, Heidelberg, 11 – 38.

Antoch J. and Hušková M. (1994). *Procedures for detection of multiple changes in series of independent observations.* In: 5th Prague Symp. on Asymptotic Statistics, Hušková M. and Mandl P. (eds.), Physica-Verlag, Heidelberg, 3 – 20.

Antoch J. and Hušková M. (1996). *Tests and estimators for epidemic alternatives.* In: Probastat'94, Witkovsky V. and Pázman A. (eds.), Tatra Mountains Math. Publ. **7**, 316 – 329.

Antoch J. and Hušková M. (1998). *Estimators of changes.* In: Nonparametrics, Asymptotics and Time Series, Ghosh S. (ed.), M. Dekker, New York, 533 – 578.

Antoch J. and Hušková M. (2000). *Bayesian type estimators for change point.* J. Stat. Plan. Infer. **91**, 195 – 208.

Antoch J. and Hušková M. (2000). *Bayesian like R- and M-estimators of change points.* Discussiones Mathematicae **20**, 115 – 134.

Antoch J. and Hušková M. (2001). *Permutation tests in change point analysis.* Statist. Probab. Letters (to appear).

Antoch J., Hušková M. and Jarušková D. (2000). In: *Lecture Notes of the 5th IASC Summer School*, Lauro N.C. et al eds., ISI, Voorburg, 1 – 75.

Antoch J., Hušková M. and Prášková Z. (1997). *Effect of dependence on statistics for determination of change.* J. Stat. Plan. Infer. **60**, 291 – 310.

Antoch J., Hušková M. and Veraverbeke N. (1995). *Change point problem and bootstrap.* J. Nonparametric Statistics **5**, 123 – 144.

Antoch J. and Víšek J. Á. (1992). *Robust estimation in linear model and its computational aspects.* Computational Aspects of Model Choice, Antoch J. (ed.), Physica – Verlag, Heidelberg, 39 – 104.

Bai J. (1991). *On the partial sums of residuals in autoregressive and moving average models.* J. Time Series Analysis **14**, 247 – 259.

Bai J. (1994). *Least squares estimation of a shift in linear processes.* J. Time Series Analysis **15**, 453 – 472.

Berman S. M. (1982). *Sojourns and extremes of stationary processes.* Ann. Probab. **10**, 1 – 46.

Bhattacharya P. K. and Brockwell P. J. (1976). *The minimum of additive process with applications to signal, estimation and storage theory.* Z. Wahrsch. Verw. Gebiete **37**, 51 – 75.

Billingsley P. (1968). *Convergence of Probability Measures.* J. Wiley, New York.

Bräker H. U. (1993). *High boundary excursions of locally stationary Gaussian processes.* Dissertation Thesis, University of Bern.

Brown R. L., Durbin J. and Evans J. M. (1975). *Techniques for testing the constancy of regression relationships over time (with discussion).* JRSS **B 37**, 149 – 192.

Carlstein E. (1988). *Nonparametric estimation of a change point.* Ann. Statist. **16**, 188 – 197.

Chen X. (1988). *Inference in a simple change point problem.* Scientia Sinica **A 31**, 654 – 667.

Chernoff H. and Zacks S. (1964). *Estimating the current mean of normal distribution which is subjected to changes in time.* Ann. Math. Statist. **35**, 999 – 1018.

Chib J. (1999). *Bayesian estimation and comparison of multiple change point models.* Manuscript.

Chow Y. S. and Teicher H. (1988). *Probability Theory.* Springer Verlag, New York.

Chung-Bow Lee (1995). *Estimation of the number of change points in exponential families distributions.* Preprint.

Chung-Bow Lee (1995). *Estimation of the number of change points in normal random variables.* Preprint.

Csörgő M. and Horváth L. (1988). *Nonparametric methods for the change point problem.* In: Handbook of Statistics **7**, Krishnaiah P. R. and Rao C. R. (eds.), North Holland, Amsterdam, 403 – 425.

Chow Yu. S. and Teicher H. (1988). *Probability Theory.* Springer Verlag, Heidelberg.

Cobb G. W. (1978). *The problem of the Nile. Conditional solution to a change point problem.* Biometrika **65**, 243 – 251.

Cramér H. and Leadbetter M. R. (1967). *Stationary and Related Stochastic Processes.* J. Wiley, New York.

Csörgő M. and Horváth L. (1993). *Weighted Approximations in Probability and Statistics.* J. Wiley, New York.

Darkhovsky B. E. (1976). *A nonparametric method for a posteriori detection of the "disorder" time of a sequence of independent random variables.* Theory Probab. Appl. **21**, 178 – 183.

Darling D. A. and Erdős P. (1956). *A limit theorem for the maximum of normalized sums of independent random variables.* Duke Math. J. **23**, 143 – 155.

Davies R. B. (1987). *Hypothesis testing when a nuisance parameter is present only under the alternative.* Biometrika **74**, 33 – 43.

Davis R. A., Huang D. and Yao Y. C. (1995). *Testing for a change in the parameter values and order of an autoregressive model.* Ann. Statist. **23**, 282-304.

De Long D. (1981). *Crossing probabilities for a square root boundary by a Bessel process.* Commun. in Statist. - Theory and Method. **A10**, 2191-2213.

Deshayes J. and Picard D. (1986). *Off-line statistical analysis of change point models using nonparametric and likelihood methods.* In: Lecture Notes in Control and Information Sciences **77**, Basseville M. et al. (eds.), Springer Verlag, New York, 103-168.

Dümbgen L. (1991). *The asymptotic behavior of some nonparametric change point estimators.* Ann. Statist. **19**, 1471-1495.

Ferger D. (1994a). *Change point estimators in case of small disorders.* J. Stat. Plan. Infer. **40**, 33-49.

Ferger D. (1994b). *On the rate of almost sure convergence of Dümbgen change point estimators.* Statist. Probab. Letters **25**, 27-31.

Ferger D. (1994c). *Asymptotic distribution theory of change point estimators and confidence intervals based on bootstrap approximation.* Math. Meth. Statist. **3**, 362-378.

Ferger D. (1995). *Nonparametric tests for nonstandard change point problems.* Ann. Statist. **23**, 1848-1861.

Ferger D. and Stute W. (1992). *Convergence of change point estimators.* Stoch. Proc. Appl. **42**, 345-351.

Gardner L. A. (1969). *On detecting changes in the mean of normal variates.* Ann. Math. Statist. **40**, 116-126.

Gombay E. and Horváth L. (1990). *Asymptotic distributions of maximum likelihood tests for change in the mean.* Biometrika **77**, 411-414.

Gombay E. and Horváth L. (1994). *Limit theorems for change in linear regression.* J. Multivariate Analysis **48**, 43-69.

Gombay E., Horváth L. and Hušková M. (1996). *Estimators and tests for change in variances.* Statistics & Decisions **14**, 145-159.

Gombay E. and Horváth L. (1996). *Approximations for the time of change and the power function in change point models.* J. Stat. Plan. Infer. **52**, 43-66.

Gombay E. and Horváth L. (1996). *On the rate of approximations for maximum likelihood tests in change point models.* J. Mult. Anal. **56**, 120-152.

Gombay E. and Hušková M. (1998). *Rank based estimators of the change point.* J. Stat. Plan. Infer. **67**, 137-154.

Hampel F. R., Ronchetti E. M., Rousseeuw P. and Stahel W. A. (1981). *Robust Statistics. The Approach Based on Influence Functions.* J. Wiley, New York.

Hawkins D. H. (1977). *Testing a sequence of observations for a shift in location.* J. American Statistical Association **72**, 180-186.

Hinkley D. V. (1969). *Inference about the intersection in two-phase regression.* Biometrika **56**, 495-504.

Hinkley D. V. (1970a). *Inference about the change point in a sequence of random variables.* Biometrika **57**, 1-17.

Hinkley D. V. and Hinkley E. A. (1970). *Inference about the change point in a sequence of binomial variables.* Biometrika **57**, 477-488.

Hinkley D. and Schechtman E. (1987). *Conditional bootstrap methods in the mean-shift model.* Biometrika **74**, 85-93.

Horváth L. (1993). *The maximum likelihood method for testing changes in the parameters of normal observations.* Ann. Statist. **21**, 671-680.

Horváth L. (1995). *Detecting changes in linear regression.* Statistics **26**, 189-208.

Horváth L. and Gombay E. (1994). *Limit theorems for change in linear regression.* J. Multivariate Analysis **48**, 43-69.

Horváth L., Hušková M. and Serbinovska M. (1997). *Estimators for the time of change in linear models.* Statistics **27**, 109-130.

Horváth L. and Kokozska P. (1997). *The effect of long-range dependence on change point estimators.* J. Stat. Plan. Infer. **64**, 57 – 81.

Horváth L. and Qi-Man Shao (1996). *Darling-Erdős type theorems for sums of Gaussian variables with long range memory.* Stoch. Proc. Appl. **63**, 117 – 137.

Huber P. J. (1981). *Robust Statistics.* J. Wiley, New York.

Hüsler J. (1990). *Extreme values and high boundary crossings of locally stationary Gaussian processes.* Ann. Probability **18**, 1141 – 1158.

Hüsler J. (1993). *A note on extreme values of locally stationary Gaussian processes.* J. Stat. Plan. Infer. **45**, 203 – 213.

Hušková M. (1994). *Nonrecursive procedures for detecting change in simple linear regression models.* In: New Progress in Probability and Statistics, Puri M. L. and Vilaplana J. P. (eds.), 105 – 126.

Hušková M. (1995). *Estimators for epidemic alternatives.* Comment. Math. Univ. Carolinae **36**, 279 – 291.

Hušková M. and Sen P. K. (1989). *Nonparametric tests for shift and change in regression at an unknown time point.* In: Statistical Analysis and Forecasting of Economic Structural Change, Hackl P. (ed.), Springer Verlag, New York, 71 – 85.

James B., James K. L. and Siegmund D. (1987). *Tests for change points.* Biometrika **74**, 71 – 84.

Jandhyala V. K. and MacNeill I. B. (1990). *Residual partial sum limit process for regression models with applications to detecting parameter changes at unknown times.* Stoch. Proc. Appl. **33**, 309 – 323.

Jandhyala V. K. and MacNeill I. B. (1991). *Tests for parameter changes at unknown times in linear regression models.* J. Stat. Plan. Infer. **27**, 291 – 316.

Jandhyala V. K. and MacNeill I. B. (1992). *On testing for the constancy of regression coefficients under random walk and change point alternatives.* Econometric Theory **8**, 501 – 517.

Jarušková D. (1996). *Change point detection in meteorological measurement.* Monthly Weather Review **124**, 1535 – 1543.

Jarušková D. (1997). *Some problems with application of change point detection methods to environmental data.* Environmetrics **8**, 469 – 483.

Jarušková D. (1998a). *Change point estimator in gradually changing sequences.* Comment. Math. Univ. Carolinae **39**, 551 – 561.

Jarušková D. (1998b). *Testing appearance of linear trend.* J. Stat. Plan. Infer. **70**, 263 – 276.

Jarušková D. (1999). *Testing appearance of polynomial trend.* Extremes **2**, 25 – 37.

Jarušková D. and Antoch J. (1994). *Detection of change in variance.* In: 5th Prague Symposium on Asymptotic Statistics, Mandl P. and Hušková M. (eds.), Physica-Verlag, Heidelberg, 297 – 301.

Jones P. D., Wigley M. L. and Briffa K. R. (1994). *Global and hemispheric temperature anomalies - land and marine instrumental records.* In: Trends'93: A Compendium of Data on Global Change, Boden T. A. et al. (eds.), ORNL/CDIAC - 65, Carbon Dioxide Information Analysis Center, Oak Ridge National Laboratory, Oak Ridge, Tenn., USA, 603 – 608.

Kander Z. and Zacks S. (1966). *Test procedures for possible changes in parameters of statistical distributions occurring at unknown time points.* Ann. Math. Statist. **37**, 1196 – 1210.

Kiefer J. (1960). *K-sample analogues of the Kolmogorov-Smirnov's and Cramér-von Mises tests.* Ann. Math. Statist. **30**, 420 – 447.

Kim Hyune-Ju (1995). *Detection of a change in longitudinal data.* In: Int. Conference on Statistical Methods and Statistical Computing for Quality and Productivity Improvement, 748 – 754.

Kim Hyune-Ju and Siegmund D. (1989). *The likelihood ratio test for a change point in simple linear regression.* Biometrika **76**, 409 – 423.

Knowles M., Siegmund D. and Zhang H. (1991). *Confidence regions in semilinear regression.* Biometrika **78**, 15 – 31.

Leadbetter M. R., Lindgren G. and Rootzén H. (1983). *Extremes and related properties of random sequences and processes.* Springer Verlag, Heidelberg.

MacNeill I. B. (1974). *Tests for change of parameter at unknown time and distribution of some related functionals of Brownian motion.* Ann. Statist. **2**, 950 – 962.

MacNeill I. B. (1978). *Properties of sequences of partial sums of polynomial regression residuals with applications to tests for change of regression at unknown times.* Ann. Statist. **6**, 422 – 433.

Michels P. and Trenkler G. (1990). *Testing the stability of regression coefficients using generalized recursive residuals.* Austr. J. Statist. **22**, 293 – 312.

Nicholson S. E. (1994). *Century-scale series of standardised annual departure of African rainfall.* In: Trends'93: A Compendium of Data on Global Change, Boden T. A. et al. (eds.), ORNL/CDIAC - 65, Carbon Dioxide Information Analysis Center, Oak Ridge National Laboratory, Oak Ridge, Tenn., USA, 952 – 962.

Ploberger W. and Krämer W. (1992). *The CUSUM test with OLS residuals.* Econometrica **60**, 217 – 285.

Ritov Y. (1990). *Asymptotic efficient estimation of the change point with unknown distribution.* Ann. Statist. **18**, 1829 – 1839.

Schechtman E. (1983). *A conservative nonparametric distribution-free confidence bound for the shift in the change point problem.* Comment. Statist. **A12**, 2455 – 2464.

Sen P. K. (1982). *Asymptotic theory of some tests for constancy of regression relationships over time.* Statistics **13**, 21 – 31.

Sen P. K. (1984). *Recursive M-tests for constancy of multivariate regression relationships over time.* Sequential Analysis **3**, 191 – 211.

Shepp L. A. (1979). *The joint density of the maximum and its location for a Wiener process with drift.* J. Appl. Probab. **16**, 423 – 427.

Shorack G. R. and Wellner J. A. (1986). *Empirical Processes with Applications to Statistics.* J. Wiley, New York.

Siegmund D. (1986). *Boundary crossing probabilities and statistical applications.* Ann. Statist. **14**, 361 – 404.

Siegmund D. (1988). *Confidence sets in change point problems.* Int. Statist. Review **56**, 31 – 48.

Siegmund D. and Zhang H. (1994). *Confidence region in broken line regression.* In: Change point problems, IMS Lecture Notes – Monograph Series **23**, Carlstein E. et al. (eds.). 292 – 316.

Steinebach J. and Eastwood V. R. (1996). *Extreme value asymptotics for multivariate renewal processes.* J. Multivariate Analysis **56**, 284 – 302.

Stryhn H. (1996). *The location of maximum of asymmetric two-sided Brownian motion with triangular drift.* Statist. Probab. Letters **29**, 279 – 284.

Tang S.M. and Mac Neill I.B. (1993). *The effect of serial correlation on tests for parameter change at unknown time.* Ann. Statist. **21**, 552 – 575.

Vostrikova L. Y. (1981). *Detection of "disorder" in multidimensional random processes.* Sov. Math. Dokl. **24**, 55 – 59.

Wolfe D. A. and Schechtman E. (1984). *Nonparametric procedures for the change point problem.* J. Stat. Plan. Infer. **9**, 389 – 396.

Worsley K. J. (1983). *Testing for a two-phase multiple regression.* Technometrics **25**, 35 – 42.

Worsley K. J. (1986). *Confidence regions and tests for a change point in a sequence of exponential family random variables.* Biometrika **73**, 91 – 104.

Yao Q. (1993). *Tests for change points with epidemic alternatives.* Biometrika **80**, 179 – 191.

Yao Y.-C. (1988). *Estimating number of change points via Schwarz criterion.* Statist. Probab. Letters **6**, 181 – 189.

Yao Y.-C. (1990). *On asymptotic behavior of a class of nonparametric tests.* Statist. Probab. Letters **19**, 173 – 177.

Yao Y.-C. (1987). *Approximating the distribution of maximum likelihood estimate of the change point in a sequence of independent random variables.* Ann. Statist. **15**, 1321 – 1328.

Yao Yi-Ching and Davis R. A. (1986). *The asymptotic behavior of the likelihood ratio statistic for testing a shift in mean in a sequence of independent normal variates.* Sankhya **48**, 339 – 353.

Zacks S. (1983). *Survey of classical and Bayesian approach to the change point problem. Fixed sample and sequential procedures of testing and sequential procedures of testing and estimation.* In: Recent Advances in Statistics, Papers in Honor of Herman Chernoff's Sixties Birthday, Rizvi M. H. (ed.), Academic Press, New York, 245 – 269.

ON-LINE STATISTICAL PROCESS CONTROL

Jaromír Antoch and Daniela Jarušková

Charles University of Prague, Department of Statistics, Sokolovská 83, CZ–186 75 Praha 8, Czech Republic; jaromir.antoch@karlin.mff.cuni.cz
Czech Technical University, Department of Mathematics, Thákurova 7, CZ–166 29 Praha 6, Czech Republic; daniela.jaruskova@fsv.cvut.cz

Abstract: Critical overview of classical and recent methods for the detection of a change in a sequence of observations based on sequential approach is presented. Attention is payed to the Shewhart, CUSUM, EWMA and Girshick-Rubin-Shiryayev procedures. Numerous applications to the normal distribution suitable for statistical process control are considered. Instead of detailed proofs only their main ideas are presented.

1. Introduction

Imagine a process which produces a potentially infinite sequence of observations $X_1, X_2, X_3, \dots$ Initially the process is *in control (in order)* in the sense that an observer is satisfied to record the observations without taking any action. At some unknown moment ν the process changes and becomes *out of control (out of order)*. The observer would like to infer from the observations that this change has taken place and take appropriate action as soon as possible after the time ν.

For simplicity, we assume that random variables $X_1, X_2, \dots$ have absolutely continuous distribution functions. Moreover, we suppose that *before the change* the variables are distributed according to the density function f_0 while *after the change* according to the another density function, say f_1. The density f_0 is supposed to be known. As to the density f_1, we discuss situations when it is both known and unknown. Throughout our applications and examples we often assume, for greater transparency of the text, that both the f_0 and f_1 are normal. This assumption is coherent with most of the software packages and "relatively simple" textbooks, where this setting is usually assumed automatically. If the distribution is different from the normal one, as is the case of so-called Koopman-Darmois family of distributions, this fact is pointed out explicitly.

Let us denote by P_0 the distribution under which $X_1, X_2, \dots$ are independent identically distributed (iid) random variables with the density function f_0 and by $\{P_\nu,\ \nu = 1, 2, \dots\}$ the distribution under which $X_1, \dots, X_{\nu-1}$ are

---

2000 *Mathematics Subject Classification.* 62-01, 62L15, 60G40, 62N10.

Key words and phrases. Statistical process control, sequential procedures, Shewhart procedure, CUSUM procedure, EWMA procedure, Girshick-Rubin-Shiryayev procedure, *ARL* function.

Acknowledgement. The work was partially supported by grants GAČR 201/00/0769, MSM 113200008 and MSM210000001.

iid with the density function f_0 and $X_\nu, X_{\nu+1}, \dots$ are iid with the density function f_1. By E_ν, $\nu = 0, 1, 2, \dots$ we denote the corresponding expectations. Our main aim is to find a stopping time τ such that if change occurs at time ν then the delay for its detection $(\tau - \nu)^+$ would be small. A reasonable measure of "quickness of detection" of change occurring at time ν is the smallest number C_ν such that for all realizations $x_1, \dots, x_{\nu-1}$ of $X_1, \dots, X_{\nu-1}$ and $\tau \geq \nu$

$$\mathsf{E}_\nu\Big(\tau - \nu + 1 \mid X_1 = x_1, \dots, X_{\nu-1} = x_{\nu-1}\Big) \leq C_\nu$$

holds. As a kind of the *worst case criterion*, let us define $\overline{\mathsf{E}}\tau = \sup_{\nu \geq 1} C_\nu$. The decision to have small $\overline{\mathsf{E}}\tau$ must be, of course, balanced against the need to have a controlled frequency of *false reactions*. In other words, when there is no change then τ would be large, hopefully infinite. It was shown, however, that in order to have $\overline{\mathsf{E}}\tau$ finite it is necessary that τ has a finite expectation even under P_0. An appropriate type of restrictions on false reactions is therefore

$$\mathsf{E}_0\tau \geq B,$$

where the constant B is to be prescribed.

REMARK: In applications we often meet the situation that the procedure is not applied to the originally observed variables $X_1, X_2, \dots$, but to the averages of m consecutive variables, i. e. to

$$\overline{X}_1 = \frac{1}{m}\sum_{i=1}^{m} X_i, \ \overline{X}_2 = \frac{1}{m}\sum_{i=m+1}^{2m} X_i, \dots$$

The averaging reduces the variance, because if the variance of X_i's is σ^2 then the variance of $\overline{X}_i$'s is σ^2/m. Some software packages enable to work also with the standardized averages of subgroup samples of different length, which can be useful if some observations are missing.

In this paper we shall deal with four methods based on different stopping times (stopping rules):

1) *Shewhart algorithm (Shewhart chart, Shewhart procedure)*

$$\tau = \inf\left\{n \,\Big|\, \log\frac{f_1(X_n)}{f_0(X_n)} \geq h_1\right\}.$$

2) *CUSUM algorithm (CUSUM procedure)*

$$\tau = \inf\left\{n \,\Big|\, \widetilde{S}_n - \min_{0 \leq j \leq n} \widetilde{S}_j \geq h_2\right\},$$

where

$$\widetilde{S}_n = \sum_{i=1}^{n} \log\frac{f_1(X_i)}{f_0(X_i)}, \qquad \widetilde{S}_0 = 0.$$

3) *Exponentially weighted moving average algorithm*

$$\tau = \inf\left\{n \,\Big|\, \overline{X}_{EWMA}(n) \geq h_3\right\},$$

where

$$\overline{X}_{EWMA}(n) = (1-\lambda)\overline{X}_{EWMA}(n-1) + \lambda X_n, \qquad 0 < \lambda \le 1.$$

4) *Girshick-Rubin-Shiryayev algorithm*

$$\tau = \inf\left\{ n \,\middle|\, W_n \ge h_4 \right\},$$

where

$$W_n = \frac{f_1(X_n)}{f_0(X_n)}(1 + W_{n-1}) = \sum_{i=1}^{n}\prod_{j=i}^{n}\frac{f_1(X_j)}{f_0(X_j)}, \qquad W_0 = 0.$$

2. Shewhart procedure

The Shewhart procedure is the oldest and the best known procedure of all sequential quality control algorithms. As noticed above, the stopping time has the form

$$\tau = \inf\left\{ n \,\middle|\, \log\frac{f_1(X_n)}{f_0(X_n)} \ge h \right\}.$$

To decide how to find the critical limit h it is necessary to evaluate $\mathsf{E}_0\tau$ and $\overline{\mathsf{E}}\tau$. As the stopping time τ for the Shewhart procedure is independent of the past of the process, the distribution of the delay does not depend upon the place where the change occurs so that $\overline{\mathsf{E}}\tau = \mathsf{E}_1\tau$. Provided P represents an arbitrary probability under which $X_1, X_2, \ldots$ are iid, the distribution of τ is geometric, i.e. for $j = 1, 2, \ldots$

$$P(\tau = j) = (1-p)^{j-1}p, \quad p = P\left(\log\frac{f_1(X_1)}{f_0(X_1)} \ge h\right), \tag{2.1}$$

so that $\mathsf{E}\tau = p^{-1}$. Especially, (2.1) holds for P_0 and P_1.

2.1. Normal distribution – one-sided Shewhart chart for shift in mean

In this section we will study independent random variables $X_1, X_2, \ldots$ distributed according to a normal distribution with the known variance σ^2. Let us denote by P_μ the distribution under which they are iid according to $N(\mu, \sigma^2)$. We suppose that the process is *in control* if the mean $\mu = 0$ and is *out of control* if the mean $\mu = \mu^* > 0$. The stopping time $\tau = \tau_+$, which depends on the parameter h, b respectively, has the form

$$\tau_+ = \inf\left\{ n \,\middle|\, \frac{1}{\sigma^2}\left(\mu^* X_n - \frac{\mu^{*2}}{2}\right) \ge h \right\} = \inf\left\{ n \,\middle|\, X_n \ge \frac{h\sigma^2}{\mu^*} + \frac{\mu^*}{2} \right\} =$$
$$= \inf\left\{ n \,\middle|\, X_n \ge b \right\}.$$

To find the optimal value of b we will study the relationship between b and $\mathsf{E}_0\tau_+$, $\mathsf{E}_{\mu^*}\tau_+$ respectively. Under P_0, the stopping time τ_+ has the geometric distribution

$$P_0(\tau_+ = j) = (1-p_0)^{j-1}p_0, \qquad j = 1, 2, \ldots,$$

where

$$p_0 = P_0(X_n \geq b) = 1 - \Phi(b/\sigma),$$

so that

$$\mathsf{E}_0\tau_+ = \frac{1}{p_0} = \frac{1}{1-\Phi(b/\sigma)}. \tag{2.2}$$

Under P_{μ^*}, the stopping time τ_+ has again the geometric distribution and

$$\mathsf{E}_{\mu^*}\tau_+ = \frac{1}{p_{\mu^*}} = \frac{1}{P_{\mu^*}(X_n \geq b)} = \frac{1}{1-\Phi\left(\frac{b-\mu^*}{\sigma}\right)}.$$

The usual rule for finding the value of b is to pick a desired value of $\mathsf{E}_0\tau_+$ and to solve (2.2) (w.r.t. b). Let us notice that in this case the choice of b, which gives τ_+ unambiguously, is independent upon the value μ^*. Therefore, we can use the same τ_+ for any $\mu > 0$.

It is evident that the mean delay interval between the change and its detection is:

$$\mathsf{E}_\mu\tau_+ = \frac{1}{1-\Phi\left(\frac{b-\mu}{\sigma}\right)}.$$

In the statistical process control a crucial notion is the *average run length (ARL) function* which (in the sequential analysis) plays similar role as the power function does in the hypotheses testing. In our particular case it is the function (in μ) given by

$$ARL_+(\mu) = \mathsf{E}_\mu\tau_+$$

with one parameter b.

In Table 1 the values of the ARL function of the one-sided Shewhart procedure for detecting a positive shift in the mean of the normal distribution, $b = 3\,\sigma$, are given. The $ARL_+(\mu)$ function is plotted in Figure 1.

	μ					
	$-\sigma$	$-\sigma/2$	0	$\sigma/2$	σ	$3\sigma/2$
b=3σ	31 574	4299	740.8	161.04	43.96	14.97
	2σ	$5\sigma/2$	3σ	$7\sigma/2$	4σ	$9/2\sigma$
b=3σ	6.30	3.24	2.00	1.45	1.19	1.07

Table 1. Selected values of ARL function of one-sided Shewhart procedure for detecting a positive shift in mean of normal distribution, $b = 3\,\sigma$.

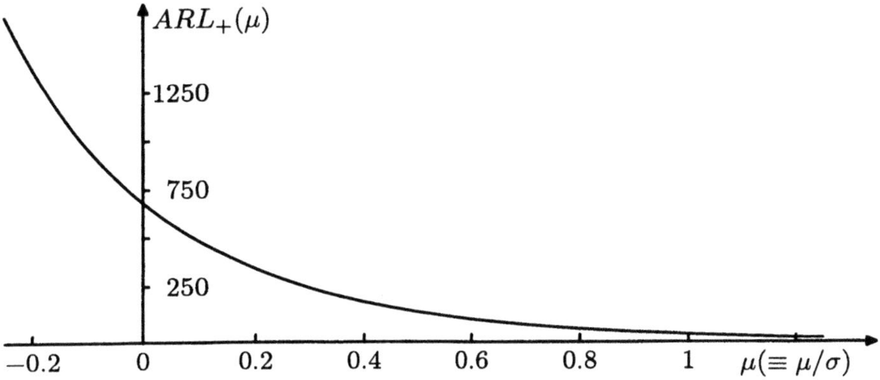

Figure 1. $ARL_+(\mu)$ function of one-sided Shewhart procedure for detecting a positive shift in mean of normal distribution, $b = 3\,\sigma$.

EXAMPLE: Independent normally distributed random variables with the known variance σ^2 are observed. The target value is zero. We would like to detect a positive shift in the mean. We wish that the mean interval between the false reactions $\mathsf{E}_0\tau_+ = 740$. The problem is to find the appropriate constant b. According to (2.2) the constant b must satisfy

$$\frac{1}{1-\Phi(b/\sigma)} = 740 \quad \equiv \quad \frac{b}{\sigma} = \Phi^{-1}\left(1-\frac{1}{740}\right) \quad \equiv \quad b = \sigma\Phi^{-1}(1-0.00135),$$

so that $b \approx 3\,\sigma$, i.e. we stop the procedure if $X_n \geq 3\,\sigma$. The mean delay in detecting, for example, the shift of the size $\mu = \sigma$ equals

$$\mathsf{E}_\mu\tau_+ = \frac{1}{1-\Phi\left(\frac{3\,\sigma-\sigma}{\sigma}\right)} = \frac{1}{1-\Phi(2)} \approx 44.$$

Analogously, we can consider the situation that the process is *in control* if the mean $\mu = 0$ and is *out of control* if the mean $\mu = \mu^* < 0$. The stopping time $\tau = \tau_-$, which depends on the parameter h, resp. b, has the form

$$\tau_- = \inf\left\{n \,\middle|\, \frac{1}{\sigma^2}\left(\mu^* X_n - \frac{\mu^{*2}}{2}\right) \geq h\right\} = \inf\left\{n \,\middle|\, X_n \leq \frac{h\sigma^2}{\mu^*} + \frac{\mu^*}{2}\right\} =$$
$$= \inf\left\{n \,\middle|\, X_n \leq -b\right\}.$$

The corresponding $ARL_-(\mu)$ function can be found similarly as $ARL_+(\mu)$.

2.2. Normal distribution – two-sided Shewhart chart for shift in mean

Let us suppose that the observations $X_1, X_2, \ldots$ are independent normally distributed with known variance σ^2. Let the process be *in control* if the mean $\mu = 0$ and be *out of control* if the mean $\mu \neq 0$. Let us define the stopping time of a two-sided symmetric Shewhart chart $\tau = \min\{\tau_+, \tau_-\}$, where

$$\tau_+ = \inf\left\{n \mid X_n \geq b\right\} \quad \text{and} \quad \tau_- = \inf\left\{n \mid X_n \leq -b\right\}.$$

The mean interval between false reactions is

$$\mathsf{E}_0\tau = \frac{1}{p_0} = \frac{1}{P_0(|X_n| \geq b)} = \frac{1}{2(1-\Phi(b/\sigma))}. \tag{2.3}$$

Similarly as in the one-sided case, the usual rule for the choice of b is to pick a desired value of $\mathsf{E}_0\tau$ and to solve the equation (2.3). The stopping time τ is again determined only by the value of b so that we can use the same stopping algorithm for all $\mu \neq 0$.

The average run length function is of the form

$$ARL(\mu) = \mathsf{E}_\mu\tau = \frac{1}{1-\left(\Phi\left(\frac{b-\mu}{\sigma}\right) - \Phi\left(\frac{-b-\mu}{\sigma}\right)\right)}.$$

EXAMPLE: The independent normally distributed random variables with known variance σ^2 are observed. The target value is zero. We would like to detect an arbitrary shift in the mean. We wish that the mean interval between false reactions $\mathsf{E}_0\tau = 370$. The problem is to find the appropriate constant b. According to (2.3), the constant b must satisfy

$$2(1-\Phi(b/\sigma)) = \frac{1}{370},$$

so that $b \approx 3\,\sigma$, i.e. we stop the procedure if $|X_n| \geq 3\,\sigma$. The mean delay in detecting, for example the shift $\mu = \sigma$, equals

$$\mathsf{E}_\mu\tau \approx 44.$$

In Table 2 the values of the ARL function of the two-sided Shewhart procedure for detecting an arbitrary shift in the mean of normal distribution, $b = 3\,\sigma$, are given. The $ARL(\mu)$ function is plotted in Figure 2.

	μ				
	0	$\pm 0.5\,\sigma$	$\pm 1\,\sigma$	$\pm 1.5\,\sigma$	$\pm 2\,\sigma$
b=3σ	370	155.22	43.89	14.97	6.30
	$\pm 2.5\,\sigma$	$\pm 3\,\sigma$	$\pm 3.5\,\sigma$	$\pm 4\,\sigma$	
b=3σ	3.24	2.00	1.45	1.19	

Table 2. Selected values of ARL function of two-sided Shewhart procedure for detecting arbitrary shift in mean of normal distribution, $b = 3\,\sigma$.

REMARK: In applications of the Shewhart procedure for the shift in the mean, instead of working with the single variables $X_1, X_2, \ldots$, one usually deals with the averages of a small number (say $m = 3 \approx 5$) of consecutive variables.

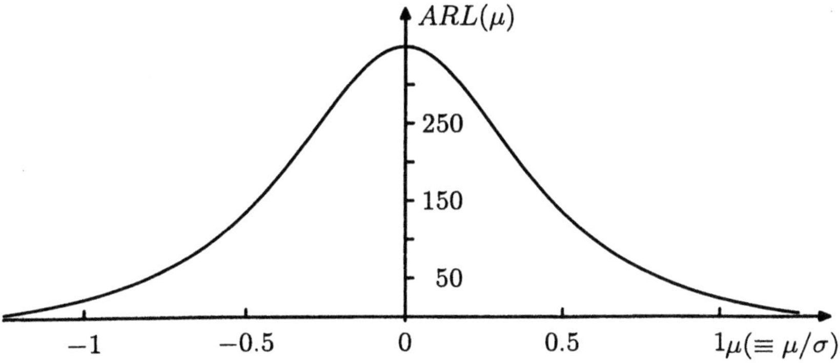

Figure 2. $ARL(\mu)$ function of two-sided Shewhart procedure for detecting arbitrary shift in mean of normal distribution, $b = 3\,\sigma$.

REMARK: As we saw, the important characteristic for determining control limits is the ARL function. However, many authors follow a long tradition and recommend, as a "rule of thumb", to take the control limits at the distance $3\,\sigma$ from the target value.

3. CUSUM procedure

CUSUM procedure is closely related to the following problem of the hypotheses testing. Suppose that $X_1, \ldots, X_n$ (n fixed for a moment) have been observed. We consider the problem of testing the null hypothesis that $X_1, \ldots, X_n$ are distributed according to the same known density function f_0 against the alternative that, at unknown time ν, $\nu \leq n$, the change occurred so that $X_1, \ldots, X_{\nu-1}$ are iid with the density function f_0 while $X_\nu, \ldots, X_n$ are iid with another *known* density function f_1. If we knew the time ν we could use the log-likelihood ratio

$$\widetilde{S}_n - \widetilde{S}_{\nu-1} = \sum_{i=\nu}^{n} \log \frac{f_1(X_i)}{f_0(X_i)}$$

as the test statistic and reject the null hypothesis if $\widetilde{S}_n - \widetilde{S}_{\nu-1} \geq K$, K being an appropriate constant. If the time of change is unknown, we reject the null hypothesis if

$$\widetilde{g}_n = \max_{0 \leq j \leq n} \left(\widetilde{S}_n - \widetilde{S}_j \right) = \widetilde{S}_n - \min_{0 \leq j \leq n} \widetilde{S}_j \geq \widetilde{h}.$$

Notice that $\widetilde{g}_n$, $n \geq 1$, can be defined also by the following recursion

$$\widetilde{g}_n = \max \left(\widetilde{g}_{n-1} + \log \frac{f_1(X_n)}{f_0(X_n)},\ 0 \right), \qquad \widetilde{g}_0 = 0.$$

Let us turn back to the situation from the introduction, i.e., that we have an infinite sequence of observations $X_1, X_2, \ldots$ available. An intuitively appealing stopping rule is

$$\tau = \inf \left\{ n \mid \widetilde{g}_n \geq \widetilde{h} \right\}.$$

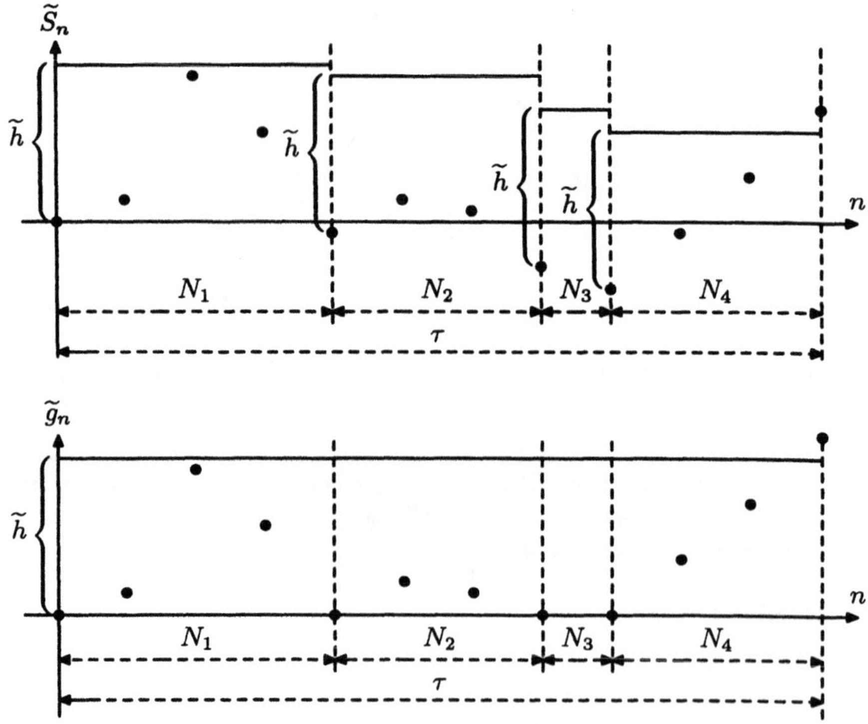

Figure 3. Example of behavior of sequences $\left\{\widetilde{S}_n, \widetilde{g}_n\right\}$.

Let us notice that $\widetilde{g}_n$ measures the current height of the random walk (see Appendix A, Section 7.1.) $\widetilde{S}_j,\ j = 0, 1, 2, \ldots$ above its minimal value. Whenever the random walk establishes a new minimum, i. e. $\widetilde{S}_n = \min_{0 \le k \le n} \widetilde{S}_k$, the process forgets its past and starts again in the sense that for all $j \ge 0$

$$\widetilde{S}_{n+j} - \min_{0 \le k \le n+j} \widetilde{S}_k = \widetilde{S}_{n+j} - \widetilde{S}_n - \min_{0 \le k \le j} \left(\widetilde{S}_{n+k} - \widetilde{S}_n\right).$$

The sequence $\widetilde{g}_n$ is in fact the random walk with one reflecting boundary.

We will study the properties of $\{\widetilde{g}_n\}$ to evaluate $\mathsf{E}_0\tau$ and $\overline{\mathsf{E}}\tau$, see the introduction. At first we show that $\mathsf{E}_1\tau = \overline{\mathsf{E}}\tau$. The inequality $\mathsf{E}_1\tau \le \overline{\mathsf{E}}\tau$ is obvious. The inequality $\mathsf{E}_1\tau \ge \overline{\mathsf{E}}\tau$ results from the following argument. The observations $X_1 = x_1, \ldots, X_n = x_n$ determine $\widetilde{g}_n = t \ge 0$ (depending on $X_1, \ldots, X_n$). Since $X_{n+1}, \ldots$ are independent upon the previous X_i's, the sequence $\widetilde{g}_{n+1}, \widetilde{g}_{n+2}, \ldots$ behaves just as $\widetilde{g}_1, \widetilde{g}_2, \ldots$ would have once started with $\widetilde{g}_0 = t \ge 0$. Since this last assertion would make any succeeding $\widetilde{g}$'s smaller, it would not increase the time required to reach $\widetilde{h}$.

Hence we have to find how $\mathsf{E}_0\tau$ and $\mathsf{E}_1\tau$ depend on $\widetilde{h}$. $\mathsf{E}_0\tau$ is the expectation of τ if all $X_1, X_2, \ldots$ are distributed according to the density f_0 and denotes the average time interval between the false reactions if the process is *in control*. Similarly, $\mathsf{E}_1\tau$ is the expectation of τ if all $X_1, X_2, \ldots$ are distributed according to the density f_1 and denotes the average delay between the change and its detection if the process is *out of control*.

To evaluate $\mathsf{E}_0\tau$ and $\mathsf{E}_1\tau$ we can use the results from the sequential analysis as the *CUSUM* procedure is in fact a sequence of Wald's sequential tests with the (log) limits $\{0, \widetilde{h}\}$. If the sum of the log-likelihood ratios is greater than or equal to $\widetilde{h}$ then the procedure definitively ends. If it is smaller than or equal to zero then a (new) sequential test begins using only the new observations. This idea can be expressed exactly as follows. First let us recall that $\tau = \inf\left\{n \mid \widetilde{g}_n \geq \widetilde{h}\right\}$ and introduce

$$N = N_1 = \inf\left\{n \geq 1 \mid \widetilde{S}_n \notin \left(0, \widetilde{h}\right)\right\}.$$

If $\widetilde{S}_{N_1} \geq \widetilde{h}$ then $\tau = N_1$. Otherwise

$$\widetilde{S}_{N_1} = \min_{0 \leq k \leq N_1} \widetilde{S}_k$$

and we put

$$N_2 = \inf\left\{n \geq 1 \,\middle|\, \widetilde{S}_{N_1+n} - \widetilde{S}_{N_1} \notin \left(0, \widetilde{h}\right)\right\}.$$

If $\widetilde{S}_{N_1+N_2} - \widetilde{S}_{N_1} \geq \widetilde{h}$ then $\tau = N_1 + N_2$. Otherwise $\widetilde{S}_{N_1+N_2} \leq \widetilde{S}_{N_1}$ and

$$\widetilde{S}_{N_1+N_2} = \min_{0 \leq k \leq N_1+N_2} \widetilde{S}_k.$$

In general, let

$$N_k = \inf\left\{n \geq 1 \,\middle|\, \widetilde{S}_{N_1+\cdots+N_{k-1}+n} - \widetilde{S}_{N_1+\cdots+N_{k-1}} \notin \left(0, \widetilde{h}\right)\right\}.$$

It is easy to see that

$$\tau = N_1 + \cdots + N_M$$

where

$$M = \inf\left\{k \mid \widetilde{S}_{N_1+\cdots+N_k} - \widetilde{S}_{N_1+\cdots+N_{k-1}} \geq \widetilde{h}\right\}.$$

For illustration see Figure 3.

Wald's identity yields

$$\mathsf{E}_i\tau = \mathsf{E}_iN \cdot \mathsf{E}_iM, \qquad i = 0, 1.$$

The random variable M is under P_0, as well as under P_1, geometrically distributed with

$$\mathsf{E}_iM = \frac{1}{P_i\left(\widetilde{S}_N \geq \widetilde{h}\right)}, \qquad i = 0, 1,$$

hence

$$\mathsf{E}_i\tau = \frac{\mathsf{E}_iN}{P_i\left(\widetilde{S}_N \geq \widetilde{h}\right)}, \qquad i = 0, 1.$$

To evaluate how $\mathsf{E}_0\tau$ depends on $\widetilde{h}$, we must find $\mathsf{E}_0 N$ and $P_0\left(\widetilde{S}_N \geq \widetilde{h}\right)$. Analogously, to evaluate how $\mathsf{E}_1\tau$ depends on $\widetilde{h}$ we must find $\mathsf{E}_1 N$ and $P_1(\widetilde{S}_N \geq \widetilde{h})$.

The first method for finding the relationship between $\widetilde{h}$ and $\mathsf{E}_0\tau$, resp. $\mathsf{E}_1\tau$, is based on the Fredholm integral equations, see Appendix A. Let us denote for $i = 1, 2$,

$$P_i(z) = P_i\left\{\widetilde{S}_N \leq 0 \,|\, \widetilde{S}_0 = z\right\},$$
$$N_i(z) = \mathsf{E}_i\left\{N \,|\, \widetilde{S}_0 = z\right\}.$$

Then

$$\mathsf{E}_i\tau = \frac{N_i(0)}{1 - P_i(0)}, \qquad i = 1, 2,$$

and the functions $P_i(z)$ and $N_i(z)$, $i = 0, 1$, satisfy the equations

$$P_i(z) = \int_{-\infty}^{0} g_i(y - z)\, dy + \int_0^{\widetilde{h}} P_i(y)\, g_i(y - z)\, dy,$$
$$N_i(z) = 1 + \int_0^{\widetilde{h}} N_i(y)\, g_i(y - z)\, dy,$$

where $g_i(y)$ is the density of $Y = \log\,(f_1(X_1)/f_0(X_1))$ under the probability P_i, $i = 1, 2$.

The second method of evaluation of $\mathsf{E}_i\tau$, $i = 0, 1$, is based on the Wald's approximations of the sequential analysis, see Appendix C, and gives

$$\mathsf{E}_0\tau \approx \frac{\widetilde{h} + 1 - e^{\widetilde{h}}}{\mathsf{E}_0 \log \dfrac{f_1(X_1)}{f_0(X_1)}} \qquad \text{and} \qquad \mathsf{E}_1\tau \approx \frac{\widetilde{h} - 1 + e^{-\widetilde{h}}}{\mathsf{E}_1 \log \dfrac{f_1(X_1)}{f_0(X_1)}}. \tag{3.1}$$

3.1. Normal distribution – one-sided CUSUM procedure for shift in mean

Here we apply the above considerations to the case of independent normally distributed random variables $X_1, X_2, \ldots$ with known variance σ^2. Let us denote again by P_μ the distribution under which they are iid according to $N(\mu, \sigma^2)$. The process is *in-control* if the mean $\mu = 0$ and is *out-of-control* if the mean $\mu = \mu^* > 0$. Applying the notation and the results covering the *CUSUM* technique (summarized in the previous lines) in the way that the density of $N(0, \sigma^2)$ stands for f_0 and the density of $N(\mu, \sigma^2)$ for f_1. Then the log-likelihood ratio $\widetilde{S}_n$ is of the form

$$\widetilde{S}_n = \sum_{i=1}^{n}\left(\mu^* X_i - \frac{\mu^{*2}}{2}\right)\frac{1}{\sigma^2} = \sum_{i=1}^{n}\left(X_i - \frac{\mu^*}{2}\right)\frac{\mu^*}{\sigma^2}.$$

Let us denote for $\mu^* > 0$

$$k = \frac{\mu^*}{2} \;\dots\; \text{the reference value,} \qquad h = \widetilde{h}\,\frac{\sigma^2}{\mu^*} \;\dots\; \text{the threshold value,}$$

and

$$S_n(k) = \sum_{i=1}^{n} (X_i - k), \qquad S_n = S_n(0) = \sum_{i=1}^{n} X_i.$$

The *CUSUM* statistics $\{\widetilde{g}_n\}$ can be expressed in the form

$$\begin{aligned}\widetilde{g}_n &= \max_{0\le j\le n} \left(\widetilde{S}_n - \widetilde{S}_j\right) = \max_{0\le j\le n} \frac{\mu^*}{\sigma^2} \sum_{i=j+1}^{n} \left(X_i - \frac{\mu^*}{2}\right) \\ &= \frac{\mu^*}{\sigma^2} \max_{0\le j\le n} \left(S_n(k) - S_j(k)\right)\end{aligned}$$

and the stopping time is

$$\begin{aligned}\tau_+ &= \inf\left\{ n \,\middle|\, \max_{0\le j\le n} \frac{\mu^*}{\sigma^2} \sum_{i=j+1}^{n} \left(X_i - \frac{\mu^*}{2}\right) \ge \widetilde{h} \right\} \\ &= \inf\left\{ n \,\middle|\, \max_{0\le j\le n} \sum_{i=j+1}^{n} (X_i - k) \ge h \right\}.\end{aligned}$$

There exists another commonly used parametrization of the problem which is more geometrically oriented and can be introduced through the new constants θ and d defined by

$$\theta = \operatorname{arctg} k, \quad d = \frac{h}{k}\cdot$$

Using this parametrization we can rewrite the stopping time as

$$\tau_+ = \inf\left\{ n \,\middle|\, \max_{0\le j\le n} \left(S_n - S_j - (n-j)\operatorname{tg}\theta\right) \ge d \operatorname{tg}\theta \right\}.$$

The validity of this inequality can be easily derived using the lower arm of so called V-mask, see Figure 4.

Let us summarize that the stopping time τ_+ characterizing the *CUSUM* procedure is determined by one of the equivalent couples of parameters, i. e. (k, h) or (d, θ).

Applying the general *CUSUM* scheme explained at the beginning of this chapter we are able to evaluate $\mathsf{E}_0\tau_+$ and $\mathsf{E}_{\mu^*}\tau_+$ for fixed μ^*, where

$$k = \frac{\mu^*}{2} > 0 \qquad \text{and} \qquad h = \widetilde{h}\,\frac{\sigma^2}{\mu^*}\cdot$$

However, we might be interested in the calculation of the average run length function $ARL_+(\mu) = \mathsf{E}_\mu\tau_+$ for arbitrary fixed parameters (k, h). We shall touch this problem in the following lines.

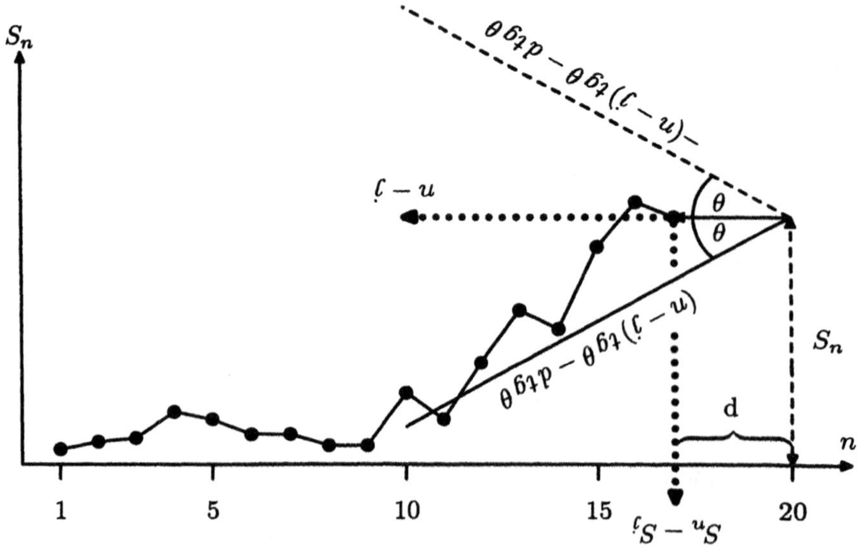

Figure 4. V-mask.

It is clear, see Appendix A, that the method of integral equations can be used also to calculate the expectation of τ_+ under an arbitrary distribution P_μ. Without loss of generality we can suppose that $\sigma^2 = 1$. The average run length function

$$\mathrm{ARL}_+(\mu) = \mathsf{E}_\mu \tau_+ = \frac{N_\mu(0)}{1 - P_\mu(0)},$$

where $P_\mu(z)$ and $N_\mu(z)$ satisfy the integral equations

$$P_\mu(z) = \int_{-\infty}^{0} \frac{1}{\sqrt{2\pi}} \exp\left\{ -\frac{(y - z - (\mu - k))^2}{2} \right\} dy + \\ + \int_0^h P_\mu(y) \frac{1}{\sqrt{2\pi}} \exp\left\{ -\frac{(y - z - (\mu - k))^2}{2} \right\} dy,$$

and

$$N_\mu(z) = 1 + \int_0^h N_\mu(y) \frac{1}{\sqrt{2\pi}} \exp\left\{ -\frac{(y - z - (\mu - k))^2}{2} \right\} dy.$$

The second method for the approximation of the $ARL_+(\mu)$ is based on the use of Wald's approximations. Let us take arbitrary μ and introduce $\widetilde{\mu} = \mu^* - \mu = 2k - \mu$ so that $k = (\mu + \widetilde{\mu})/2$. If we suppose that $\mu - \widetilde{\mu} = 2(\mu - k) > 0$, then

$$\tau_+ = \inf\left\{ n \,\middle|\, \max_{0\le j\le n} \sum_{i=j+1}^{n} (X_i - k) > h \right\} =$$

$$= \inf\left\{ n \,\middle|\, \max_{0\le j\le n} (\mu - \widetilde{\mu}) \sum_{i=j+1}^{n} \left(X_i - \frac{\mu + \widetilde{\mu}}{2} \right) > h(\mu - \widetilde{\mu}) \right\}.$$

Therefore, τ_+ is also the stopping time for detecting the shift from $\widetilde{\mu}$ to μ. Using the second formula of (3.1), where the density of $N(\widetilde{\mu}, \sigma^2)$ stands for f_0, the density of $N(\mu, \sigma^2)$ for f_1 and $\widetilde{h} = h(\mu - \widetilde{\mu})$, it follows

$$ARL_+(\mu) = \mathsf{E}_\mu \tau_+ \approx \frac{\widetilde{h} - 1 + e^{-\widetilde{h}}}{\mathsf{E}_\mu\left((\mu - \widetilde{\mu})X_1 - \left(\frac{\mu^2}{2} - \frac{\widetilde{\mu}^2}{2} \right) \right)} \tag{3.2}$$

$$= \frac{h(\mu - \widetilde{\mu}) - 1 + e^{-h(\mu - \widetilde{\mu})}}{\frac{(\mu - \widetilde{\mu})^2}{2}}$$

$$= \frac{2h(\mu - k) - 1 + e^{-2h(\mu - k)}}{2(\mu - k)^2}, \qquad \mu \neq k.$$

Analogously we get the same result if $\mu - \widetilde{\mu} = 2(\mu - k) < 0$. Taking the limit we obtain

$$ARL_+(\mu) \approx h^2, \qquad \mu = k.$$

REMARK: Let us remind that the same result was obtained by Reynolds (1975) if the sequence $\{\widetilde{g}_n\}$ is approximated by the Wiener process with one reflecting boundary.

Unfortunately, the approximation (3.2) is not accurate enough. Therefore, Siegmund (1985) suggested the following improvement

$$\mathrm{ARL}_+(\mu) \approx \begin{cases} \dfrac{2(h + 1.166)(\mu - k) - 1 + e^{-2(h+1.166)(\mu - k)}}{2(\mu - k)^2}, & \mu \neq k, \\ (h + 1.166)^2, & \mu = k. \end{cases} \tag{3.3}$$

The usual recommendations for the choice of k and h of $CUSUM$ procedure are as follows:

- Choose k such that $\mu = 2k$ is the value of the process where it becomes important to detect shift of given magnitude;
- Pick a desired mean time between false reactions ($MTBFR$) and choose, for already fixed k, such h that $ARL_+(0) = \mathsf{E}_0\tau_+ = MTBFR$.

EXAMPLE: The independent normally distributed random variables with the variance σ^2 are observed. The target value is zero. We would like to detect a positive shift in the mean provided the mean interval between false reactions

$\mathsf{E}_0\tau_+ = 740$. Values of the *ARL* function for selected values of the parameters k, h and $\sigma = 1$ are summarized in the following table. For the calculation Siegmund's approximation (3.3) was used.

		μ					
h	k	-1	-0.5	0	0.5	1	1.5
8.006	0.25	10^9	838807	740	28.77	11.34	7.02
4.766	0.50	10^7	71032	740	35.19	9.87	5.43
2.487	1.00	276883	12757	740	67.85	13.34	5.36
1.538	1.50	59677	6237	740	108.50	22.48	7.31
1.006	2.00	25366	4162	740	148.49	35.83	11.21
h	k	2	2.5	3	3.5	4	4.5
8.006	0.25	5.08	3.98	3.27	2.77	2.41	2.13
4.766	0.50	3.73	2.84	2.29	1.92	1.65	1.45
2.487	1.00	3.15	2.21	1.70	1.38	1.16	1.00
1.538	1.50	3.54	2.21	1.58	1.23	1.00	0.85
1.006	2.00	4.72	2.57	1.68	1.23	0.96	0.79

Table 3. Values of *ARL* function of one-sided *CUSUM* procedure for detecting a positive shift in mean of normal distribution with $\sigma^2 = 1$ for selected values of parameters k, h.

Let us suppose that it is for us important to detect the shift of the size $\mu = \sigma$. In that case we stop the process as soon as

$$\max_{0\le j\le n} \sum_{i=j+1}^{n} \left(\frac{X_i}{\sigma} - 0.5\right) > 4.766.$$

Using this stopping rule, the mean delay in detecting $\mu = \sigma$ is approximately 10 observations.

If in reality the shift would be of size $2\,\sigma$, while our procedure is based on assumptions $\mathsf{E}_0\,\tau_+ = 740$, and a desire is to detect (preferably) the shift of the size $\mu = \sigma$, we get immediately from Table 3. that the mean delay to detect such a shift is 3.73 observations (in mean).

On the other hand, provided $\mathsf{E}_0\tau_+ = 740$ and we want to detect the shift $\mu = 2\,\sigma$ preferably, we should use $(h, k) = (1, 2.487)$. The mean delay to detect $\mu = 2\sigma$ would be then 3.15 observations.

Up to now the setting of the parameters k and h (θ and d, respectively) for the *CUSUM* procedure was based on the *ARL* function. Aside that, many authors follow (usually without noticing that fact) the ideas used in the sequential hypotheses testing and derive these parameters from the error probabilities α and β. Due to the fact that most of the software packages offer also this option, it is worth paying attention to this approach for a moment.

Let us assume we observe independent random variables following $N(\mu, 1)$, the target value is zero and we would like to detect the positive shift in the process, i. e. the process is *in-control* if $\mu = 0$ and *out-of-control* if $\mu = \mu^* > 0$. At a fixed time moment n, let us have observations $X_1, X_2, \ldots, X_n$ available and we wish to decide whether the process is *in-control* or not. Let us define new variables $Y_1 = X_n, Y_2 = X_{n-1}, \ldots, Y_n = X_1$ and consider the nth step of the sequential test

$$H_0 : \mu = 0 \quad \text{against} \quad H_1 : \mu = \mu^* > 0$$

with the (fixed) probabilities α and β. Due to the fact that the "harm" can be caused if H_1 holds we shall concentrate only on this hypothesis. It follows from the Appendix B that if we accept H_1 at the time moment n, then the *CUSUM* procedure with parameters

$$d = \frac{2}{\mu^{*2}} \log \frac{1-\beta}{\alpha} \quad \text{and} \quad \theta = \operatorname{arctg} \frac{\mu^*}{2}$$

signals that the process is *out-of-control*. This is illustrated in Figures 3 and 5.

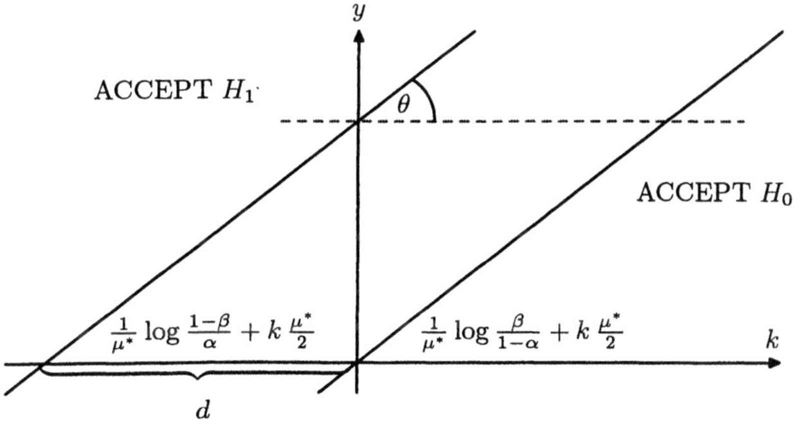

Figure 5. Setting of parameters for *CUSUM* procedure from error probabilities α and β.

REMARK: If the series of observations extends far enough back, it is possible that *an out-of-control* will be signalled eventually, even though the corresponding sequential procedure would be terminated previously with acceptance of H_0.

For the negative shift, the stopping time has the form $(k, h > 0)$

$$\tau_- = \inf \left\{ n \,\middle|\, \min_{0 \le j \le n} \sum_{i=j+1}^{n} (X_i + k) \le -h \right\},$$

respectively for the parameters θ and d

$$\tau_- = \inf\left\{ n \,\middle|\, \min_{0\le j\le n}\left(S_n - S_j + (n-j)\,\mathrm{tg}\,\theta\right) \le -d\,\mathrm{tg}\,\theta \right\}.$$

The validity of this inequality can be discovered using the upper arm of the V-mask. The average run length function for negative alternative with the stopping time τ_- can be found analogously as for τ_+.

3.2. One-sided CUSUM procedure for shift in parameter of Koopman-Darmois family of distributions

The same procedure for deriving the *CUSUM* stopping time and its properties that we used for the normal distribution can be applied to the one-parameter Koopman-Darmois family of density functions defined by

$$f_\theta(x) = h(x)\cdot\exp\left\{\theta T(x) - b(\theta)\right\}, \qquad \theta\in\Theta\subset R_1,\ x\in R_1, \tag{3.4}$$

where $b(0)=0$ and b is twice differentiable. Let us suppose that the process is *in-control* if $\theta = 0$ and *out-of-control* if $\theta = \theta^* > 0$. The *CUSUM* stopping time τ can be expressed as follows:

$$\begin{aligned}
\tau &= \inf\left\{ n \,\middle|\, \max_{0\le j\le n}\left(\widetilde{S}_n - \widetilde{S}_j\right) > \widetilde{h} \right\} = \\
&= \inf\left\{ n \,\middle|\, \max_{0\le j\le n}\sum_{i=j+1}^{n}\left(\theta^* T(X_i) - b(\theta^*)\right) > \widetilde{h} \right\} = \\
&= \inf\left\{ n \,\middle|\, \max_{0\le j\le n}\sum_{i=j+1}^{n}\left(T(X_i) - k\right) > h \right\},
\end{aligned}$$

where

$$h = \frac{\widetilde{h}}{\theta^*} \qquad\text{and}\qquad k = \frac{b(\theta^*)}{\theta^*}.$$

It means that, instead of studying the behavior of the random walk created by

$$\log\frac{f_{\theta^*}(X_1)}{f_0(X_1)},\ \log\frac{f_{\theta^*}(X_2)}{f_0(X_2)},\dots$$

we study the random walk created by

$$T(X_1) - k,\ T(X_2) - k,\dots$$

The *CUSUM* procedure has two parameters k and h and we are, as in the previous cases, interested in the average run length function

$$\mathrm{ARL}_+(\theta) = \mathsf{E}_\theta\tau_+,$$

where $\mathsf{E}_\theta\tau_+$ is the expectation of τ_+ corresponding to the probability under which the observations are iid with the same density f_θ given by (3.4).

The average run length function can be again calculated from the integral equations of the renewal theory, see Appendix A, so that

$$\mathrm{ARL}_+(\theta) = \frac{N_\theta(0)}{1 - P_\theta(0)},$$

where

$$P_\theta(z) = \int_{-\infty}^{0} g_\theta(y-z)\,dy + \int_0^h P_\theta(y)\,g_\theta(y-z)\,dy,$$

$$N_\theta(z) = 1 + \int_0^h N_\theta(y)\,g_\theta(y-z)\,dy$$

and $g_\theta(y)$ denotes the density of $T(X_1) - k$. Another possibility is to use Wald's approximations resulting in

$$\mathrm{ARL}_+(\theta) = \mathsf{E}_\theta \tau_+ \approx \frac{h(\theta - \widetilde{\theta}) - 1 + \mathrm{e}^{-(\theta-\widetilde{\theta})h}}{(\theta - \widetilde{\theta})\left(b'(\theta) - \dfrac{b(\theta) - b(\widetilde{\theta})}{\theta - \widetilde{\theta}}\right)},$$

where $\widetilde{\theta}$ is defined by the relation

$$\frac{b(\theta) - b(\widetilde{\theta})}{\theta - \widetilde{\theta}} = \frac{b(\theta^*)}{\theta^*} = k.$$

REMARK: The density of $N(\mu, \sigma^2)$ with known σ^2 belongs to the Koopman-Darmois family as

$$\theta = \frac{\mu}{\sigma}, \quad T(x) = x \quad \text{and} \quad b(\theta) = \frac{\theta^2}{2}.$$

3.3. Normal distribution – two-sided CUSUM procedure for arbitrary shift in mean

A two-sided symmetric *CUSUM* procedure for detecting an arbitrary shift in mean from the target value $\mu = 0$ with the parameters k and h is given by the stopping time

$$\tau = \min\left(\tau_+, \tau_-\right),$$

where

$$\tau_+ = \inf\left\{ n \,\middle|\, \max_{0 \le j \le n} \sum_{i=j+1}^{n} (X_i - k) > h \right\}$$

and

$$\tau_- = \inf\left\{ n \,\middle|\, \min_{0 \le j \le n} \sum_{i=j+1}^{n} (X_i + k) < -h \right\}.$$

For discovering the validity of these equations we can use symmetrical V-mask with $\theta = \operatorname{arctg} k$ and $d = h/k$.

It was shown by Kemp (1961) that

$$\frac{1}{\mathsf{E}_\mu \tau} = \frac{1}{\mathsf{E}_\mu \tau_+} + \frac{1}{\mathsf{E}_\mu \tau_-},$$

so that for $ARL(\mu) = \mathsf{E}_\mu \tau$, $ARL_+(\mu) = \mathsf{E}_\mu \tau_+$, $ARL_-(\mu) = \mathsf{E}_\mu \tau_-$ and

$$ARL(\mu) = \frac{ARL_+(\mu) ARL_-(\mu)}{ARL_+(\mu) + ARL_-(\mu)}$$

holds.

The usual recommendations for the choice of k and h are as follows:

(1) Choose k such that $\mu = |2k|$ is the magnitude of the shift which we wish to detect.

(2) Pick a desired mean time between false reactions ($MTBFR$) and choose, for already fixed k, such h that $ARL(0) = \mathsf{E}_0 \tau = MTBFR$.

EXAMPLE: Let the independent normally distributed random variables with variance $\sigma^2 = 1$ are observed and the target value is zero. We would like to detect an arbitrary shift in the mean provided the mean interval between false reactions $\mathsf{E}_0 \tau = 370$. Values of the ARL function for selected values of the parameters k, h are summarized in the following table. For the calculation we used again Siegmund's approximation (3.3).

		μ				
h	k	0	± 0.5	± 1	± 1.5	± 2
8.006	0.25	370	28.77	11.34	7.02	5.08
4.766	0.50	370	35.17	9.87	5.43	3.73
2.487	1.00	370	67.49	13.34	5.36	3.15
1.546	1.50	370	106.64	22.48	7.31	3.54
1.006	2.00	370	143.38	35.78	11.21	4.72
h	k	0	± 2.5	± 3	± 3.5	± 4
8.006	0.25	370	3.98	3.27	2.77	2.41
4.766	0.50	370	2.84	2.29	1.92	1.65
2.487	1.00	370	2.21	1.70	1.38	1.16
1.546	1.50	370	2.21	1.58	1.23	1.00
1.006	2.00	370	2.57	1.68	1.23	0.96

Table 4. Values of the ARL function of two-sided $CUSUM$ procedure for detecting an arbitrary shift in mean of normal distribution with $\sigma^2 = 1$ for selected values of parameters k, h.

If the variance σ^2 is known but different from 1 we can proceed analogously as in the previous example.

3.4. Methods for estimating the standard deviation

In the applications of the considered methods to the detection of a shift in mean of normally distributed data, we assumed that the standard deviation σ, resp. the variance σ^2, is known from the previous experience. However, if the standard deviation is not known, it must be estimated from the observed data at the same time as we record the process. The usual estimator

$$\widehat{\sigma}_{n,1} = \sqrt{\frac{1}{n-1}\sum_{i=1}^{n}\left(X_i - \overline{X}\,\right)^2}$$

does not perform well in the case that there is a shift in the mean. Therefore, other estimators are recommended.

If our method is based on the individual observations, the suitable estimator of σ at the nth stage of the sampling is

$$\widehat{\sigma}_{n,2} = \sqrt{\frac{1}{2(n-1)}\sum_{i=1}^{n-1}\left(X_{i+1} - X_i\right)^2}.$$

If the method works with the subgroup means $\overline{X}_1, \overline{X}_2, \ldots$ calculated from m consecutive observations, then at the nth stage of subgroup sampling, i.e. we have already observed $n \times m$ individual observations, we can estimate σ using two different methods:

(1) average of n unbiased estimators of consecutive subgroup standard deviations

$$\widehat{\sigma}_{n,3} = \frac{1}{c(m)}\,\frac{s_1 + \cdots + s_n}{n};$$

(2) root-mean-square estimator

$$\widehat{\sigma}_{n,4} = \frac{1}{c(mn-n+1)}\sqrt{\frac{s_1^2 + \cdots + s_n^2}{n}},$$

where s_i^2 is the best unbiased estimator of σ^2 calculated from the ith subgroup, i. e.

$$s_i^2 = \frac{1}{m-1}\sum_{j=im+1}^{(i+1)m}\left(X_j - \overline{X}_i\right)^2, \qquad i = 1, \ldots, m,$$

and $c(m)$ is the unbiasing factor defined by

$$c(m) = \frac{\Gamma\left(\frac{m}{2}\right)\sqrt{\frac{2}{m-1}}}{\Gamma\left(\frac{m-1}{2}\right)}.$$

3.5. Two-sided CUSUM procedure for shift in parameter of Koopman-Darmois family of distributions

Two-sided general procedures for detecting an arbitrary shift from the target value $\theta = 0$ is given by

$$\tau = \min\,(\tau_+, \tau_-),$$

where

$$\tau_+ = \inf\left\{ n \,\middle|\, \max_{0\le j\le n} \sum_{i=j+1}^{n} (T(X_i) - k_+) > h_+ \right\},$$

$$\tau_- = \inf\left\{ n \,\middle|\, \min_{0\le j\le n} \sum_{i=j+1}^{n} (T(X_i) + k_-) < -h_- \right\}.$$

Siegmund (1985) gives the conditions (that can be interpreted as a measure of symmetry) under which

$$\frac{1}{\mathsf{E}_\theta \tau} = \frac{1}{\mathsf{E}_\theta \tau_+} + \frac{1}{\mathsf{E}_\theta \tau_-}.$$

REMARK: Nadler &Robbins (1971) recommended the application of *CUSUM* scheme for the detection of an arbitrary shift in location parameter θ of the distribution functions $F(x-\theta), \theta \in \Theta$ providing the target value $\theta = 0$. They showed that the symmetric *CUSUM* procedure with $k = 0$ based on the stopping time

$$\tau = \min\,(\tau_+, \tau_-),$$

where

$$\tau_+ = \inf\left\{ n \,\middle|\, \max_{0\le j\le n} \sum_{i=j+1}^{n} X_i > h \right\},$$

$$\tau_- = \inf\left\{ n \,\middle|\, \min_{0\le j\le n} \sum_{i=j+1}^{n} X_i < -h \right\}$$

is equivalent to the procedure based on the stopping time

$$\tau = \left\{ n \,\middle|\, \max_{1\le j\le n} \sum_{i=1}^{j} X_i - \min_{1\le j\le n} \sum_{i=1}^{j} X_i > h \right\}.$$

They approximated the average length function using the distribution of a range of a Wiener process.

3.6. Lorden's CUSUM procedure

We shall deal again with the Koopman-Darmois family of densities. If the real *out of control value* θ is too far from the value θ^* on which we have based our *CUSUM* stopping time, than the procedure looses its power. Therefore Lorden (1970) suggested (for arbitrary shift in θ) to use the procedure based on the stopping time

$$\tau = \inf \left\{ n \,\middle|\, \max_{1\le j\le n} \sup_{|\theta|\ge\theta_1} \sum_{i=j}^{n} \left(\theta T(X_i) - b(\theta)\right) > h \right\}.$$

For the shift in mean of normally distributed random variables with the unit variance the alarm is raised at time n if at least for one j

$$S_n - S_j \ge c_j \qquad \text{or} \qquad S_n - S_j \le -c_j,$$

where

$$c_j = \begin{cases} \sqrt{2jh} & j \le \dfrac{2h}{\theta_1^2} \\[2ex] \dfrac{h}{\theta_1} + \dfrac{j\theta_1}{2} & j > \dfrac{2h}{\theta_1^2} \end{cases}$$

The limits are so called *U-shaped*. Unfortunately, in the literature we did not found any suitable approximation of the ARL function except of the Lorden's suggestion to apply Monte-Carlo method.

3.7. Principal differences between the CUSUM and Shewhart procedures

Shewhart procedure	CUSUM procedure
• The decision whether to stop the process in each time moment is based only on the information about the current single subgroup sample.	• The decision whether to stop the process in each time moment is based on the information from all samples up to and including the current subgroup.
• In the Shewhart chart the horizontal control limits are used to decide whether a point signals an out-of-control behavior.	• In the *CUSUM* chart the limits can be either in the form of V-mask (U-mask) or a horizontal decision interval.

4. Exponentially weighted moving average

The *Exponentially Weighted Moving Average* algorithm (*EWMA*) is usually applied if we wish to detect the change in the mean of the variables $\{X_i\}$ with common density f. Supposing the target value is zero and the variance σ^2 is known, the exponentially weighted moving average is defined by the recursion

$$\overline{X}_{EWMA}(n) = (1-\lambda)\overline{X}_{EWMA}(n-1) + \lambda X_n, \qquad \lambda \in (0,1],\ n = 1,2,\dots$$

For discovering a positive shift one sided algorithm raises alarm if

$$\overline{X}_{EWMA}(n) \geq h,$$

and the two-sided algorithm raises alarm if

$$\left| \overline{X}_{EWMA}(n) \right| \geq h.$$

The average run length function ARL has two parameters h and λ. The second constant λ is a smoothing constant. If $\lambda = 1$ then we get the Shewhart chart. On the other hand, if λ is small the $EWMA$ procedure resembles $CUSUM$.

Robinson &Ho (1970) suggested a numerical approximation for the evaluation of the average run length function. Crowder (1987a) used Fredholm integral equation whose derivation is based on the same idea as the one for integral equation for $CUSUM$ procedure. If $L(u)$ is ARL of two sided procedure, given that the weighted moving average starts in u $\left(\overline{X}_{EWMA}(0) = u\right)$, then

$$\begin{aligned} L(u) &= 1.P\Big(\big|\,(1-\lambda)u + \lambda X_1\,\big| > h\Big) + \\ &+ \int_{\{|(1-\lambda)u+\lambda y| \leq h\}} \Big(1 + L((1-\lambda)u + \lambda y)\Big)\, f(y)\, dy = \\ &= 1 + \frac{1}{\lambda}\int_{-h}^{h} L(y)\, f\left(\frac{y-(1-\lambda)u}{\lambda}\right)\, dy. \end{aligned}$$

If $L(u)$ is ARL of the one-sided procedure for the detection of a positive shift, given that the weighted moving average starts in u, then $L(u)$ satisfies

$$L(u) = 1 + \frac{1}{\lambda}\int_{-\infty}^{h} L(y)\, f\left(\frac{y-(1-\lambda)u}{\lambda}\right)\, dy.$$

Usually the control limits for the exponentially weighted moving average are specified to be $h = L\sigma_{EWMA}$, where

$$\sigma^2_{EWMA} = \left(\frac{\lambda}{2-\lambda}\sigma^2\right).$$

EXAMPLE: Let us suppose that the observations are normally distributed with the unit variance and let the target value be zero. We would like to detect a positive as well as negative shift in the mean and we wish that the mean interval between false reactions $\mathsf{E}_0\tau = ARL(0) = 370$. Several values of ARL function are given in the following table.

	μ				
	0	± 0.5	± 1	± 1.5	± 2
$L = 2.50,\quad \lambda = 0.50$	370	26.6	10.8	6.8	5
$L = 2.75,\quad \lambda = 0.12$	370	29.6	9.6	5.6	4
	± 2.5	± 3	± 3.5	± 4	
$L = 2.50,\quad \lambda = 0.50$	4.0	3.4	2.9	2.6	
$L = 2.75,\quad \lambda = 0.12$	3.2	2.6	2.3	2.0	

Table 5. Values of the ARL function of two-sided $EWMA$ procedure for detection of arbitrary shift in the mean of normal distribution with $\sigma^2 = 1$ for selected values of L, λ, μ (reprinted from Crowder (1987a)).

5. Girshick – Rubin – Shiryayev's algorithm

Girshick &Rubin (1952) studied the problem in a Bayesian framework. They supposed that the time of change is a random variable θ distributed according to the geometric distribution

$$P(\theta = k) = (1-q)^{k-1}q, \qquad k = 1, 2, \ldots$$

Let us denote again f_0 the density function of $\{X_i\}$'s before the change and f_1 the density after the change. Observing the process up to the time k the posterior probability q_k that a change occurs later then in the time k is given by the following recursion

$$q_k = P\Big(\theta > k \,\Big|\, X_k = x_k, \ldots, X_1 = x_1\Big) = \frac{f_0(X_k)q_{k-1}(1-q)}{q_{k-1}f_0(X_k) + \big(1 - q_{k-1}\big)f_1(X_k)}.$$

Let $p_k = 1 - q_k$ and Z_k is a monotone transformation of q_k of the form

$$Z_k = \frac{1-g}{g}\left(\frac{1}{q_k} - \frac{1}{1-q}\right),$$

where $g \in (0, 1)$ is fixed constant. (For the role of g see paper of Girshick &Rubin (1952), where more general model than ours is considered.) Then the relationship $p_k > p$ is equivalent to the relationship $Z_k > A$. Moreover, the variables Z_k satisfy the following recursion

$$Z_0 = 0, \qquad Z_k = y_k\left(1 + Z_{k-1}\right),$$

where

$$y_k = \frac{f_1(X_k)}{(1-g)f_0(X_k)}.$$

Shiryayev (1963a) showed that the stopping time

$$\tau = \inf\left\{n \,\middle|\, p_n \geq p(q,c)\right\}$$

is the Bayesian solution for the loss function

$$L(\theta,\tau) = \begin{cases} \text{one loses one unit if } \tau < \theta, \\ \text{one loses } c \text{ units for each observation taken after } \theta \text{ if } \tau \geq \theta. \end{cases}$$

Letting $g \searrow 0$ we get so called Girshick-Rubin-Shiryayev's algorithm with the stopping time

$$\tau = \inf\left\{n \,|\, W_n \geq A\right\},$$

where

$$W_0 = 0 \quad \text{and} \quad W_k = w_k\big(1 + W_{k-1}\big) \quad \text{with} \quad w_k = \frac{f_1(X_k)}{f_0(X_k)}.$$

Pollack (1987) studied this procedure in non-Bayesian setting for the one-parameter Koopman-Darmois family of distributions given by the density functions

$$f_\theta(x) = h(x)\exp\left\{\theta T(x) - b(\theta)\right\}, \qquad \theta \in \Theta \subset R_1,\ x \in R_1.$$

He supposed that $b(0) = b'(0) = 0$ and the process being in control if $\theta = 0$. Denoting the *out of control value* θ^*, the stopping time is

$$\tau = \inf\left\{n \,|\, W_n(\theta^*) \geq A\right\},$$

where

$$W_k(\theta^*) = \sum_{j=1}^{k} \exp\left\{\theta^* \sum_{i=j}^{k} T(X_i) - (k-j+1)\, b\,(\theta^*)\right\}.$$

The average run length function $ARL(\theta)$ has two parameters θ^* and A and is very uneasy to compute. Pollack (1987) studied the function $ARL(\theta)$ if $A \to \infty$ and showed that provided $I(\theta^*) < \infty$ and distribution of

$$\log \frac{f_{\theta^*}(X_i)}{f_0(X_i)} = \theta^* T(X_i) - b(\theta^*)$$

is non lattice, we have

$$ARL(0) = \frac{A}{\nu(\theta^*)}\big(1 + o(1)\big),$$

where $o(1) \to 0$ as $A \to \infty$. The function $\nu(\theta^*)$ is given by

$$\nu(\theta^*) = \lim_{a\to\infty} \mathsf{E}_{\theta^*} \exp\left\{-\left[\sum_{i=1}^{N_a}\Big(\theta^* T(X_i) - b(\theta^*)\Big) - a\right]\right\},$$

where

$$N_a = \min\left\{n \,\middle|\, \sum_{i=1}^{n}\Big(\theta^* T(X_i) - b(\theta^*) Big) \geq a\right\}.$$

Pollack (1987) also found the expression for $ARL(\theta)$ for arbitrary θ, but his expression seems to be too complicated for the routine calculation.

For detecting the shift in mean of normally distributed random variables with the known variance σ^2 the function $\nu(\mu^*)$ can be expressed in the form

$$\nu(\mu^*) = \lim_{a\to\infty} \mathsf{E}_{\mu^*} \exp\left\{-\left[\mu^*\left(S_{N_a} - N_a\frac{\mu^*}{2}\right) - a\right]\right\},$$

where

$$S_n = \sum_{i=1}^{n} X_i \qquad \text{and} \qquad N_a = \inf\left\{n \,\middle|\, \mu^* S_{N_a} - N_a\frac{mu^{*2}}{2} > a\right\}.$$

The function $\nu(\mu^*)$ was tabulated by Woodroofe (1982). For small values of μ^* Siegmund (1985) suggested the approximation

$$\nu(\mu^*) = \exp\{-0.583\mu^*\}.$$

6. Comparison of studied procedures

Only very few books and papers really try to compare the methods considered above. Roberts (1966) compares all four methods using simulations. He appreciates highly Girshick-Rubin-Shiryayev procedure. This method gives especially good results if we know (at least approximately) the *out-of-control* value. Montgomery (1990) gives the comparison of the two most frequently applied procedures – Shewhart and *CUSUM* – as follows

6.1. Advantages of CUSUM procedure over Shewhart procedure

- The *CUSUM* procedure is more efficient for small permanent shifts in the process mean, in particular, shifts of $0.5 \sim 2$ standard deviations from the target mean.
- Shifts in the process mean are visually easy to detect on a *CUSUM* chart since they produce a change in a slope of the plotted points. The point at which the slope changes is the point at which the shift has occurred.

6.2. Advantages of Shewhart procedure over CUSUM procedure

- The Shewhart procedure is more efficient to detect large abrupt shifts in the process.
- It can be easier to interpret patterns on a Shewhart chart since the points are uncorrelated.
- The Shewhart procedure is not so much affected by the possible dependence of the original data as the *CUSUM* procedure.

REMARK: Nelson (1984) and some other authors recommend to use some special tests to complement Shewhart procedure to improve its performance of smaller permanent shifts detection.

7. Appendix

Throughout the Appendix we provide main ideas of the proofs only, not their detailed derivation. Details can be found in referenced papers and books.

7.1. A. Random walk

Let $X_1, X_2, \ldots$ be iid real-valued random variables. Define

$$S_0 = z \quad \text{and} \quad S_n = S_0 + \sum_{i=1}^{n} X_i, \quad n = 1, 2, \ldots$$

The sequence S_n is called a random walk on the real line starting at the point z.

The generalized (improper) random variable N is called the stopping time if the event $[N = n]$ depends only on $X_1, \ldots, X_n$, i. e. $[N = n] \in \sigma(X_1, \ldots, X_n)$.

Lemma 7.1. *(Wald (1947)) Let $X_1, X_2, \ldots$ be iid real-valued random variables with $\mathsf{E}X_1 < \infty$ and let $S_n = \sum_{i=1}^{n} X_i$, $n = 1, 2, \ldots$ If N is a stopping time with $\mathsf{E}N < \infty$, then*

$$\mathsf{E}S_N = \mathsf{E}X_1 \mathsf{E}N. \tag{7.1}$$

Let a, b be arbitrary real numbers such that $b < a$. In the sequel we shall consider the stopping time N defined by

$$N = \inf\Big\{n \geq 1 \,\big|\, S_n \geq a \ \text{ or } S_n \leq b\Big\}. \tag{7.2}$$

Lemma 7.2. *(Stein (1946)) Let $X_1, X_2, \ldots$ be iid real-valued random variables with $P(X_1 = 0) < 1$. Let a, b and z be arbitrary numbers such that $b < z < a$ and $S_n = z + \sum_{i=1}^{n} X_i$. Then there exist constants c and $0 < \varrho < 1$ such that the stopping time N defined by (7.2) satisfies*

$$P(N < n) \leq c\,\varrho^n, \quad n = 1, 2, \ldots$$

Consequently, all moments of N are finite and $P(N < \infty) = 1$.

Let us suppose that the variables $X_1, X_2, \ldots$ are continuous with the density function $g(x)$. Then $S_1, S_2, \ldots$ are also continuous and for all $i = 1, 2, \ldots$ the conditional density of S_{i+1} given $S_i = s$ is $g(x - s)$.

Lemma 7.3. *Let a, b and z be arbitrary numbers such that $b < z < a$ and let the stopping time N is defined by (7.2). The functions $P(z) = P(S_N \leq b \mid S_0 = z)$ and $N(z) = \mathsf{E}(N \mid S_0 = z)$ satisfy integral equations*

$$P(z) = \int_{-\infty}^{b} g(x - z)\,dx + \int_{b}^{a} P(x)\,g(x - z)\,dx,$$

$$N(z) = 1 + \int_{a}^{b} N(x)\,g(x - z)\,dx.$$

Proof.

$$P(z) = P(S_N \leq b \,|\, S_0 = z) = \sum_{n=1}^{\infty} P\Big(S_n \leq b \cap N = n \,\Big|\, S_0 = z\Big) =$$

$$= P\Big(S_1 \leq b \,\Big|\, S_0 = z\Big) + \sum_{n=2}^{\infty} P\Big(S_n \leq b \cap N = n \,\Big|\, S_0 = z\Big) =$$

$$= \int_{-\infty}^{b} g(x-z)\,dx + \sum_{n=2}^{\infty} \int_{b}^{a} P\Big(S_n \leq b \cap N = n \,\Big|\, S_1 = x, S_0 = z\Big) g(x-z)\,dx =$$

$$= \int_{-\infty}^{b} g(x-z)\,dx + \sum_{n=2}^{\infty} \int_{b}^{a} P\Big(S_n \leq b \cap N = n \,\Big|\, S_1 = x\Big) g(x-z)\,dx =$$

$$= \int_{-\infty}^{b} g(x-z)\,dx + \sum_{n=2}^{\infty} \int_{b}^{a} P\Big(S_{n-1} \leq b \cap N = n-1 \,\Big|\, S_0 = x\Big) g(x-z)\,dx =$$

$$= \int_{-\infty}^{b} g(x-z)\,dx + \int_{b}^{a} \sum_{n=1}^{\infty} P\Big(S_n \leq b \cap N = n \,\Big|\, S_0 = x\Big) g(x-z)\,dx =$$

$$= \int_{-\infty}^{b} g(x-z)\,dx + \int_{b}^{a} P(x)\, g(x-z)\,dx.$$

Analogously

$$N(z) = \mathsf{E}(N \,|\, S_0 = z) = \sum_{n=1}^{\infty} P\Big(N \geq n \,\Big|\, S_0 = z\Big) =$$

$$= 1 + \sum_{n=2}^{\infty} \int_{b}^{a} P(N \geq n \,|\, S_1 = x, S_0 = z)\, g(x-z)\,dx =$$

$$= 1 + \sum_{n=2}^{\infty} \int_{b}^{a} P\Big(N \geq n \,\Big|\, S_1 = x\Big)\, g(x-z)\,dx =$$

$$= 1 + \sum_{n=2}^{\infty} \int_{b}^{a} P\Big(N \geq n-1 \,\Big|\, S_0 = x\Big)\, g(x-z)\,dx =$$

$$= 1 + \int_{b}^{a} \sum_{n=1}^{\infty} P\Big(N \geq n \,\Big|\, S_0 = x\Big)\, g(x-z)\,dx =$$

$$= 1 + \int_{b}^{a} N(x)\, g(x-z)\,dx.$$

□

7.2. B. Sequential probability ratio test

Wald's sequential probability ratio test is designed to test a simple hypothesis against a simple alternative. Let $X_1, X_2, \dots$ be iid with the unknown distribution P and let P_0 and P_1 be given distinct distributions with densities f_0 and f_1 with respect to the dominating σ-finite measure. The two hypotheses are $H_i: \ P = P_i, \ i = 0, 1$. Define the likelihood ratio (the probability ratio) at the nth stage of sampling

$$l_n = \prod_{j=1}^{n} \frac{f_1(X_j)}{f_0(X_j)}.$$

The test consists of choosing real positive numbers A, B $(B < A)$ and defining the stopping time

$$N = \inf \left\{ n \,\middle|\, l_n \geq A \text{ or } l_n \leq B \right\}.$$

The hypothesis H_0 is accepted if $l_N \leq B$ and the hypothesis H_1 is accepted if $l_N \geq A$. The two error probabilities are

$\alpha = P_0\,(l_N \geq A)$ probability that we accept H_1 providing H_0 holds,

$\beta = P_1\,(l_N \leq B)$ probability that we accept H_0 providing H_1 holds.

It is often more convenient to express the test in terms of the logarithms of the likelihood ratios. Put

$$\widetilde{S}_n = \log\, l_n = \sum_{j=1}^{n} \log \frac{f_1(X_j)}{f_0(X_j)}, \quad n = 1, 2, \dots, \quad \widetilde{S}_0 = 0.$$

Then $\{\widetilde{S}_n\}$ is a random walk starting at zero with

$$\mathsf{E}_0\, \frac{f_1(X_1)}{f_0(X_1)} < 0 < \mathsf{E}_1\, \frac{f_1(X_1)}{f_0(X_1)}.$$

Put $a = \log A$ and $b = \log B$. The stopping time N can be expressed in the form

$$N = \inf \left\{ n \,\middle|\, \widetilde{S}_n \geq a \text{ or } \widetilde{S}_n \leq b \right\}.$$

Further we shall suppose (as is usual) $B < 1 < A$ so that $b < 0 < a$.

The behavior of N is of interest not only under P_0 and P_1, but under arbitrary P as well. It follows from Stein (1946) that if $P\{ \log\, (f_1(X_1)/f_0(X_1)) = 0\} < 1$ then $P(N < \infty) = 1$ and N has moments of all orders under P. This conclusions hold in particular for P_0 and P_1 since $\log\, (f_1(X_j)/f_0(X_j))$, $j = 1, 2, \dots$ are nondegenerate under both P_i, $i = 0, 1$. Thus sequential probability ratio test terminates with probability 1 under both P_i.

Consider the problem how to relate $\alpha = P_0\big(l_N \geq A\big)$ and $\beta = P_1\big(l_N \leq B\big)$ to A and B. Let for $n = 1, 2, \dots$ A_n denote the subset of n-dimensional space where $B < l_k(x_1, \dots, x_k) < A$ for $k = 1, \dots, n-1$ and $l_n(x_1, \dots, x_n) \geq A$. Hence $\{N = n \cap l_n \geq A\} = \{(X_1, \dots, X_n) \in A_n\}$. By direct calculation

$$\alpha = P_0(l_N \geq A) = \sum_{n=1}^{\infty} P_0\left(N = n, l_n \geq A\right) =$$
$$= \sum_{n=1}^{\infty} \int_{A_n} f_0(x_1) \dots f_0(x_n)\, dx_1 \dots dx_n =$$
$$= \sum_{n=1}^{\infty} \int_{A_n} \frac{f_0(x_1) \dots f_0(x_n)}{f_1(x_1) \dots f_1(x_n)} \cdot f_1(x_1) \dots f_1(x_n)\, dx_1 \dots dx_n =$$
$$= \sum_{n=1}^{\infty} \int_{A_n} \frac{1}{l_n(x_1, \dots, x_n)} \cdot f_1(x_1) \dots f_1(x_n)\, dx_1 \dots dx_n \leq$$
$$= \frac{1}{A} \sum_{n=1}^{\infty} P_1\{N = n, l_n \geq A\} = \frac{1}{A} P_1(l_N \geq A) = \frac{1}{A}(1 - \beta\ ),$$

so that

$$\frac{1-\beta}{\alpha} \geq A \quad \text{and similarly} \quad \frac{\beta}{1-\alpha} \leq B. \tag{7.3}$$

If one assumes that the excess over the boundaries is negligible when the test ends (in the other words that there is no *overshoot*) the inequalities (7.3) become the approximate equalities

$$\frac{1-\beta}{\alpha} \approx A \quad \text{and} \quad \frac{\beta}{1-\alpha} \approx B,$$

so that

$$\alpha \approx \frac{1-B}{A-B} \quad \text{and} \quad \beta \approx \frac{B(A-1)}{A-B}$$

or, equivalently in a and b,

$$\alpha \approx \frac{1-e^b}{e^a - e^b} \quad \text{and} \quad \beta \approx \frac{e^b(e^a-1)}{e^a - e^b}.$$

Finally, let us turn to the approximation of $\mathsf{E}_0 N$ and $\mathsf{E}_1 N$. Neglecting again "overshoot" we have approximately $\widetilde{S}_N = b$ or a and, moreover,

$$P_0\left(\widetilde{S}_N = b\right) \approx 1 - \alpha \qquad \text{and} \qquad P_0\left(\widetilde{S}_N = a\right) \approx \alpha,$$
$$P_1\left(\widetilde{S}_N = b\right) \approx \beta \qquad \text{and} \qquad P_1\left(\widetilde{S}_N = a\right) \approx 1 - \beta.$$

Hence

$$\mathsf{E}_0 \widetilde{S}_N \approx \alpha a + (1-\alpha) b \quad \text{and} \quad \mathsf{E}_1 \widetilde{S}_N \approx (1-\beta) a + \beta b$$

or

$$\mathsf{E}_0 \widetilde{S}_N \approx \frac{1-e^b}{e^a-e^b} a + \left(1 - \frac{1-e^b}{e^a-e^b}\right) b,$$
$$\mathsf{E}_1 \widetilde{S}_N \approx \left(1 - \frac{e^b\left(e^a-1\right)}{e^a-e^b}\right) a + \frac{e^b\left(e^a-1\right)}{e^a-e^b} b.$$

From Wald's identity (7.1)

$$\mathsf{E}_0 N \approx \frac{\mathsf{E}_0 \widetilde{S}_N}{\mathsf{E}_0 \log \dfrac{f_1(X_1)}{f_0(X_0)}} = \frac{1}{\mathsf{E}_0 \log \dfrac{f_1(X_1)}{f_0(X_0)}} \left[\frac{1-e^b}{e^a - e^b} a + \left(1 - \frac{1-e^b}{e^a - e^b}\right) b \right]$$

and

$$\mathsf{E}_1 N \approx \frac{\mathsf{E}_1 \widetilde{S}_N}{\mathsf{E}_1 \log \dfrac{f_1(X_1)}{f_0(X_0)}} = \frac{1}{\mathsf{E}_1 \dfrac{f_1(X_1)}{f_0(X_0)}} \left[\left(1 - \frac{e^b(e^a-1)}{e^a - e^b}\right) a + \frac{e^b(e^a-1)}{e^a - e^b} b \right].$$

The above approximations are called *Wald's approximations*.

EXAMPLE: Let $X_1, X_2, \dots$ be iid copies of the random variable X distributed according to $N(\mu, 1)$. Let us consider the sequential test

$$H_0 : \ \mu = 0 \quad \text{against} \quad H_1 : \ \mu = \mu^* > 0.$$

It is evident that corresponding likelihood ratio at the kth stage of sampling is of the form

$$l_k = \prod_{i=1}^{k} \frac{f_1(X_i)}{f_0(X_i)} = \exp\left\{ \mu^* \sum_{i=1}^{k} X_i - \mu^{*2} \frac{k}{2} \right\}.$$

Let us fix probability errors α and β. Following the general rules of the sequential analysis we:

- accept H_0 iff $\quad \log l_k \le \log \dfrac{\beta}{1-\alpha} \quad \equiv \quad \mu^* \sum_{i=1}^{k} X_i - \dfrac{k}{2} \mu^{*2} \le \log \dfrac{\beta}{1-\alpha}$
- accept H_1 iff $\quad \log l_k \ge \log \dfrac{1-\beta}{\alpha} \quad \equiv \quad \mu^* \sum_{i=1}^{k} X_i - \dfrac{k}{2} \mu^{*2} \ge \log \dfrac{1-\beta}{\alpha}$
- take another observation iff $\quad \log \dfrac{\beta}{1-\alpha} < \log l_k < \log \dfrac{1-\beta}{\alpha}$

The boundary of the region for acceptance of H_1 is given by the line

$$y = \frac{1}{\mu^*} \log \frac{1-\beta}{\alpha} + k \frac{\mu^*}{2}. \tag{7.4}$$

and, analogously, the boundary of the region for acceptance H_0 is given by the line

$$y = \frac{1}{\mu^*} \log \frac{\beta}{1-\alpha} + k \frac{\mu^*}{2}.$$

These lines are evidently parallel and their slopes correspond to the angle

$$\theta = \operatorname{arctg} \frac{\mu^*}{2}.$$

Moreover, the distance d between the intersection of the line (7.4) with the horizontal axis and the origin is

$$d = \frac{2}{\mu^{*2}} \log \frac{1-\beta}{\alpha}.$$

As $\log(1-\beta)$ is usually small, some authors and developers of the software packages recommend to calculate d according to

$$d \approx \frac{2}{\mu^{*2}} \log \frac{1}{\alpha}.$$

For illustration see Figure 5.

7.3. C. Approximations of $\mathsf{E}_0\tau$ and $\mathsf{E}_1\tau$ for CUSUM procedure

Further, we would like to apply Wald's approximations for deriving $\mathsf{E}_i, i = 0, 1$, for the *CUSUM* stopping time

$$\tau = \inf\left\{n \,\middle|\, \widetilde{S}_n - \min_{0 \le j \le n} \widetilde{S}_j \ge \widetilde{h}\right\}.$$

Let us denote

$$N = \inf\left\{n \ge 1 \,\middle|\, \widetilde{S}_n \in \left(0, \widetilde{h}\right)\right\}.$$

Unfortunately, we cannot immediately use Wald's approximations to evaluate $P_i\left(\widetilde{S}_N \ge h\right)$ and $\mathsf{E}_i N, i = 0, 1$, as the lower bound of the interval $\left(0, \widetilde{h}\right)$ is zero. However, we can define

$$N(\varepsilon) = \inf\left\{n \ge 1 \,\middle|\, \widetilde{S}_n \in \left(-\varepsilon, \widetilde{h}\right)\right\}$$

and let $\varepsilon \searrow 0$. Applying Wald's approximations it follows that

$$P_0\left(\widetilde{S}_{N(\varepsilon)} \ge \widetilde{h}\right\} \approx \frac{1-e^{-\varepsilon}}{e^{\widetilde{h}} - e^{-\varepsilon}}, \qquad P_1\left(\widetilde{S}_{N(\varepsilon)} \ge \widetilde{h}\right\} \approx e^{\widetilde{h}} \frac{1-e^{-\varepsilon}}{e^{\widetilde{h}} - e^{-\varepsilon}},$$

and

$$\mathsf{E}_0 \widetilde{S}_{N(\varepsilon)} \approx \widetilde{h}\, \frac{1-e^{-\varepsilon}}{e^{\widetilde{h}} - e^{-\varepsilon}} + (-\varepsilon)\, \frac{e^{\widetilde{h}} - 1}{e^{\widetilde{h}} - e^{-\varepsilon}},$$

$$\mathsf{E}_1 \widetilde{S}_{N(\varepsilon)} \approx \widetilde{h}\, e^{\widetilde{h}}\, \frac{1-e^{-\varepsilon}}{e^{\widetilde{h}} - e^{-\varepsilon}} + (-\varepsilon)\, e^{-\varepsilon}\, \frac{e^{\widetilde{h}} - 1}{e^{\widetilde{h}} - e^{-\varepsilon}}.$$

Letting $\varepsilon \searrow 0$ we get (provided the limit exists)

$$\mathsf{E}_0\tau = \lim_{\varepsilon \searrow 0} \frac{\mathsf{E}_0\, N(\varepsilon)}{P_0\left(\widetilde{S}_{N(\varepsilon)} \ge \widetilde{h}\right)} = \lim_{\varepsilon \searrow 0} \frac{\mathsf{E}_0 \widetilde{S}_{N(\varepsilon)}}{\left(\mathsf{E}_0 \log \dfrac{f_1(X)}{f_0(X)}\right) P_0\left(\widetilde{S}_{N(\varepsilon)} > \widetilde{h}\right)} \approx$$

$$\approx \lim_{\varepsilon \searrow 0} \frac{\widetilde{h}\, \dfrac{1-e^{-\varepsilon}}{e^{\widetilde{h}} - e^{-\varepsilon}} + (-\varepsilon)\, \dfrac{e^{\widetilde{h}} - 1}{e^{\widetilde{h}} - e^{-\varepsilon}}}{\left(\mathsf{E}_0 \log \dfrac{f_1(X)}{f_0(X)}\right) \dfrac{1-e^{-\varepsilon}}{e^{\widetilde{h}} - e^{-\varepsilon}}} = \frac{\widetilde{h} + 1 - e^{\widetilde{h}}}{\mathsf{E}_0 \log \dfrac{f_1(X)}{f_0(X)}}$$

and

$$\mathsf{E}_1\tau = \lim_{\varepsilon\searrow 0} \frac{\mathsf{E}_1 N(\varepsilon)}{P_1\left(\widetilde{S}_{N(\varepsilon)} \geq \widetilde{h}\right)} = \lim_{\varepsilon\searrow 0} \frac{\mathsf{E}_1 \widetilde{S}_{N(\varepsilon)}}{\left(\mathsf{E}_1 \log \dfrac{f_1(X)}{f_0(X)}\right) P_1\left(\widetilde{S}_{N(\varepsilon)} > \widetilde{h}\right)} \approx$$

$$\approx \lim_{\varepsilon\searrow 0} \frac{\widetilde{h}\, e^{\widetilde{h}} \dfrac{1-e^{-\varepsilon}}{e^{\widetilde{h}}-e^{-\varepsilon}} + (-\varepsilon)\, e^{-\varepsilon} \dfrac{e^{\widetilde{h}}-1}{e^{\widetilde{h}}-e^{-\varepsilon}}}{\left(\mathsf{E}_1 \log \dfrac{f_1(X)}{f_0(X)}\right) e^{\widetilde{h}} \dfrac{1-e^{-\varepsilon}}{e^{\widetilde{h}}-e^{-\varepsilon}}} = \frac{\widetilde{h} - 1 + e^{-\widetilde{h}}}{\mathsf{E}_1 \log \dfrac{f_1(X)}{f_0(X)}}.$$

References

Anderson T. W. (1960). *A modification of the sequential probability ratio test to reduce the sample size.* Annals of Mathematical Statistics **31**, 165 – 197.

Antoch J. and Hušková M. (1993). *Change-point problem.* In: Computational Aspects of Model Choice, Antoch J. (ed.), 11 – 38, Physica-Verlag, Heidelberg.

Antoch J. and Hušková M. (1994). *Procedures for detection of multiple changes in series of independent observations.* In: 5th Prague Symp. on Asymptotic Statistics, Hušková M. and Mandl P. eds., 3 – 20, Physica-Verlag, Heidelberg.

Antoch J. and Hušková M. (1996). *Tests and estimators for epidemic alternatives.* In: Probastat'94, Witkovsky V. and Pázman A. eds., 316 – 329, Tatra Mountains Math. Publ. **7**.

Antoch J. and Hušková M. (1998). *Estimators of changes.* In: Nonparametrics, Asymptotics and Time Series, Ghosh S. ed., 533 – 578, M. Dekker, New York.

Antoch J., Hušková M. and Prášková Z. (1997). *Effect of dependence on statistics for determination of change.* J. Stat. Plan. Infer. **60**, 291 – 310.

Antoch J., Hušková M. and Veraverbeke N. (1995). *Change-point problem and bootstrap.* J. Nonparametric Statistics **5**, 123 – 144.

Armitage P. (1950). *Sequential analysis with more than two alternative hypotheses and its relation to discriminant analysis.* J. Royal Statistical Society **B 12**, 137 – 144.

Assaf D. and Ritov Ya. (1988). *A double sequential procedure for detecting a change in distribution.* Biometrika **75**, 715 – 722.

Bagshaw M. and Johnson R. A. (1975a). *The influence of reference values and estimated variance on the ARL of CUSUM tests.* J. Royal Statistical Society **37**, 413 – 419.

Bagshaw M. and Johnson R. A. (1975b). *The effect of serial correlation on the performance of CUSUM tests II.* Technometrics **17**, 73 – 80.

Bagshaw M. and Johnson R. A. (1977). *Sequential procedures for detecting parameter changes in a time series model.* J. American Statistical Association **72**, 593 – 597.

Baker S. V. (1977). *The Numerical Treatment of Integral Equations.* Clarendon Press. Oxford.

Barnard G. A. (1959). *Control charts and stochastic processes (with discussion).* J. Royal Statistical Society **B 21**, 239 – 271.

Basseville M. and Benveniste A. (eds.) (1986). *Detection of abrupt changes in signals and dynamic systems.* Springer, New York.

Bather J. A. (1963). *Control charts and the minimization of costs.* J. Royal Statistical Society **B 25**, 49 – 80.

Bather J. A. (1976). *A control chart model and a generalized stopping problem for Brownian motion.* Mathematical Operational Research **1**, 209 – 224.

Bauer P. and Hackl P. (1982). *A parabolic mask for continuous sampling inspection.* Zeitschrift für Operational Research **26**, 47 - 57.

Bhattacharya G. K. and Frierson F. Jr. (1981). *A nonparametric control chart for detecting small disorders.* Annals of Statistics **9**, 544 – 554.

Bhattacharya G. K. and Johnson R. A. (1968). *Non-parametric tests for shift at an unknown time point.* Annals of Mathematical Statistics **39**, 1731 – 1743.

Billingsley P. (1968). *Convergence of Probability Measures.* J. Wiley, New York.

Bissel A. F. (1969). *CUSUM techniques for quality control.* Applied Statistics **18**, 1 – 30.

Bissel A. F. (1979). *A semi-parabolic mask for CUSUM charts.* Statistician **28**, 1 – 7.

Brook D. and Evans D. A. (1972). *An approach to the probability distribution of CUSUM run length.* Biometrika **59**, 539 – 549.

Burr I. W. (1969). *Control charts for measurements with varying sample sizes.* J. Quality Technology **1**, 163 – 167.

Burr I. W. (1976). *Quality Control Methods.* M. Dekker, New York.

Champ C. W. and Woodal W. H. (1987). *Exact results for Shewhart control chart with supplementary run rules.* Technometrics **29**, 393 – 400.

Charboneau H. and Webster G. L. (1978). *Industrial Quality Control.* Prentice Hall, Englewood Cliffs, New Jersey.

Chernoff H. and Zacks S. (1964). *Estimating the current mean of a normal distribution which is subject to changes in time.* Annals of Mathematical Statistics **35**, 999 – 1018.

Chiu W. (1974). *The economic design of CUSUM charts for controlling normal means.* Applied Statistics **23**, 420 – 433.

Crosier R. B. (1986). *A new two-sided cumulative sum quality control scheme.* Technometrics **28**, 187 – 194.

Crowder S. V. (1987a). *Average run lengths of exponentially weighted moving average charts.* J. Quality Technology **19**, 161 – 164.

Crowder S. V. (1987b). *A simple method for studying run-length distributions for exponentially weighted moving average charts.* Technometrics **29**, 401 – 407.

Csörgő M. and Horváth S. (1987). *Detection of changes in random sequences.* J. Multivariate Analysis **47**, 119 – 130.

Darkhovskii B. S. and Brodskii B. E. (1987). *A nonparametric method for fastest detection of a change in the mean of a random sequence.* Theory of Probability and its Applications **32**, 640 – 648.

Darkhovskii B. S. and Brodskii B. E. (1993). *Nonparametric Methods in Change-Point Problems.* Kluver, Dordrecht.

Darling D. A. and Siegert A. J. F. (1953). *The first passage problem for a continuous Markov process.* Annals of Mathematical Statistics **24**, 624 – 639.

Deming W. E. (1983). *Quality, Productivity and Competitive Position.* MIT Press, Cambridge MA.

Devor R. E., Chang T. and Sutherland J. W. (1992). *Statistical Quality Design and Control.* MacMillan, New York.

Duncan A. J. (1974). *Quality Control and Industrial Statistics, 4th ed.* Irwin, New York.

Dvoretzky A., Kiefer J. and Wolfowitz J. (1950). *Sequential decision problems for processes with continuous time parameter.* Annals of Mathematical Statistics **24**, 254 – 264.

Eisenberg B. and Ghosh B. K. (1991). *The sequential probability ratio test.* Handbook of Sequential Analysis, Ghosh B. K. and Sen P. K. eds., 47 – 66, M. Dekker, New York.

Ewan D. E. (1963). *When and how to use CUSUM charts.* Technometrics **5**, 1 – 22.

Ewan W. D. and Kemp K. W. (1960). *Sampling inspection of continuous processes with no autocorrelation between successive results.* Biometrika **47**, 555 – 573.

Feller W. (1951). *An Introduction to the Probability Theory and its Applications, Vol. I.* J. Wiley, New York.

Feller W. (1966). *An Introduction to the Probability Theory and its Applications, Vol. II.* J. Wiley, New York.

Ghosh B. K. and Sen P. K. (1991). *Handbook of Sequential Analysis.* M. Dekker, New York.

Girshick M. A. and Rubin H. (1952). *A Bayes approach to a quality control model.* Annals of Mathematical Statistics **23**, 114 – 125.

Goel A. L. and Wu S. M. (1971). *Determination of ARL and contour nomogram for CUSUM charts to control normal mean.* Technometrics **13**, 221 – 230.

Goldsmith R. L. and Whitfield H. (1961). *Average run lengths in cumulative chart quality control schemes.* Technometrics **3**, 11 – 22.

Grant E. L. (1988). *Statistical Quality Control, 5th ed.* McGraw-Hill, New York.

Healy J. D. (1987). *A note on multivariate CUSUM procedures.* Technometrics **29**, 409 – 412.

Hinkley D. V. (1969). *Inference about the intersection in two-phase regression.* Biometrika **56**, 495 – 504.

Hinkley D. V. (1970). *Inference about the change-point in a sequence of random variables.* Biometrika **57**, 1 – 17.

Hinkley D. V. (1971). *Inference about the change-point from cumulative sum tests.* Biometrika **58**, 509 – 523.

Hudson D. J. (1966). *Fitting segmented curves whose joint points have to be estimated.* J. American Statistical Association **61**, 1097 – 1129.

Hunter J. S. (1986). *The exponentially weighted moving average.* J. Quality Technology **18**, 239 – 250.

Hušková M. (1994). *Nonrecursive procedures for detecting change in simple linear regression models.* In: New Progress in Probability and Statistics, Puri M. L. and Vilaplana J. P. eds., 105 – 126.

Hušková M. (1995). *Estimators for epidemic alternatives.* Comment. Math. Univ. Carolinae **36**, 279 – 291.

Hušková M. and Sen P. K. (1989). *Nonparametric tests for shift and change in regression at an unknown time point.* In: Statistical Analysis and Forecasting of Economic Structural Change, Hackl P. ed., 71 – 85, Springer Verlag, New York.

Ishikawa K. (1982). *A Guide to Quality Control.* Asian Productivity Organization, Tokyo.

Jackson J. E. (1985). *Multivariate quality control.* Communications in Statistics **A 14**, 2657 – 2688.

James B., James K. L. and Siegmund D. (1987). *Tests for a change point.* Biometrika **74**, 71 – 83.

Jarušková D. (1996). *Change-point detection in meteorological measurements.* Monthly Weather Review **124**, 1535 – 1543.

Jarušková D. (1997). *Some problems with application of change-point detection methods to environmental data.* Environmetrics **8**, 469 – 483.

Jarušková D. (1998a). *Change-point estimator in gradually changing sequences.* Comment. Math. Univ. Carolinae.

Jarušková D. (1998b). *Testing appearance of linear trend.* Stat. Plann. Infer. **70**, 263 – 276.

Jarušková D. and Antoch J. (1994). *Detection of change in variance.* In: 5th Prague Symposium on Asymptotic Statistics Mandl P. and Hušková M. eds., 297 – 301, Physica-Verlag, Heidelberg.

Johnson N. L. (1961). *A simple theoretical approach to cumulative sum control charts.* J. American Statistical Association **56**, 835 - 840.

Johnson R. A. and Bagshaw M. (1974). *The effect of serial correlation on the performance of CUSUM tests I.* Technometrics **16**, 103 – 112.

Johnson N. L. and Leone F. C. (1962). *Cumulative control charts: Mathematical principles applied to their construction and use I.* Industrial Quality Control **18**, 15 – 21.

Johnson N. L. and Leone F. C. (1963). *Cumulative control charts: Mathematical principles applied to their construction and use II.* Industrial Quality Control **19**, 29 – 36.

Johnson N. L. and Leone F. C. (1964). *Cumulative control charts: Mathematical principles applied to their construction and use III.* Industrial Quality Control **20**, 22 – 28.

Juran J. M. (ed.) (1988). *Quality Control Handbook, 4th ed.* McGraw-Hill, New York.

Kantorowich L. V. and Krylov V. I. (1958). *Approximate Methods of Higher Analysis.* Interscience Publishers, New York.

Keats J. B. and Hubele N. F. (eds.) (1989). *Statistical Process Control in Automated Manufacturing.* M. Dekker, New York.

Kemp K. W. (1961). *The average run length of the cumulative sum chart when a V-mask is used.* J. Royal Statistical Society **B 23**, 149 – 153.

Kemp K. W. (1962). *The use of cumulative sums for sampling inspections schemes.* Applied Statistics **11**, 16 – 31.

Kennedy D. P. (1971). *Some martingales related to cumulative sum tests and single server queues.* Stochastic Processes and their Applications **4**, 261 – 269.

Khan R. A. (1978). *Wald's approximations to the average run length in CUSUM procedures.* J. Statistical Planning and Inference **2**, 63 – 77.

Khan R. A. (1979). *A sequential detection procedure and the related CUSUM procedure.* Sankhyā **B 40**, 146 – 162.

Khan R. A. (1984). *On cumulative sum procedures and the SPRT with applications.* J. Royal Statistical Society **B 46**, 79 – 85.

Kiefer J. and Sacks J. (1963). *Asymptotically optimum sequential inference and design.* Annals of Mathematical Statistics **34**, 705 – 750.

Krylov V. I. (1963). *Approximate Calculation of Integrals.* MacMillan, New York.

Kume H. (1985). *Statistical Methods for Quality Improvement AOTS.* Chosakai, Tokyo.

Lai T. L. (1974). *Control charts based on weighted sums.* Annals of Statistics **2**, 134 – 147.

Lalley S. (1984). *Limit theorems for first passage times in linear and non-linear renewal theory.* Advances in Applied Probability **16**, 766 – 803.

Lehmann E. L. (1983). *Theory of Point Estimation.* J. Wiley, New York.

Lehmann E. L. (1986). *Testing Statistical Hypotheses, 2nd ed.* J. Wiley, New York.

Lombard F. (1987). *Rank tests for change-point problem.* Biometrika **74**, 615 – 624.

Lombard F. (1988). *Detecting change-points by Fourrier analysis.* Technometrics **30**, 305 – 310.

Lorden G. (1970). *On excess over the boundary.* Annals of Mathematical Statistics **41**, 520 – 527.

Lorden G. (1971). *Procedures for reacting to a change in distribution.* Annals of Mathematical Statistics **42**, 1897 – 1908.

Lorden G. and Eisenberger I. (1973). *Detection of failure rate increases.* Technometrics **15**, 167 – 175.

Lucas J. M. (1973). *A modified V-mask control scheme.* Technometrics **15**, 833 – 847.

Lucas J. M. (1976). *The design and use of V-mask control schemes.* J. Quality Technology **8**, 1 – 12.

Lucas J. M. (1982). *Combined Shewhart-CUSUM quality control schemes.* J. Quality Technology **14**, 51 – 59.

Lucas J. M. and Crosier R. B. (1982). *Fast initial response for CUSUM quality control schemes: Give your CUSUM a head start.* Technometrics **24**, 199 – 205.

Maeder R. (1991). *Programming in Mathematica, 2nd ed.* Addison – Wesley, Redwood City, California.

MATLAB®. The MathWorks Inc., 24 Prime Park Way, Natick, MA 01 760, USA.

Montgomery D. C. (1990). *Introduction to Statistical Quality Control, 2nd ed.* J. Wiley, New York.

Moustakides G. V. (1986). *Optimal stopping times for detecting changes in distribution.* Annals of Statistics **14**, 1379 – 1387.

Mundorf A. G. (1980). *A control chart based on cumulative scores.* Applied Statistics **29**, 252 – 258.

Nadler J. and Robbins N. B. (1971). *Some characteristics of Page's two-sided procedure for detecting a change in a location parameter.* Annals of Mathematical Statistics **42**, 538 – 551.

Nelson L. (1982a). *Control charts for individual measurements.* J. Quality Technology **14**, 172 – 174.

Nelson L. (1982b). *Control charts for medians.* J. Quality Technology **14**, 226 – 227.

Nelson L. (1983). *The deceptiveness of moving averages.* J. Quality Technology **15**, 99 – 100.

Nelson L. (1984). *The Shewhart control chart tests for special causes.* J. Quality Technology **16**, 237 – 239.

Nelson L. (1985). *Interpreting Shewhart Control $\overline{X}$-charts.* J. Quality Technology **17**, 114 – 116.

Neuhaus W. (1987). *Early warning.* Scandinavian Actuarial J. **57**, 128 – 156.

Nikiforov I. V. (1983). *Sequential Detection of Changes in Characteristics of Time Series.* Nauka, Moscow.

Ott E. R. (1975). *Process Quality Control.* McGraw-Hill, New York.

Page E. S. (1954a). *Continuous inspection schemes.* Biometrika **41**, 100 – 115.

Page E. S. (1954b). *A test for a change in a parameter occurring at a unknown point.* Biometrika **41**, 523 – 527.

Page E. S. (1955). *A test for a change in parameter occurring at an unknown point.* Biometrika **42**, 523 – 527.

Page E. S. (1957). *On problems in which a change in a parameter occurs at an unknown point.* Biometrika **44**, 248 – 252.

Page E. S. (1961). *Cumulative sum charts.* Technometrics **3**, 1 – 9.

Page E. S. (1963). *Controlling the standard deviation by CUSUM and warning lines.* Technometrics **5**, 307 – 315.

Pall G. A. (1987). *Quality Process Management.* Prentice-Hall, Englewood Cliffs, NJ.

Pollack M. (1985). *Optimal detection of a change in distribution.* Annals of Statistics **13**, 206 – 227.

Pollack M. (1987). *Average run lengths of an optimal method of detecting a change in distribution.* Annals of Statistics **15**, 749 – 779.

Pollack M. and Siegmund D. (1975). *Approximations to the expected sample size of certain sequential tests.* Annals of Statistics **3**, 1267 – 1282.

Reynolds M. R. Jr. (1975). *Approximations to the average run length in cumulative sum control charts.* Technometrics **17**, 65 – 71.

Roberts S. W. (1959). *Control chart tests based on geometric moving averages.* Technometrics **1**, 239 – 250.

Roberts S. W. (1966). *A comparison of some control charts procedures.* Technometrics **8**, 411 – 430.

Robins H. and Siegmund D. (1970). *Boundary crossing probabilities for the Wiener process and sample sums.* Annals of Mathematical Statistics **41**, 1410 – 1429.

Robinson P. B. and Ho T. Y. (1978). *Average run lengths of geometric moving average charts by numerical methods.* Technometrics **20**, 85 – 93.

Rowlands R. J. et al. (1982). *Snub-nosed V-mask control schemes.* Statistician **31**, 1 – 10.

Rowlands R. J. and Wetherill G. B. (1991). *Quality control.* Handbook of Sequential Analysis, Ghosh B. K. and Sen P. K. eds., 563 – 580, M. Dekker, New York.

S+®. Statistical Sciences Inc., 1700 Westlake Avenue North, Suite 500, Seatle, WA 98 109, USA.

SAS®. SAS Institute Inc., SAS Campus Drive, Cary, NC 27 513, USA

SAS Institute Inc. (1991). *SAS/QC® Software*: Reference Version 6, 1st ed., 3rd printing SAS Institute Inc. Cary.

Shaban S. A. (1980). *Change-point problem and two-phase regression: An annotated bibliography.* International Statistical Review **48**, 83 – 93.

Shewhart W. A. (1928). *The application of statistics as an aid in maintaining quality of a manufactured products.* J. American Statistical Association **23**, 546 – 548.

Shewhart W. A. (1931). *The Economic Control of Quality of a Manufactured Product.* Van Nostrand, New York (Republished in 1980 by The American Society for Quality Control).

Shirland L. E. (1993). *Statistical Quality Control with Microcomputer Applications.* J. Wiley, New York.
Shiryayev A. N. (1963a). *On optimal methods in the quickest detection problems.* Theory of Probability and its Applications **8**, 26 – 51.
Shiryayev A. N. (1963b). *To disorder detection for a technological process I.* Theory of Probability and its Applications **8**, 264 – 281.
Shiryayev A. N. (1963c). *To disorder detection for a technological process II.* Theory of Probability and its Applications **8**, 431 – 443.
Shiryayev A. N. (1963d). *Some precise formulas in change-point problems.* Theory of Probability and its Applications **10**, 380 – 385.
Shiryayev A. N. (1976). *Statistical Sequential Analysis.* Nauka, Moscow.
Shiryayev A. N. (1978). *Optimal Stopping Rules.* Springer, Heidelberg.
Siegmund D. (1975). *Error probability and average sample number of the sequential probability ratio test.* J. Royal Statistical Society **B 37**, 394 – 401.
Siegmund D. (1985). *Sequential Analysis: Test and Confidence Intervals.* J. Wiley, New York.
Siegmund D. (1986). *Boundary crossing probabilities and statistical applications.* Annals of Statistics **14**, 361 – 404.
Siegmund D. (1988). Confidence sets in change-point problems Inter. Statist. Review **56**, 31 – 48.
SPSS®. SPSS Inc., 444 N Michigan Avenue, Chicago, IL 60611, USA
STATGRAPHICS®. STSC Inc., 2125 East Jefferson Street, Rockville, MD 20852, USA
STATISTICA®. StatSoft, 2325 E. 13th St., TULSA, OK 74104, USA
Stein C. (1946). *A note on cumulative sums.* Annals of Mathematical Statistics **17**, 498 – 499.
Telksnys L. (1986). *Detection of Changes in Random Processes.* Optimization Software, New York.
Tukey J. W. (1977). *Exploratory Data Analysis.* Addison-Wesley, Massachusetts.
Vance L. C. (1983). *A bibliography of statistical quality control chart techniques.* J. Quality Technology **15**, 59 – 62.
van Dobben de Bruyn C. S. (1968). *Cumulative Sum Tests: Theory and Practice.* Hafner, New York.
Vasilopoulos A. V. and Stamboulis A. P. (1978). *Modification of control chart limits in the presence of data correlation.* J. Quality Technology **10**, 20 – 30.
Wadsworth H. M., Stephens K. S. and Godfrey A. B. (1986). *Modern Methods for Quality Control and Improvement.* J. Wiley, New York.
Wald A. (1947). *Sequential Analysis.* J. Wiley, New York.
Weiler K. (1954). *A new type of control chart, limits for mean, ranges and sequential runs.* J. American Statistical Association **49**, 298 – 314.
Wetherill G. B. (1977). *Sampling Inspection and Quality Control, 2nd ed.* Chapman and Hall, London.
Wheeler D. F. (1983). *Detecting a shift in process average: tables of the power function for $\overline{X}$-charts.* J. Quality Technology **15**, 155 – 170.
Wijsman R. A. (1991). *Stopping times: termination, moments, distribution.* Handbook of Sequential Analysis, Ghosh B. K. and Sen P. K. eds., M. Dekker, New York, 67 – 120.
Wolfram S. (1991). *Mathematica: A System for Doing Mathematics by Computer, 2nd ed.* Addison – Wesley Redwood City, California.
Woodall W. H. (1983). *The distribution of the run length of one-sided CUSUM procedures for continuous random variables.* Technometrics **25**, 295 – 300.
Woodall W. H. (1984). *On the Markov chain approach to the two-sided CUSUM procedure.* Technometrics **26**, 41 – 46.
Woodall W. H. (1985a). *The statistical design of quality control charts.* The Statistician **34**, 155 – 160.

Woodall W. H. and Ncube M. M. (1985b). *Multivariate CUSUM quality-control procedures.* Technometrics **27**, 285 – 292.

Woodroofe M. (1982). *Nonlinear Renewal Theory in Sequential Analysis.* SIAM, Philadelphia.

Woodroofe M. (1991). *The role of renewal theory in sequential analysis.* Handbook of Sequential Analysis, Ghosh B. K. and Sen P. K. eds, 145 – 168, M. Dekker, New York.

Woodward R. H. and Goldsmith R. L. (1964). *Cumulative Sum Techniques.* Oliver and Boyd, London.

Yaschin E. (1985). *On a unified approach to the analysis of two-sided cumulative sum schemes with headstarts.* Advances in Applied Probability **17**, 562 – 593.

Zacks S. (1981). *The probability distribution and the expected value of a stopping variable associated with one-sided CUSUM procedures for non-negative integer valued random variables.* Communications in Statistics **A 10**, 2245 – 2258.

Zacks S. (1983). *Survey of classical and Bayesian approaches to the change-point problem.* Recent Advances in Statistics, Rizvi M. H. ed., 245 – 269, Academic Press, New York.

Zacks S. (1991). *Detection and change-point problems.* Handbook of Sequential Analysis, Ghosh B. K. and Sen P. K. eds., 531 – 562, M. Dekker, New York.

Zacks S. and Kander Z. (1966). *Test procedures for possible changes in parameters of statistical distributions occurring at unknown time points.* Annals of Mathematical Statistics **37**, 1196 – 1210.

MULTIVARIATE CONTROL CHARTS FOR COMPLEX PROCESSES

Luan Jaupi

Département de Mathématiques, Conservatoire National des Arts et Métiers, 292 Rue Saint Martin, F – 751 41 PARIS CEDEX 03, France; jaupi@cnam.fr

Abstract: This paper deals with multivariate control problems and the use of influence function is proposed to distinguish between chance and special causes of variation. The influence function of mean is proposed to monitor the process mean. To investigate process variability, the control charts based on the influence functions of eigenvalues are suggested. Finally, in order to describe process orientation, control charts based on the influence functions of eigenvectors are employed. A real application illustrating the proposed control charts is presented.

1. Introduction

Frequently, manufactured items need the values of several different quality characteristics for an adequate description of their quality. Each of a number of these quality characteristics must satisfy certain specifications, but the quality of the product depends on the combined effect of these characteristics, rather than on their individual values. When assignable causes are present in a multivariate process they may affect different process parameters: the process mean, and/or the process variability, and/or the process orientation. Indeed, the quality characteristic vector has both a magnitude and a direction. Therefore control charts to monitor process mean-variability-orientation are necessary. A diversity of multivariate control charts for the process mean and dispersion have been proposed recently to distinguish between random and assignable causes of process variability. Most of the proposed control charts are based on T^2 or χ^2 statistics. This paper deals with multivariate control problems and the use of influence functions is proposed to distinguish between chance and special causes of variation. When the process has reached a state of statistical control, the process mean and the structure of dispersion matrix should be stable over the time. The effect of observations or subgroups on these parameters may be evaluated, among others, by the means of influence functions. Hence, special causes of variation could be identified by an unusual influence of observations or subgroups on the process mean and/or dispersion parameters. To monitor the process mean the influence function of mean is proposed, to investigate process variability control charts based on the influence functions of eigenvalues are suggested, and to

2000 *Mathematics Subject Classification.* 62P30.

Key words and phrases. Multivariate control charts, influence function, PCA, CUSUM, T^2-statistic, process mean, process variability, process orientation.

describe process orientation control charts based on the influence functions of eigenvectors may be employed.

With complex products we usually find many different types of quality characteristics. Generally, not all of these quality characteristics are equally important. Some of them may be very critical to describe each item quality, and some of them may be moderate ones. Moreover, when the number of quality characteristics is high, the use of multivariate control charts based on T^2 or χ^2 statistics may be very inappropriate, because the upper control limit of these control charts raises steeply with the number of variables. Therefore it is convenient to employ control procedures that are based on leading quality characteristics but enables to identify special causes that may affect the overall process. Our approach to build up such control charts consists in monitoring the stable level of variability of the process according to the directions settled by a generalized PCA which is based on the leading variables.

In Section 2 we derive the influence function of a multivariate location-scale parameter and those of each eigenvalue and eigenvector of a covariance matrix. Multivariate control charts for process mean, variability and orientation are given in Section 3. In Section 4 control charts to monitor a complex process are developed. In Section 5 an application from the field of automobiles is presented aiming at illustrating the proposed control charts. Some remarks on the use of influence functions in process control are given in Section 6.

2. Influence measures to monitor multivariate processes

2.1. Influence function of mean and covariance matrix

Let $T = T(F)$ be a statistical functional. The influence function $IF(\boldsymbol{x}; T, F)$ of the statistical functional T at F is defined as the limit, when $\varepsilon \to 0$, of

$$\left\{T[(1-\varepsilon)F + \varepsilon\delta_x] - T(F)\right\}/\varepsilon, \tag{2.1}$$

where $\delta_{\boldsymbol{x}}$ denotes the distribution giving unit mass to the point $\boldsymbol{x} \in \mathcal{R}_p$. The perturbation of F by $\delta_{\boldsymbol{x}}$ is denoted as

$$F_{\varepsilon \boldsymbol{x}} = (1-\varepsilon)F + \varepsilon\delta_{\boldsymbol{x}}, \qquad 0 \leq \varepsilon \leq 1. \tag{2.2}$$

As such, the influence function measures the rate of change of T as F is shifted infinitesimally in the direction of $\delta_{\boldsymbol{x}}$, for details see Hampel, Ronchetti, Rousseeuw and Stahel (1986). For the sake of simplicity, we denote the influence function of T by attaching a superscript $\tilde{}$, i.e. $\widetilde{T}$.

Let $\boldsymbol{X}$ be a p-dimensional random variable defined on $\mathcal{R}_p$, let $(\boldsymbol{\mu}, \boldsymbol{\Sigma})$ denote the location-scale parameter defined by

$$\begin{aligned} \mathsf{E}\,(\boldsymbol{x}-\boldsymbol{\mu}) &= 0 \\ \mathsf{E}\,\left[(\boldsymbol{x}-\boldsymbol{\mu})\,(\boldsymbol{x}-\boldsymbol{\mu})'\right] &= 0 \end{aligned} \tag{2.3}$$

In order to calculate the influence function of mean $\boldsymbol{\mu}$ and covariance matrix $\boldsymbol{\Sigma}$, we substitute F by $F_{\varepsilon \boldsymbol{x}}$ in (2.3) and take the derivative with respect to ε at $\varepsilon = 0$. The differentiation of the mean equation gives

$$\widetilde{\boldsymbol{\mu}} = \boldsymbol{x} - \boldsymbol{\mu} \tag{2.4}$$

and the differentiation of the covariance matrix equation gives

$$\widetilde{\boldsymbol{\Sigma}} = (\boldsymbol{x} - \boldsymbol{\mu})(\boldsymbol{x} - \boldsymbol{\mu})' - \boldsymbol{\Sigma}, \tag{2.5}$$

cf. Huber (1981), Critchley (1985), Hampel, Ronchetti, Rousseeuw and Stahel (1986) or Jaupi and Saporta (1993).

2.2. Influence function of eigenvalues and eigenvectors

We assume that $\boldsymbol{\Sigma}$ has distinct eigenvalues $\lambda_1 > \lambda_2 > \cdots > \lambda_p$ and we denote by $\boldsymbol{\alpha}_1, \boldsymbol{\alpha}_2, \ldots, \boldsymbol{\alpha}_p$ the associated eigenvectors. Under regularity conditions and by using Sibson's lemma [1979], we find the following expressions for the influence function of the j-th eigenvalue and the associated eigenvector of the covariance matrix respectively

$$\widetilde{\lambda_j} = \boldsymbol{\alpha}_j{}' \widetilde{\boldsymbol{\mu}} \boldsymbol{\alpha}_j \tag{2.6}$$

$$\widetilde{\boldsymbol{\alpha}_j} = \sum_{k=1, k \neq j}^{p} (\lambda_j - \lambda_k)^{-1} \boldsymbol{\alpha}_k \boldsymbol{\alpha}_k{}' \widetilde{\boldsymbol{\Sigma}} \boldsymbol{\alpha}_j, \tag{2.7}$$

cf. Sibson (1979), Critchley (1985), Tanaka (1988) or Jaupi (1992). Generally, in the applications of the influence function, the unknown distribution function F has to be estimated by $\widehat{F}$, the empirical distribution function based on a random sample from F. Replacing F by $\widehat{F}$ and taking $x = x_i$, the i-th observation, we have a sample version of the influence function. The importance about the influence function lies in its heuristic interpretation: it describes the effect of an infinitesimal contamination at point x_i on the estimate.

3. Multivariate control charts

Consider a manufacturing process where each item is characterized by a p quality characteristics vector $\boldsymbol{X} = (x_1, x_2, \ldots, x_p)'$. Because of interdependency between the quality characteristics, the random variables are correlated. The problem therefore requires multivariate control procedures to distinguish between random and assignable causes of variability. When the process has reached a state of statistical control, that is the process is operating within its limits of expected variation, the process mean and the structure of the dispersion matrix should be stable over time. Special causes of variation could be identified by an unusual influence of observations or subgroups over process parameter estimators. When special causes are present in a multivariate process they may affect different process parameters the process mean, and/or orientation and/or variability. Generally special causes that affect one of these parameters do not necessarily affect the others. Therefore special causes detection techniques to monitor different process parameters are necessary. To monitor the process mean the influence function of mean will be employed, to investigate process variability control charts based on the influence functions of eigenvalues will be used, and to describe process orientation control charts based on the influence functions of eigenvectors will be employed. Detailed applications of multivariate control charts for process mean, variability and orientation are given in Jaupi and Saporta (1994), Jaupi and Saporta (1997a), Jaupi and Saporta (1997b).

3.1. Control chart for process mean

When one is monitoring the process mean, what is calculated and plotted on a chart is the value of a quadratic form. In general, a quadratic form can be written as $\boldsymbol{X}'\boldsymbol{M}\boldsymbol{X}$, where $\boldsymbol{M}$ is a definite positive matrix and $\boldsymbol{X} \in \mathcal{R}_p$. Taking

$$\boldsymbol{M} = \widehat{\boldsymbol{\Sigma}}^{-1} \tag{3.1}$$

with $\widehat{\boldsymbol{\Sigma}}^{-1}$ the inverse of sample covariance matrix, we find a first sample version of a dignified statistic, Hotelling's $\boldsymbol{T}^2$. Thus, for the i-th subgroup, $i = 1, \ldots, k$, one would calculate and plot in the control chart the quantity

$$\boldsymbol{T}_i^2 = \widetilde{\mu}' \widehat{\boldsymbol{\Sigma}}^{-1} \widetilde{\mu} = (x_i - \bar{x})' \, \widehat{\boldsymbol{\Sigma}}^{-1} \, (x_i - \bar{x}) \, . \tag{3.2}$$

Each of $\boldsymbol{T}_i^2$ values, $i = 1, \ldots, k$, of (3.2) would be compared with

$$UCL = \frac{(k-1)\,p}{k-p} F_{\alpha,p,k-p}, \tag{3.3}$$

cf. Jackson (1959).

Points lying outside the control limit indicate the time at which the potential assignable causes occured in the process. Knowing the time of such highly influential occurrence is the best evidence for searching out and eliminating a process disturbance.

3.2. Control charts for process variability

The total variance of $\boldsymbol{X}$ is equal to the sum of the eigenvalues of the covariance matrix $\boldsymbol{\Sigma}$

$$\mathsf{Var}\,(\boldsymbol{X}) = \lambda_1 + \lambda_2 + \cdots + \lambda_p . \tag{3.4}$$

Assignable causes that affect the variability of the output do not increase significantly each component of total variance of $\boldsymbol{X}$. Instead, they may have a large influence in the variability of some components and small effect in the remaining directions. Therefore, an approach to design control charts for variability consists in detecting any significant departure from the stable level of the variability of each component. That is, if one wants to monitor the process variability according to the j-th direction, $j = 1, \ldots, p$, what would be calculated and plotted on a control chart is the value of the influence function of the j-th eigenvalue of the covariance matrix $\boldsymbol{\Sigma}$. Thus, for the i-th subgroup, $i = 1, \ldots, n$, one would calculate and plot

$$\widetilde{\lambda}_j = IF\left(x_i, \lambda_j, \widehat{F}\right) = C_{ij}^2 - \widehat{\lambda}_j, \tag{3.5}$$

where C_{ij} is the j-th principal component, $j = 1, \ldots, p$, of the i-th observation. The control limits are three sigma control limits as in any Shewhart control chart. The decision procedure is the same as in all control charts. It means that points lying outside the control limits indicate the time at which special causes that affect the variability are present in the process.

3.3. Control charts for process orientation

It is important to underline a key feature of the multivariate processes. A multivariate quality vector $\boldsymbol{X}$ has both a magnitude and a direction. Therefore, it is possible that a special cause may have a small effect on process mean and/or variability, and its presence is only revealed in relation with the orientation structure of the process. The process orientation may be described by means of the influence function of the eigenvectors. The influence functions of eigenvectors are p-dimensional vectors too. Therefore, two different approaches may be used to monitor the process orientation. The first strategy consists in setting up control charts to monitor each component of these p-dimensional vectors. The second approach consists in calculating and plotting the value of a quadratic form based on the influence functions of eigenvectors. These values should be compared with control limits which are three sigmas control limits as in any Shewhart control chart. The decision procedure is the same as in all control charts.

4. Control charts to monitor multivariate complex processes

4.1. Motivation

Consider a manufacturing process where the values of p-quality characteristics are necessary to describe each item quality. Let denote the quality vector by $\boldsymbol{X} = (x_1, x_2, \ldots, x_p)'$. Generally, not all of these quality characteristics are equally important. Suppose that the first l quality characteristics are leading ones and the last $(p - l)$ quality characteristics are moderate ones, noted respectively as $\boldsymbol{X}_1 = (x_1, x_2, \ldots, x_l)'$ and $\boldsymbol{X}_2 = (x_{l+1}, x_{l+2}, \ldots, x_p)'$. Then the quality vector $\boldsymbol{X}$ may be written as $\boldsymbol{X} = (\boldsymbol{X}_1', \boldsymbol{X}_2')'$. Each of these quality characteristics must satisfy certain specifications. The quality of the product depends on the combined effect of these characteristics, rather than on their individual values. Our aim is to detect quickly the presence of assignable causes in the process. When the number of quality characteristics is large the use of conventional methods may be very inappropriate. Therefore it will be convenient to employ a control procedure that is based on the leading quality characteristics but enables to identify special causes that may affect the overall process. An approach to build up such control charts consists in monitoring the stable level of variability of the process according to the directions settled by a generalized PCA, which is based on the leading variables. To find such principal components Rao (1964)'s approach on principal components of instrumental variables will be used. Then the influence functions of eigenelements of the generalized PCA may be evaluated by using the perturbation theory of eigenvalue problems, Tanaka and Mori (1995). Finally, these influence functions are used to detect the presence of assignable causes in the process.

4.2. Generalized PCA

Let $\boldsymbol{X}$ be a p-dimensional random variable defined on $\mathcal{R}_p$, let $(\boldsymbol{\mu}, \boldsymbol{\Sigma})$ denote the location-scale parameter defined by the classical estimators. Let $\boldsymbol{X}_1$ and $\boldsymbol{X}_2$ be two sub-vectors of $\boldsymbol{X}$ with l and $(p-l)$ dimensions respectively, such as $\boldsymbol{X} = (\boldsymbol{X}_1', \boldsymbol{X}_2')'$. Let $(\boldsymbol{\mu}_1, \boldsymbol{\Sigma}_1)$ and $(\boldsymbol{\mu}_2, \boldsymbol{\Sigma}_2)$ denote the location scale parameters of $\boldsymbol{X}_1$ and $\boldsymbol{X}_2$. We have the following expressions for $\boldsymbol{\mu}$ and $\boldsymbol{\Sigma}$:

$$\boldsymbol{\mu} = (\boldsymbol{\mu}_1', \boldsymbol{\mu}_2')' \qquad \text{and} \qquad \boldsymbol{\Sigma} = \begin{pmatrix} \boldsymbol{\Sigma}_1 & \boldsymbol{\Sigma}_{12} \\ \boldsymbol{\Sigma}_{21} & \boldsymbol{\Sigma}_2 \end{pmatrix} \tag{4.1}$$

We wish to generate l linear combinations $\boldsymbol{Y} = \boldsymbol{A}'\boldsymbol{X}_1$, in such way that the predictive efficiency of $\boldsymbol{Y}$ for $\boldsymbol{X}$ is maximum. It is well known that the solution depends on the measure of predictive efficiency. Using as predictive efficiency the trace operator we have the following solution: the columns of matrix $\boldsymbol{A}$ consist of eigenvectors of the following equation

$$\left|(\boldsymbol{\Sigma}_1^2 + \boldsymbol{\Sigma}_{12}\boldsymbol{\Sigma}_{21}) - \lambda\boldsymbol{\Sigma}_1\right| = 0. \tag{4.2}$$

Assuming that the eigenvalues $\lambda_1, \lambda_2, \ldots, \lambda_l$ are ordered from the largest to the smallest and denoting by $\boldsymbol{\alpha}_1, \boldsymbol{\alpha}_2, \ldots, \boldsymbol{\alpha}_l$ the associated eigenvectors, the solution is given as follows

$$\boldsymbol{A}' = (\boldsymbol{\alpha}_1, \boldsymbol{\alpha}_2, \ldots, \boldsymbol{\alpha}_l), \tag{4.3}$$

cf. Rao (1964). We refer this type of PCA as generalized PCA.

Under regularity conditions and by using the perturbation theory of eigenvalue problems we have the following expressions for the influence function of the j-th eigenvalue of a generalized PCA

$$\widetilde{\lambda}_j = \widehat{\alpha}_j \left(\widetilde{\boldsymbol{\Pi}} - \widehat{\lambda}_j\widetilde{\boldsymbol{\Sigma}}_1\right)\widehat{\alpha}_j, \qquad j = 1, \ldots, l. \tag{4.4}$$

where $\widetilde{\boldsymbol{\Pi}} = \widetilde{\boldsymbol{\Sigma}}_1\boldsymbol{\Sigma}_1 + \boldsymbol{\Sigma}_1\widetilde{\boldsymbol{\Sigma}}_1 + \widetilde{\boldsymbol{\Sigma}}_{12}\boldsymbol{\Sigma}_{21} + \boldsymbol{\Sigma}_{12}\widetilde{\boldsymbol{\Sigma}}_{21}$, cf. Tanaka and Mori (1995).

4.3. Control charts for process variability

Assignable causes that affect the variability of the output do not increase significantly each component of total variance. Instead, they may have a large influence on the variability of some components and small effect in the remaining directions. Therefore one way to design control charts for variability consists in detecting any significant departure from the stable level of the variability of each component. That is, if one wants to monitor the process variability according to the j-th direction, settled by a generalized PCA, $j = 1, \ldots, l$, what would be calculated and plot on a control chart is the value of the influence function of the j-th eigenvalue. Thus, for the i-th subgroup, $i = 1, \ldots, n$, one would calculate and plot the value of

$$\widetilde{\lambda}_j = \widehat{\alpha}_j \left(\widetilde{\mathbf{\Pi}} - \widehat{\lambda}_j \widetilde{\mathbf{\Sigma}}_1 \right) \widehat{\alpha}_j . \tag{4.5}$$

This value should be compared with the control limits, which are three sigma control limits as in any Shewhart control chart. The decision procedure is the same as in all control charts. It means that, points lying outside the control limits indicate the time at which special causes that affect the variability are present in the process.

5. Numerical example

To illustrate the proposed control charts, a fifteen-variables problem dealing with an automobile bumper manufacturing plant will be discussed. For confidentiality reasons, the variables are coded and will simply be referred to as $\boldsymbol{X}_1$ to $\boldsymbol{X}_{15}$. Table 1 presents data for a particular car bumper. There are five leading quality characteristics which are given by the first five variables and ten moderate ones. For each of the measurements there are eighteen original observations. Three additional points, observations n° 19–21, are included to indicate different cases of out of control conditions in the process.

Nr.	X1	X2	X3	X4	X5	X6	X7	X8	X9	X10	X11	X12	X13	X14	X15
1	-0.39	0.34	1.09	1.04	0.38	-0.52	0.57	-0.10	1.80	0.13	-0.20	-0.50	0.58	0.48	-0.02
2	-0.32	0.91	1.81	0.69	0.07	0.17	0.03	0.66	1.00	0.07	0.05	-0.33	0.27	0.13	0.21
3	-0.50	0.49	1.50	0.47	-0.34	-0.48	0.00	0.00	1.81	-0.05	-0.39	-0.56	0.51	0.12	-0.43
4	-0.54	0.32	1.18	0.22	0.10	-1.02	0.48	-0.28	-0.18	-0.08	0.42	-0.42	-0.12	0.28	0.70
5	-0.23	0.80	1.65	0.06	-0.30	0.17	-0.28	0.90	1.27	0.08	-0.10	-0.24	0.38	0.17	0.12
6	-0.08	0.90	1.08	0.05	-0.66	-0.28	-0.50	0.26	0.66	0.10	0.35	-0.34	0.34	0.10	0.21
7	-0.15	0.82	1.57	0.35	-0.37	0.16	0.00	0.81	1.38	0.08	-0.24	-0.59	0.66	0.17	-0.17
8	-0.20	0.67	1.39	-0.65	-0.17	-0.05	-0.12	0.78	1.96	0.10	0.09	-0.23	0.66	0.73	0.24
9	-0.43	0.60	1.95	-0.41	-1.08	0.10	-0.81	1.11	0.98	-0.22	-0.75	-0.07	0.56	-0.15	-0.55
10	-0.42	0.55	1.82	-0.20	-1.05	-0.18	-0.67	0.49	0.73	0.24	-0.28	-0.47	0.60	0.35	0.07
11	-0.10	0.91	0.93	0.05	-0.65	0.29	-0.42	1.30	0.97	-0.09	-0.40	-0.21	0.41	0.03	-0.20
12	-0.56	0.87	1.95	0.64	-0.12	0.30	-0.03	1.17	1.02	-0.28	-0.11	-0.37	0.15	-0.34	-0.09
13	-0.38	0.62	1.63	0.20	-0.31	-0.21	-0.06	0.37	1.77	-0.02	-0.38	-0.36	1.04	0.08	-0.30
14	-0.04	1.05	1.49	0.60	-0.11	0.49	0.15	0.77	0.94	0.05	-0.04	-0.50	0.05	0.08	0.08
15	-0.37	0.63	1.89	0.07	-0.17	-0.23	-0.02	0.49	1.86	0.16	-0.17	-0.67	0.78	0.20	-0.06
16	0.15	0.95	1.78	-0.10	-1.65	0.42	-0.78	1.53	0.75	0.27	-0.53	0.06	0.74	0.35	-0.15
17	-0.38	0.57	1.53	0.34	-0.30	-0.32	-0.08	0.38	1.18	-0.10	-0.25	-0.42	1.10	0.10	-0.42
18	-0.08	1.15	1.17	-0.13	-1.10	0.60	0.36	1.40	1.52	0.05	-0.33	-0.17	0.67	0.10	-0.02
19	-0.26	1.08	1.37	-0.90	-0.93	0.48	-1.23	-0.29	-0.40	-0.24	0.44	0.40	-0.15	-0.28	0.80
20	0.50	1.50	1.62	0.12	-0.48	1.29	-0.97	2.50	-0.63	-0.27	0.53	0.19	-0.39	-0.31	0.81
21	-0.10	0.91	0.93	0.05	-0.65	0.29	-0.42	1.30	0.97	-0.09	-0.75	-0.21	0.41	-0.40	0.48

Table 1. Car bumper data.

The fifteen univariate control charts set up one for each variable show that the process is under control. The same message is obtained by multivariate control charts using Hotelling's T^2 statistic based on the fifteen variables and the moderate ones. The graphical displays of these charts are not given here.

Table 2 shows the values of the empirical influence functions for the five eigenvalues of a generalized PCA, based on the five leading variables and the last column shows the values of T^2 statistic based only on the leading variables.

Figures 1–5 give the graphical displays of Shewhart control charts for process variability using the empirical influence functions of the eigenvalues. All these control charts have three sigma control limits. Figure 6 shows the multivariate control chart for process mean based on the T^2 statistic which is calculated from the five leading variables. For these control charts the control limits are

	$IF(x_i, \lambda_1)$	$IF(x_i, \lambda_2)$	$IF(x_i, \lambda_3)$	$IF(x_i, \lambda_4)$	$IF(x_i, \lambda_5)$	T^2
λ_i	0.221	0.060	0.042	0.008	0.001 89	
LCL	-0.663	-0.180	-0.126	-0.024	-0.005 67	0.0
UCL	0.663	0.180	0.126	0.024	0.005 67	17.8

Nr.	$IF(x_i, \lambda_1)$	$IF(x_i, \lambda_2)$	$IF(x_i, \lambda_3)$	$IF(x_i, \lambda_4)$	$IF(x_i, \lambda_5)$	T^2
1	0.4924	-0.0057	-0.0189	0.0017	-0.0005	7.8684
2	-0.1524	0.0257	-0.0194	0.0008	-0.0011	4.3000
3	0.1261	-0.0328	-0.0112	-0.0091	-0.0002	2.0780
4	-0.0375	-0.0627	0.0786	-0.0086	-0.0013	4.3761
5	-0.1698	-0.0434	-0.0192	-0.0048	-0.0004	0.7917
6	-0.1889	-0.0458	0.0008	-0.0057	-0.0008	1.9449
7	-0.1708	-0.0442	-0.0043	-0.0086	-0.0006	0.6333
8	-0.1690	-0.0335	-0.0448	-0.0073	0.0067	9.4916
9	-0.1473	0.0926	0.0000	-0.0016	-0.0007	4.9272
10	-0.1780	0.0526	-0.0297	-0.0061	-0.0007	4.0446
11	-0.1326	-0.0452	-0.0372	0.0168	-0.0009	3.4384
12	-0.1982	0.0128	0.0025	0.0090	0.0019	8.9336
13	-0.0536	-0.0353	-0.0138	-0.0084	-0.0007	0.5723
14	-0.1896	0.0516	-0.0168	-0.0086	-0.0007	2.7487
15	-0.0714	-0.0425	-0.0173	0.0038	0.0001	2.9041
16	0.1289	0.0420	0.0237	-0.0028	0.0006	11.0100
17	-0.0582	-0.0386	-0.0115	-0.0086	-0.0008	0.7625
18	-0.1875	-0.0449	-0.0266	0.0175	-0.0030	5.3879
19	0.1318	-0.0052	0.1994	0.0030	0.0041	8.9120
20	1.1777	0.2016	-0.0237	-0.0049	-0.0013	11.4362
21	-0.1219	-0.0440	-0.0282	0.0240	-0.0003	3.4384

Table 2. Values of the empirical influence functions for the eigenvalue and T^2 statistic.

An inspection of these control charts shows that:

(1) The same message is given by the control charts for variability based on the influence function of the first and the second eigenvalue. These charts show that special causes were present in the process at the time when observation n° 20 was obtained.

(2) The control chart used to monitor the variability according to the third eigenvector shows that assignable causes of variation were present in the process at the time when the subgroup number n° 19 was obtained.
(3) The graphical display of the control chart for the fourth eigenvalue shows that special causes of variation were also present at the time when the observation n° 21 was obtained.
(4) Conversely no out-of-control message is received at the observations n° 19–21 from the control chart for the fifth eigenvalue and the control charts using the T^2 statistic.

Thus, it would be raisonable to suspect that something is wrong in the process at the time when the the observations n° 19, 20 and 21 were obtained.

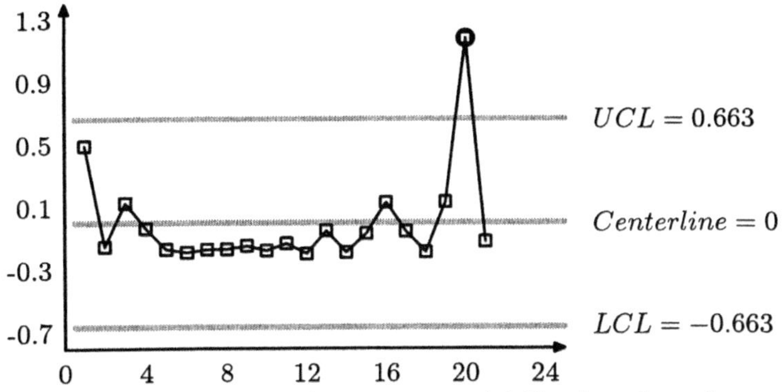

Figure 1. Control chart for process variability based on the empirical influence function of the first eigenvalue.

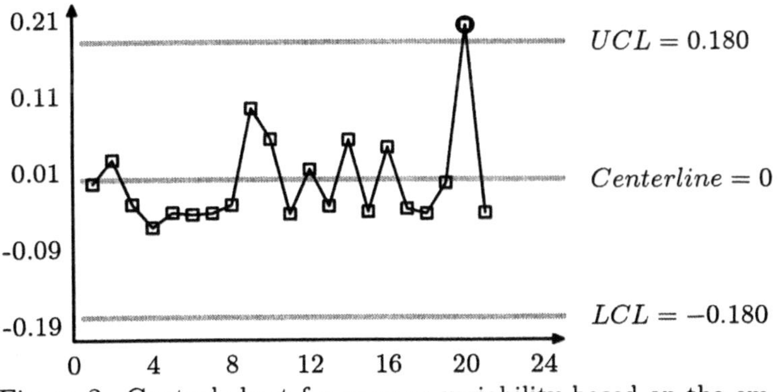

Figure 2. Control chart for process variability based on the empirical influence function of the second eigenvalue.

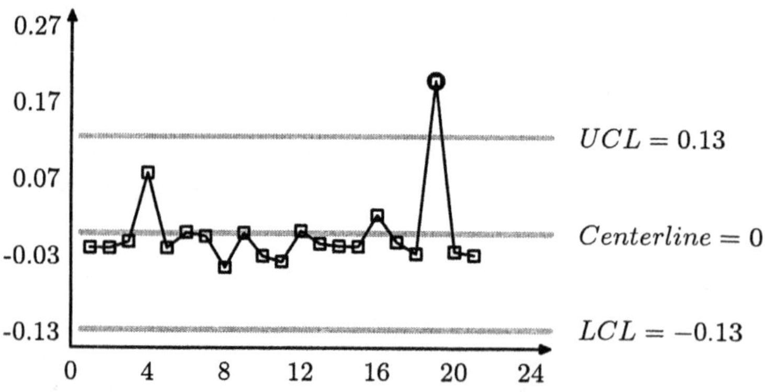

Figure 3. Control chart for process variability based on the empirical influence function of the third eigenvalue.

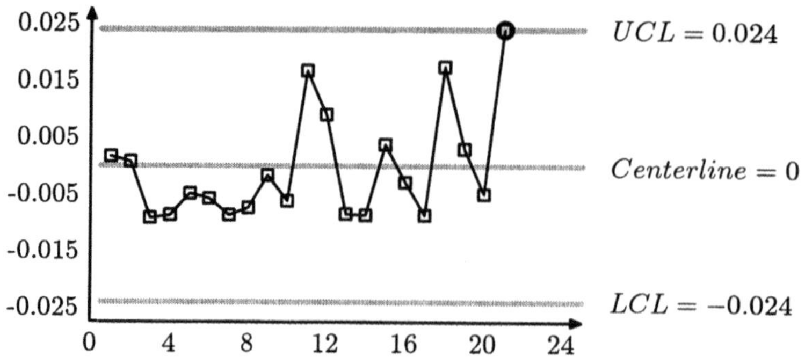

Figure 4. Control chart for process variability based on the empirical influence function of the fourth eigenvalue.

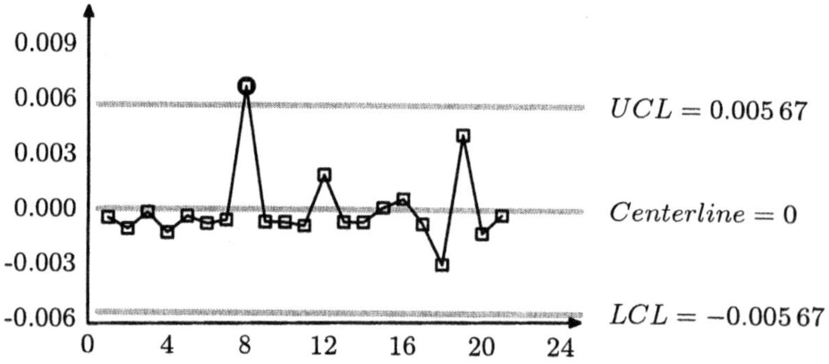

Figure 5. Control chart for process variability based on the empirical influence function of the fifth eigenvalue.

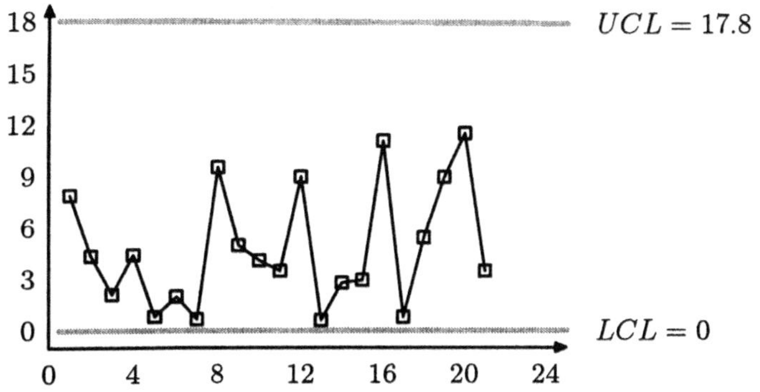

Figure 6. T^2 Control chart for process mean based only on the leading variables.

6. Remarks

When special causes are present in a multivariate process, they may affect different process parameters, e.g. the process mean and/or the process orientation and/or the process variability. Moreover, the special causes that affect one of these parameters do not necessarily affect the others. Therefore on-line methods to monitor different process parameters are necessary. An approach that can lead to build up control charts for all consists in exploiting the influence of the subgroups or individuals over the monitored parameters. The influence functions may be calculated almost in all real situations. Therefore they may be used to set up:

- Control charts for different process parameters, such as: mean-variability-orientation.
- Shewhart control charts. It means control charts that are based on the natural variability of the process. The control limits are three sigma control limits.
- Control charts with probabilistic control limits. It means that the control limits are set up based on the distribution of the statistic that is used to monitor the process and the first type error.
- Control charts for attributes, or for broader applicabilities for countable data.
- CUSUM charts for measurable data as well as for attributes.

For more information see, e.g., Jaupi and Saporta (1994), Jaupi and Saporta (1995), Jaupi and Saporta (1997a) or Jaupi and Saporta (1997b).

References

Alt F. B. and Smith N. D. (1988). *Multivariate process control.* Handbook of Statistics Vol. **7**, 333 – 351, North-Holland.

Critchley F. (1985). *Influence in principal components analysis.* Biometrika **72**, 627-636.

Hampel Fr., Ronchetti E. M., Rousseeuw P. J. and Stahel W. A. (1986). *Robust Statistics: The Approach Based on Influence Functions.* J. Wiley New-York.

Hotelling H. (1947). *Multivariate quality control.* In C. Eisenhart, M. W. Hastay and W. A. Wallis eds, Techniques of Statistical Analysis, McGraw-Hill, New York.

Huber P. (1977). *Robust Covariances in Statistical Decision Theory and Related Topics, II.* Academic Press.

Huber P. (1981). *Robust Statistics.* J. Wiley, New-York.

Jackson J. E. (1959). *Quality control methods for several related variables.* Technometrics **1**, 359 – 377.

Jaupi L. (2000). *Méthodes multivariées pour le pilotage et la maîtrise de processus et produits complexes.* XXXII - ème Journée de Statistique, Fès, Maroc, 15 – 19 mai 2000.

Jaupi L. (1999). *Comparison of multivariate control charts for complex processes.* First International Symposium on Industrial Statistics, Linköping, Suède, 19 – 21 août 1999.

Jaupi L. (1998). *Méthodes multidimensionnelles pour la maîtrise statistique des procédés basées sur l'ACP et l'ACPVI.* XXX - ème Journée de Statistique, Rennes, 24 – 29 mai 1998.

Jaupi L. and Saporta G. (1997a). *Cumulative sum charts for monitoring multivariate processes.* Industrial Statistics, Ch. Kitsos and L. Edler eds., Physica Verlag, Heidelberg.

Jaupi L. and Saporta G. (1997b). *Control charts for multivariate process based on influence functions.* Proceedings of the Conference on Statistical Science, Monte Verità, Zwitzerland, Birkhäuser, 193 – 201.

Jaupi L. and Saporta G. (1995). *A simple approach to design cusum quality control schemes for noconformities.* International Conference on Statistical Methods and Statistical Computing for Quality and Productivity Improvement, Seoul, Korea, 1995, 565 – 569.

Jaupi L. and Saporta G. (1994). *Multivariate process control through the means of influence functions.* Proceedings in Computaional Statistics, COMPSTAT'94, 195 – 200, Physica-Verlag.

Jaupi L. and Saporta G. (1993). *Using the influence function in robust principal components analysis.* In S. Morgenthaler et al. eds., New Directions in Statistical Data Analysis and Robustness, 147 – 156, Birkhäuser, Basel.

Jaupi L. (1992). *Méthodes robustes en analyse en composantes principales.* Thèse CNAM.

Maronna R. A. (1976). *Robust M-estimators of multivariate location and scatter.* Ann. Statist. **4**, 51 – 67.

Pignatello J. J. and Runger G. C. (1990). *Comparison of multivariate CUSUM charts.* Journal of Quality Technology **22**, 173 – 186.

Rao C. R. (1964). *The use and interpretation of principal components in applied research.* Sankhya textbfA 26, 329 – 358.

Sibson R. (1979). *Studies in the robustness of multidimensional scaling - perturbation analysis of classical scaling.* J. Statist. Soc. **B 41**, 217 – 229.

Tanaka J. and Mori Y. (1995). *Principal component analysis based on a subset of variables and its application.* Proceedings of International Conference on Statistical Methods and Statistical Computing for Quality and Productivity Improvement, Seoul, Korea, 1995, 308 – 317.

Tanaka Y. (1988). *Sensitivity analysis in PCA: Influence on the subspace spanned by principal components.* Comm. Statist. Theory-Methods **17**, 3157 – 3175.

MULTIDIMENSIONAL METHODS FOR STATISTICAL PROCESS CONTROL: SOME CONTRIBUTIONS OF ROBUST STATISTICS

Ndèye Niang

Département de Mathématiques, Conservatoire National des Arts et Métiers, 292 Rue Saint Martin, F-75141 PARIS CEDEX 03, France; niang@cnam.fr

Abstract: We deal with the problem of multivariate process control. The aim of this chapter is to contribute to the development of methodological and computational tools for a simultaneous consideration of several characteristics. First a critical study of the literature about multivariate Shewhart charts will be done and two new multivariate moving average charts are proposed. Afterwards we point out a number of limitations of the Shewhart chart for the first stage of the process control. These limitations make this first stage inapplicable. In order to remedy this, a set of methods based on robust estimations of the process mean and dispersion is proposed. Simulated examples are given to illustrate the methods.

1. Introduction

The general multivariate statistical quality control problem considers a repetitive process represented by p interdependent characteristics $X^1, X^2, \ldots, X^p$. The distribution of these p variables is assumed to be multivariate normal: $\mathcal{N}_p(\boldsymbol{\mu}, \boldsymbol{\Sigma})$.

When controlling such a process, the usual industrial methods consist in the construction of p univariate charts, one for each parameter.

These techniques ignore correlations between the variables and can lead to false alarms or non-detection of out of control situations as it will be illustrated in the following example.

EXAMPLE 1.1. Let us consider the case of two quality characteristics, which are jointly distributed as a bivariate normal with parameters

$$\boldsymbol{\mu}_0 = \begin{pmatrix} \mu_{01} \\ \mu_{02} \end{pmatrix} \qquad \text{and} \qquad \boldsymbol{\Sigma}_0 = \begin{pmatrix} \sigma_{01}^2 & \sigma_{012} \\ \sigma_{012} & \sigma_{02}^2 \end{pmatrix} \tag{1.1}$$

Using two Shewhart charts for each mean with control limits equal to

$$\mu_{0i} \pm 3\frac{\sigma_{0i}}{\sqrt{n}}, \qquad i = 1, 2,$$

leads to the rectangular control region (Figure 1.1). Thus if a pair of sample means falls within this region, the process is considered to be in control.

2000 *Mathematics Subject Classification.* 62P30.

Key words and phrases. Statistical process control, multivariate control charts, Hotelling's statistic, multidimensional methods, principal component analysis, robust estimations, multivariate outliers, variance-covariance matrix.

Each of the previous Shewhart control charts is equivalent to univariate hypothesis testing. But for the bivariate test: $H_0 : \boldsymbol{\mu} = \boldsymbol{\mu}_0$ vs $H_1 : \boldsymbol{\mu} \neq \boldsymbol{\mu}_0$ the critical region corresponding to the likelihood ratio test is:

$$W = \left\{ \boldsymbol{X}_i, \; i = 1, \ldots, n; \quad n \left(\overline{\boldsymbol{X}} - \boldsymbol{\mu}_0 \right)' \boldsymbol{\Sigma}_0{}^{-1} \left(\overline{\boldsymbol{X}} - \boldsymbol{\mu}_0 \right) > \chi^2_{2,\alpha} \right\}, \qquad (1.2)$$

where $\boldsymbol{X}_i$, $i = 1, \ldots, n$, is a sample of n observations of $\boldsymbol{X} = \left(X^1_, X^2_, \ldots, X^p \right)'$.

This region corresponds to the out-of-control area on the chart; the control region for the multivariate mean process is elliptical (Figure 1.1). The bivariate process must be judged out of control only if the pair of means falls outside the ellipse.

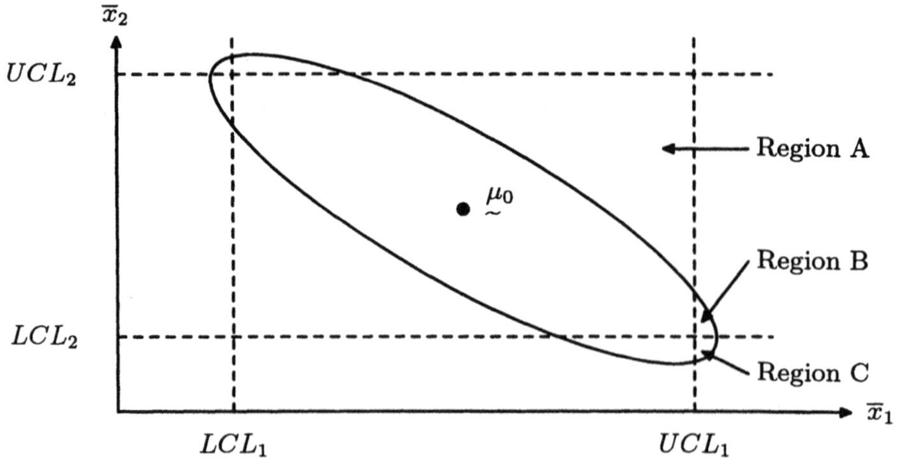

Figure 1.1.

Then the use of separate control charts or the equivalent rectangular region can be very misleading. Two situations may appear; first a non detection of a process mean shift: Region A, where it is erroneously concluded that both process means are in control, and secondly a false alarm situation: one process mean is out of control and the other is in control (Region B), and both process means are out of control (Region C).

The size of these regions depends on the correlation between the characteristics. But the probability that a sample mean vector will fall within the ellipse when the process is in control is equal to $1 - \alpha$. The rectangular region corresponds to a simultaneous test and even choosing correctly (Bonferroni inequalities) the type I error of each univariate test the probability that a sample mean vector will fall within the ellipse when the process is in control is at least $1 - \alpha$. The type I error for this simultaneous test is greater than for bivariate test and increase, with the number of variables. When the variables are uncorrelated it is possible to calculate it exactly but even in that case these regions are still existing.

Then the global approach using multivariate charts must be the only one applicable. However, once a mean shift of the multivariate process is detected, the univariate charts with adequate type I errors can be sometimes used in determining which characteristic mean is out of control; that can be helpful for the assignable causes research.

Various multivariate control charts, cf. Alt (1973), Alt (1985), Crosier (1988), Lowry et al. (1992), Tracy et al. (1992) or Niang (1994), have been proposed to take into account the dependencies between the characteristics. They are generally based on quadratic transformations of the $\mathcal{R}_p$ vector $\boldsymbol{X} = (X^1_, X^2_, \ldots, X^p)'$ into a scalar: Mahalanobis or Hotelling distances.

They can be considered as multivariate outliers detection method since a shift in the process parameters leads to statistically unreasonable discordant or extreme observations of the Mahalanobis or Hotelling distances given the assumed distribution.

The problem of detecting multivariate outliers is a complex one (Barnett and Lewis (1984)). The idea of extremeness arises from some form of ordering the data but with p greater than 1 there isn't any notion of total ordering; what it isn't the case in a univariate situation. A multivariate outlier needs not to be extreme in any of its p components; it can be the result of an extreme observation in only one direction or the result of systematic small errors in several directions. The later case of outliers corresponds to a problem of orientation (correlation) and not of location (mean) or scale (variance). This will be illustrated in the simulated example later in this paper.

When the presence of outliers is suspected, two opposite attitudes are possible: detection and rejection or accommodation. In quality control the two attitudes are to be conjointly used. Accommodation of outliers is the basis of the robust methods proposed in Section 3.

Usually a process control comprises both scenarios: the process dispersion is first controlled and if it is deemed to be in control the process mean can be controlled. Distinction is also made between phase II of known standard values $\boldsymbol{\mu}_0$, $\boldsymbol{\Sigma}_0$ and phase I when the values are unknown. This paper is focused on the Shewhart case and especially on phase I for mean control charts.

Multivariate Shewhart and cumulative sum (CUSUM) control charts have been widely studied in the literature. Several textbooks and papers have been published, see, e.g., Alt (1973), Alt (1985), Crosier (1988), Pignatiello and Runger (1990), Montgomery (1985) and Ryan (1989). A review of these multivariate control charts can be found in Niang (1994).

Unlike Shewhart and CUSUM charts, Exponentially Weighted Moving Average (EWMA) and Uniformly Weighted Moving Average (UWMA) for multivariate process are less present in the quality control literature.

In the Section 2, classical multivariate Shewhart control charts for controlling the process mean will be presented for both phase I and phase II with a brief presentation for phase II being first to facilitate the phase I presentation.

After, multivariate EWMA and UWMA charts are proposed that are natural extensions of the univariate case.

Principal component analysis as an outlier detection method can be useful in quality control. Control chart, based on PCA, will be presented.

Examples with simulated data are given to illustrate the phase I. It will be shown that, for specific large mean shifts, the first stage of the classical Shewhart charts presents some limitations due to the fact that the classical estimations for $\boldsymbol{\mu}$ and $\boldsymbol{\Sigma}$ are not optimal in the case of large shifts. Their non-robustness makes the whole analysis polluted by the out of control observations.

In Section 3, methods based on simple robust estimations of the mean and dispersion are proposed to remove these limitations. Illustrations of the methods on simulated data are given.

2. Multivariate control charts for the mean

2.1. Notations

One fundamental assumption required by standard methods is that the process characteristics are interdependent random variables distributed as a multinormal $\mathcal{N}_p(\boldsymbol{\mu}, \boldsymbol{\Sigma})$ with $\boldsymbol{\mu}$ a $\mathcal{R}_p$ vector and $\boldsymbol{\Sigma}$ a $p \times p$ matrix. The data are observed samples $\boldsymbol{X}_1, \boldsymbol{X}_2, \ldots, \boldsymbol{X}_n$ described in a $p \times n$ matrix $\boldsymbol{X}$:

$$\boldsymbol{X} = (\boldsymbol{X}_1, \boldsymbol{X}_2, \ldots, \boldsymbol{X}_n) = \begin{pmatrix} X_1^1 & X_2^1 & \cdots & X_n^1 \\ \vdots & \vdots & & \vdots \\ X_1^p & X_2^p & \cdots & X_n^p \end{pmatrix} \tag{2.1}$$

$\overline{\boldsymbol{X}}$ is the p-empirical vector mean:

$$\overline{\boldsymbol{X}} = \begin{pmatrix} \overline{X}^1 \\ \vdots \\ \overline{X}^p \end{pmatrix} = \begin{pmatrix} \frac{1}{n}\sum_{i=1}^n X_i^1 \\ \vdots \\ \frac{1}{n}\sum_{i=1}^n X_i^p \end{pmatrix} = \frac{1}{n}\sum_{i=1}^n \boldsymbol{X}_i \tag{2.2}$$

$\boldsymbol{S}$ is the empirical variance-covariance matrix:

$$\boldsymbol{S} = \frac{1}{n-1}\sum_{i=1}^n (\boldsymbol{X}_i - \overline{\boldsymbol{X}})(\boldsymbol{X}_i - \overline{\boldsymbol{X}})' \tag{2.3}$$

2.2. Multivariate Shewhart control charts

2.2.1. Phase II control chart: nominal values $\boldsymbol{\mu}_0$,$\boldsymbol{\Sigma}_0$

The aim of this phase II is to maintain the process mean at $\boldsymbol{\mu}_0$. $\boldsymbol{\mu}_0$ and $\boldsymbol{\Sigma}_0$ are referred to as standard, target or nominal values. They may be derived from a large amount of past data or could also have been selected by management to attain certain objectives.

As in the univariate case it is tested if the process mean is at $\boldsymbol{\mu}_0$. A likelihood ratio test is used:

$$\boldsymbol{\mu} = \boldsymbol{\mu}_0 \quad vs \quad \boldsymbol{\mu} \neq \boldsymbol{\mu}_0 \,.$$

Let $\boldsymbol{X}_i$, $i = 1, \ldots, n$, be a sample of n observations of $\mathcal{N}_p(\boldsymbol{\mu}, \boldsymbol{\Sigma})$ with the density function:

$$\begin{aligned} f(\boldsymbol{X}) &= f\left(X^1_, X^2_, \ldots, X^p\right) = \\ &= (2\pi)^{-p/2} |\Sigma|^{-1/2} \exp\left\{-\frac{1}{2}(\boldsymbol{X}-\boldsymbol{\mu})' \boldsymbol{\Sigma}^{-1}(\boldsymbol{X}-\boldsymbol{\mu})\right\} \end{aligned} \quad (2.4)$$

The likelihood function is :

$$\begin{aligned} L(\boldsymbol{X}_1, \boldsymbol{X}_2, \ldots, \boldsymbol{X}_n) &= (2\pi)^{-np/2} |\Sigma|^{-n/2} \times \\ &\times \exp\left\{-\frac{1}{2}\sum_{i=1}^{n}(\boldsymbol{X}_i-\boldsymbol{\mu})' \boldsymbol{\Sigma}^{-1}(\boldsymbol{X}_i-\boldsymbol{\mu})\right\} \end{aligned} \quad (2.5)$$

and the corresponding likelihood ratio statistic is:

$$\lambda = \exp\left\{-\frac{1}{2} n \left(\overline{\boldsymbol{X}}-\boldsymbol{\mu}_0\right)' \boldsymbol{\Sigma}^{-1}\left(\overline{\boldsymbol{X}}-\boldsymbol{\mu}_0\right)\right\} \quad (2.6)$$

It is well known that if $\overline{\boldsymbol{X}}$ is distributed as $\mathrm{N}_p\left(\boldsymbol{\mu}, n^{-1}\boldsymbol{\Sigma}\right)$ then $n\left(\overline{\boldsymbol{X}}-\boldsymbol{\mu}_0\right)' \boldsymbol{\Sigma}^{-1}\left(\overline{\boldsymbol{X}}-\boldsymbol{\mu}_0\right)$ has a $\chi^2_{p,d}$ distribution with $d = n(\boldsymbol{\mu}-\boldsymbol{\mu}_0)' \boldsymbol{\Sigma}^{-1}(\boldsymbol{\mu}-\boldsymbol{\mu}_0)$ (Anderson (1984)). The likelihood ratio statistic (2.6) gives the critical region:

$$W = \left\{\boldsymbol{X}_i,\ i = 1, \ldots, n;\quad n\left(\overline{\boldsymbol{X}}-\boldsymbol{\mu}_0\right)' \Sigma_0{}^{-1}\left(\overline{\boldsymbol{X}}-\boldsymbol{\mu}_0\right) > \chi^2_{p,\alpha}\right\}, \quad (2.7)$$

where $\chi^2_{p,\alpha}$ is the α-percentile of χ^2_p.
The Shewhart chart for the mean control is then defined by:

$$\begin{aligned} UCL &= \chi^2_{p,\alpha} \\ LCC &= 0 \\ Stat &= \mathrm{D}^2(\overline{\boldsymbol{X}}_i; \boldsymbol{\mu}_0) = n\left(\overline{\boldsymbol{X}}_i-\boldsymbol{\mu}_0\right)' \Sigma_0{}^{-1}\left(\overline{\boldsymbol{X}}_i-\boldsymbol{\mu}_0\right) \end{aligned} \quad (2.8)$$

Note: LSC is a probability limit:

$$P\left(\mathrm{D}^2(\overline{\boldsymbol{X}}_i; \boldsymbol{\mu}_0) < UCL\right) = P\left(\mathrm{D}^2(\overline{\boldsymbol{X}}_i; \boldsymbol{\mu}_0) < \chi^2_{p,\alpha}\right) = 1-\alpha$$

It consists in plotting the values of *Stat*, the Mahalanobis distances between the nominal vector $\boldsymbol{\mu}_0$ and $\overline{\boldsymbol{X}}_i$ the mean vectors of samples i. If this statistic plots above the upper control limit the process mean is deemed to be out of control and assignable causes of variation are sought. The control chart is essentially a graphical visualization of the likelihood ratio test for successive samples.

In the bivariate case (Example 1.1) an ellipse can be used instead of horizontal limits; but thus the time and successive samples notions are lost.

2.2.2. Phase I control chart: unknown values μ, Σ

In some cases, for example during the initial stages of process surveillance, μ_0 and Σ_0 are usually unknown and must be estimated from preliminary samples taken when the process is believed to be in control. These preliminary samples are referred to as rational subgroups, in this paper the expression "grouped observations" will be also used.

As in the univariate case, when the mean μ and the dispersion matrix Σ of the process to control are unknown a multivariate control chart is generally set up according to the following two-stage procedure: replace the nominal value in Phase II by unbiased estimates from the rational subgroups. These limits are considered as provisional limits.

With these limits it is retrospectively tested if the process was in control when the sample measurements were taken, this is called the first stage or preliminary phase. If there isn't any rational subgroup mean out of control, the provisional limits become definitive and are used to control future observations.

Grouped observations

For the first stage of a process control, a set of m samples of size n: $\{(X_{i1}, X_{i2}, \ldots, X_{in}),\ i = 1, \ldots, m\}$ is considered. It is assumed that the covariance matrix Σ is common to all subgroups. Let $\overline{\boldsymbol{X}}_i$ be the mean vector and $\boldsymbol{S}_i$ the empirical covariance matrix of the sample i. For each subgroup we have:

$$\overline{\boldsymbol{X}}_i = \begin{pmatrix} \overline{X}_i^1 \\ \vdots \\ \overline{X}_i^p \end{pmatrix} = \begin{pmatrix} \frac{1}{n}\sum_{j=1}^n X_j^{1(i)} \\ \vdots \\ \frac{1}{n}\sum_{j=1}^n X_j^{p(i)} \end{pmatrix}$$

$$\boldsymbol{S}_i = \frac{1}{n-1}\sum_{j=1}^{n}\left(\boldsymbol{X}_j^{(i)} - \overline{\boldsymbol{X}}_i\right)\left(\boldsymbol{X}_j^{(i)} - \overline{\boldsymbol{X}}_i\right)'$$

Unbiased estimates of μ and Σ are given by:

$$\overline{\overline{\boldsymbol{X}}} = \frac{1}{m}\sum_{i=1}^{m}\overline{\boldsymbol{X}}_i\,, \qquad \bar{\boldsymbol{S}} = \frac{1}{m}\sum_{i=1}^{m}\boldsymbol{S}_i$$

$\bar{\boldsymbol{S}}$ is the within variance matrix.

By replacing μ_0 in (2.8) with $\overline{\overline{\boldsymbol{X}}}$ and $\bar{\boldsymbol{S}}$ respectively, the control statistic for the first stage becomes:

$$T_i^2 = n\left(\overline{\boldsymbol{X}}_i - \overline{\overline{\boldsymbol{X}}}\right)'\bar{\boldsymbol{S}}^{-1}\left(\overline{\boldsymbol{X}}_i - \overline{\overline{\boldsymbol{X}}}\right) \tag{2.9}$$

T_i^2 is the estimated Mahalanobis distance between $\overline{\boldsymbol{X}}_i$ and $\overline{\overline{\boldsymbol{X}}}$ relatively to $\bar{\boldsymbol{S}}$ the estimate of Σ.

T_i^2 has a Hotelling distribution which can be identified to:

$$\frac{p(n-1)(m-1)}{mn-m-p+1}F_{p,mn-m-p+1}$$

(Alt (1973)).

The stability of the process mean during the first stage is tested by plotting the value of T_i^2 for $i = 1, \ldots, m$ on a control chart with

$$\begin{aligned} UCL &= \frac{p(n-1)(m-1)}{mn-m-p+1} F_{p,mn-m-p+1,\alpha} \\ LCL &= 0. \end{aligned}$$

These limits are provisional and are used for a retrospective control: if one or more of the initial m subgroups is out of control, they are excluded and the limits are recomputed on the basis of the remaining subgroups. More precisely, suppose that k of the m initial samples are discarded, the upper limit becomes:

$$UCL = \frac{(m-k-1)\,(n-1)\,p}{n(m-k)-(m-k)-p+1} F_{p,n(m-k)-(m-k)-p+1,\alpha} \tag{2.10}$$

This procedure is repeated as often as necessary until all the remaining subgroups are in control. When this occurs, the second stage can be introduced to test if the process remains in control when future subgroups are analysed; this stage is equivalent to phase II.

Let $\overline{\boldsymbol{X}}_f$ be the mean vector of a future subgroup. The corresponding control statistic is:

$$T_f^2 = n\left(\overline{\boldsymbol{X}}_f - \overline{\overline{\boldsymbol{X}}}\right)' \bar{S}^{-1}\left(\overline{\boldsymbol{X}}_f - \overline{\overline{\boldsymbol{X}}}\right) \tag{2.11}$$

This Hotelling's statistic T_f^2 is distributed as $\frac{(m+1)(n-1)}{mn-m-p+1} F_{p,mn-m-p+1}$ (Alt (1973)).

Consequently the control of the process mean with future subgroups consist in comparing T_f^2 to the control limits

$$\begin{aligned} UCL &= \frac{(m+1)(n-1)}{mn-m-p+1} F_{p,mn-m-p+1,\alpha} \\ LCL &= 0. \end{aligned}$$

If k samples have been excluded the upper control limit is

$$UCL = \frac{(m-k+1)\,(n-1)\,p}{n(m-k)-(m-k)-p+1} F_{p,n(m-k)-(m-k)-p+1,\alpha}.$$

In the usual practice, the first stage provisional UCL becomes definitive and is used to control future observations. But this upper control limit is not exactly the same as the first stage UCL. The difference is due to the fact that here both provisional and definitive upper control limits are exact probability limits (Niang (1994)).

Individual observations

In some applications individual measurements can be more suitable. It is the case, for example, for a slow process of individual pieces production where it can take too much time to obtain a subgroup or for chemical process where the nature of the characteristic like temperature or pressure can make useless to take several measures since they can be quite the same.

The multivariate Shewhart control charts for individual measurements have been presented in Jackson(1956, 1957, 1959) especially through principal component analysis.

In the case of individual measurements, the within variance matrix $\bar{\boldsymbol{S}}$ can no longer be computed. The method usually used consists in estimating the process dispersion matrix by the total variance-covariance matrix $\boldsymbol{S}$.

The control statistic is then $\mathrm{D}_i^2(\boldsymbol{X}_i; \overline{\boldsymbol{X}}) = (\boldsymbol{X}_i - \overline{\boldsymbol{X}})' \boldsymbol{S}^{-1} (\boldsymbol{X}_i - \overline{\boldsymbol{X}})$.

Unlike what it is found in the literature, $\mathrm{D}_i^2(\boldsymbol{X}_i; \overline{\boldsymbol{X}})$ hasn't an Hotelling distribution (Alt (1973)). Several approximations (Ryan (1989), Alt (1973)) can be made; but the approximate limits are efficient to detect out of control situations only if m is very large. $\mathrm{D}_i^2(\boldsymbol{X}_i; \overline{\boldsymbol{X}})$ is exactly distributed as $\frac{(m-1)^2}{m}\mathrm{Beta}\left(p/2, (m-p-1)/2\right)$ (Gnanadesikan and Kettenring (1972)).

Then a control chart for the mean has the following limits (Tracy et al. (1992)):

$$UCL = \frac{(m-1)^2}{m}\mathrm{Beta}\left(\frac{\alpha}{2}; \frac{p}{2}, \frac{(m-p-1)}{2}\right)$$

$$LCL = \frac{(m-1)^2}{m}\mathrm{Beta}\left(1-\frac{\alpha}{2}; \frac{p}{2}, \frac{(m-p-1)}{2}\right)$$

When controlling future observations $\boldsymbol{X}_f$ the control statistic $(\boldsymbol{X}_f - \overline{\boldsymbol{X}})' \boldsymbol{S}^{-1} (\boldsymbol{X}_f - \overline{\boldsymbol{X}})$ has a Hotelling distribution: $\frac{p(m+1)(m-1)}{m(m-p)} F_{p,m-p}$.

The second stage control of future observations $\boldsymbol{X}_f$ consists in plotting the values $(\boldsymbol{X}_f - \overline{\boldsymbol{X}})' \boldsymbol{S}^{-1} (\boldsymbol{X}_f - \overline{\boldsymbol{X}})$ on a control chart with

$$UCL = \frac{p(m+1)(m-1)}{m(m-p)} F_{\alpha;p,m-p}.$$

This probability limit is exact.

2.3. Exponentially weighted and uniformly weighted moving average charts

In the univariate case, EWMA and UWMA charts are as efficient as CUSUM charts and have better performances than the Shewhart one for detecting small shifts. But, as it was said in the introduction, EWMA and UWMA charts for more than one characteristic have not been really studied (Lowry et al. (1992)). For the multivariate case, natural extensions of EWMA and UWMA univariate charts are proposed. They are based on the same transformation used for the multivariate Shewhart chart.

2.3.1. EWMA control charts for mean process

As for the Shewhart chart, the p-vector of process characteristics is transformed into univariate statistic $\mathrm{D}^2(\overline{\boldsymbol{X}}_i; \boldsymbol{\mu}_0) = n\,(\overline{\boldsymbol{X}}_i - \boldsymbol{\mu}_0)' \boldsymbol{\Sigma}_0{}^{-1} (\overline{\boldsymbol{X}}_i - \boldsymbol{\mu}_0)$.

For each sample i the following W statistic is computed:

$$\begin{aligned} W_i &= r\mathrm{D}^2(\overline{\boldsymbol{X}}_i;\boldsymbol{\mu}_0) + (1-r)\,W_{i-1} \qquad r\in[0,1] \\ W_0 &= 0 \end{aligned} \tag{2.12}$$

Note: $W_0 = 0$ *corresponds to* $\boldsymbol{\mu} = \boldsymbol{\mu}_0$.

W_i is a weighted sum of *iid* (independent and identically distributed) chi square variables with p degree of freedom. Its exact distribution is not obvious to derive but it is easy to compute its two first moments.

W_i can be rewritten as:

$$W_i = r\sum_{j=0}^{i-1}(1-r)^j\mathrm{D}^2(\overline{\boldsymbol{X}}_{i-j};\boldsymbol{\mu}_0) \tag{2.13}$$

As $\quad r\sum_{j=0}^{i-1}(1-r)^j = r\left[\frac{1-(1-r)^i}{1-(1-r)}\right] = 1-(1-r)^i$

It comes: $\mathsf{E}\,W_i = p$ and $\mathsf{Var}\,W_i = 2rp/(2-r)$.

The proposed chart is a kind of classical Shewhart: $\mathsf{E}\,W_i \pm 3\sqrt{\mathsf{Var}\,W_i}$. It is defined by:

$$UCL = p + 3\sqrt{\frac{2rp}{2-r}},\quad CCL = p,\quad LCL = p - 3\sqrt{\frac{2rp}{2-r}},\quad Stat = W_i.$$

REMARK 2.1. The W_i are positive values, then LCL should be in fact equal to

$$\sup\left(0; p - 3\sqrt{\frac{2rp}{2-r}}\right).$$

Obviously, these limits are not probability limits. But considering the central-limit theorem, the W_i distribution can be assumed asymptotically normal. Furthermore, the following approximations of the chi square distribution can be used but the chart formulas become too complex.

Fisher approximation:

$$\chi^2_p \approx \frac{\left(U + \sqrt{2p-1}\right)^2}{2}$$

or Wilson-Hilferty approximation:

$$\chi^2_p \approx p\left(U\sqrt{\frac{2}{9p}} + 1 - \frac{2}{9p}\right)^3$$

The later one is very precise and good even for small values of p.

2.3.2. UWMA control chart for mean process

As in the univariate case the UWMA control chart is based on the conjoint using of the w newer measures. The p-vector of process characteristics is transformed into a univariate statistic $\mathrm{D}^2(\overline{\boldsymbol{X}}_i;\boldsymbol{\mu}_0)$. The proposed chart consists in a kind of Shewhart with 3-sigma limits for a uniformly weighted moving average of distances $\mathrm{D}^2(\overline{\boldsymbol{X}}_i;\boldsymbol{\mu}_0)$. The control statistic is then:

$$M_i = \frac{1}{w} \sum_{j=i-w+1}^{i} \mathrm{D}^2(\overline{\boldsymbol{X}}_i; \boldsymbol{\mu}_0) \qquad \text{if } i > w$$

$$M_i = \frac{1}{i} \sum_{j=1}^{i} \mathrm{D}^2(\overline{\boldsymbol{X}}_i; \boldsymbol{\mu}_0) \qquad \text{if } i \leq w$$

Let $w_{\text{inf}} = \inf(i; w)$; $w_{\text{inf}} M_i$ is a sum of w_{inf} random variables independent and identically distributed as χ^2 with p degrees of freedom. $w_{\text{inf}} M_i$ is thus distributed as a χ^2 with $p \cdot w_{\text{inf}}$ degrees of freedom. The two first moments are then: $\mathsf{E}\, M_i = p$ and $\mathsf{Var}\, M_i = 2p/w_{\text{inf}}$.

And thus the control chart with limits at $\mathsf{E}\, M_i \pm 3\sqrt{\mathsf{Var}\, M_i}$ is:

$$UCL = p + 3\sqrt{\frac{2p}{w_{\text{inf}}}}, \quad CCL = p, \quad LCL = p - 3\sqrt{\frac{2p}{w_{\text{inf}}}}, \quad Stat = M_i.$$

REMARK 2.2. Using the central-limit theorem allows considering these limits as approximate probability limits. The M_i are always positive values, then LCL should be in fact equal to $\sup\left(0; p - 3\sqrt{2p/w_{\text{inf}}}\right)$.

It is also possible to compute exact probability limits using the distribution of M_i:

$$UCL = \frac{1}{w_{\text{inf}}} \chi^2_{p w_{\text{inf}}, \alpha}, \quad LCL = 0, \quad Stat = M_i.$$

It could be interesting to compare these chart performances in terms of average run length ARL. The ARL computations will be done by simulation or using exact formulas in the case of UWMA chart (Niang (1994), Niang and Saporta (1995)).

2.4. Principal component analysis and control charts

Using principal component analysis (PCA) allows the transformation of a set of correlated initial variables $\boldsymbol{X}^i$ in new uncorrelated principal components c_k, combination of the $\boldsymbol{X}^i$ with maximal variance. Simultaneous univariate charts based on these new variables can be used to study the multivariate process control: the problems of false alarms or non detection of out of control situations noticed in the introduction are then attenuated because of the non correlation between the principal components.

Furthermore, if the principal components are interpretable this can be very helpful for the determination of assignable causes.

Generally, PCA is used as a technique for reducing dimension. Only the first q principal components corresponding to the q greatest eigenvalues are retained. In quality control, PCA should be considered rather as a method of detecting outliers: the last principal components may be as interesting as the first ones. Jolliffe (1986) shows that the first principal components allow the detection of outliers that inflate the variances and covariances, but those outliers can be detected with the initial variables. On the other hand, outliers that perturbate the correlation between variables will be detected on the last principal components.

Several methods of outliers detection, based on PCA, have been proposed (Barnett and Lewis (1984), Hawkins (1974), Gnanadesikan and Kettenring (1972), Jackson (1991)). They consist if normal or gamma probability plots and of more formal methods based on statistical tests. These methods can be applied to components individually or to residual statistics calculated using conjointly several components. The most widely used residual statistics are :

$$R_{1i}^2 = \sum_{k=p-q+1}^{p} c_{ik}^2 = \sum_{k=p-q+1}^{p} \lambda_k y_{ik}^2 , \qquad R_{2i}^2 = \sum_{k=p-q+1}^{p} \frac{c_{ik}^2}{\lambda_k} = \sum_{k=p-q+1}^{p} y_{ik}^2$$

Since the p process characteristics are assumed to be multinormal, in the case of individual data with $\boldsymbol{\mu}$ and $\boldsymbol{\Sigma}$ known, the principal components of the standardized PCA y_k, have an exact gaussian distribution $\mathrm{N}(0;1)$. If there aren't any outliers in the data, the residual statistics R_{1i}^2 and R_{2i}^2 have χ_{p-k}^2 distribution.

When data are grouped, or if $\boldsymbol{\mu}$ and $\boldsymbol{\Sigma}$ are estimated, we have to deal with pseudo principal components that do not have exactly the classical PCA properties. They still have a mean equal to 0 but the variance will be approximately equal to 1. The distributions of the residual statistics are also approximate.

However quality control methods based on these components are still efficient. The for each component y_i a control chart can be defined as following:

$$UCL = +3, \quad CCL = 0, \quad LCL = -3, \quad Stat = y_i.$$

Or if data are grouped:

$$UCL = +\frac{3}{\sqrt{n}}, \quad CCL = 0, \quad LCL = -\frac{3}{\sqrt{n}}, \quad Stat = \overline{y}_i.$$

The presence of outliers can be tested with a control chart on the R_{2i}^2:

$$UCL = \chi_{p-k,\alpha}^2, \quad LCL = 0, \quad Stat = R_{2i}^2.$$

These methods are to be used conjointly with the multivariate charts and are efficient. But it should be noticed that if there are many variables, these PCA methods may be time consuming because both the k first and the q last components have to be studied since the kind of shift is a priori unknown. On an other hand, even if it is possible to summarize the p initial variables by a little number k of components, these components depend on all initial variables since they are combination of these p variables. Methods for simplifying the interpretation of principal components have been proposed (Flury (1988), Jackson and Morris (1957), Jackson (1959), Jackson (1991)).

REMARK 2.3. The T^2 statistic can be computed using principal components (Jackson (1959)). In fact PCA is used as a matrix decomposition: the standardized principal components $\boldsymbol{Y}$ are linear combination of $\boldsymbol{X}$: $\boldsymbol{Y} = \boldsymbol{W}'(\boldsymbol{X} - \overline{\boldsymbol{X}})$ where $\boldsymbol{W}$ is the $\boldsymbol{S}$ standardized eigenvectors matrix.

Then:

$$T^2 = (\boldsymbol{X} - \overline{\boldsymbol{X}})' \boldsymbol{S}^{-1} (\boldsymbol{X} - \overline{\boldsymbol{X}}) = (\boldsymbol{X} - \overline{\boldsymbol{X}})' \boldsymbol{W}\boldsymbol{W}' (\boldsymbol{X} - \overline{\boldsymbol{X}}) = \boldsymbol{Y}'\boldsymbol{Y}$$

2.5. Numerical examples

Grouped observations (30 samples of 5 observations) are simulated from a multinormal distribution $N_3(\boldsymbol{\mu}, \boldsymbol{R})$ where $\boldsymbol{R}$, the correlation matrix, is:

$$\boldsymbol{R} = \begin{pmatrix} 1 & & \\ 0.9 & 1 & \\ 0.1 & 0.3 & 1 \end{pmatrix}$$

The variances are $\sigma_1^2 = 0.5$, $\sigma_2^2 = 0.3$, $\sigma_3^2 = 0.1$.

Three out of control situations are illustrated: the first 25 samples have a **0**-vector mean and for the last 5 samples the mean is supposed to have shifted from **0** to $\boldsymbol{\mu}$ with:

EXAMPLE 2.1. $\boldsymbol{\mu} = (\frac{\sigma_1}{2}, -\frac{\sigma_2}{2}, 0)'$

Here we have small shifts of the two first variables without respect of the correlation. Figures 2.1 to 2.6 show that the univariate charts does not detect the shift while the five out of control points are detected on the multivariate chart and on the last principal component.

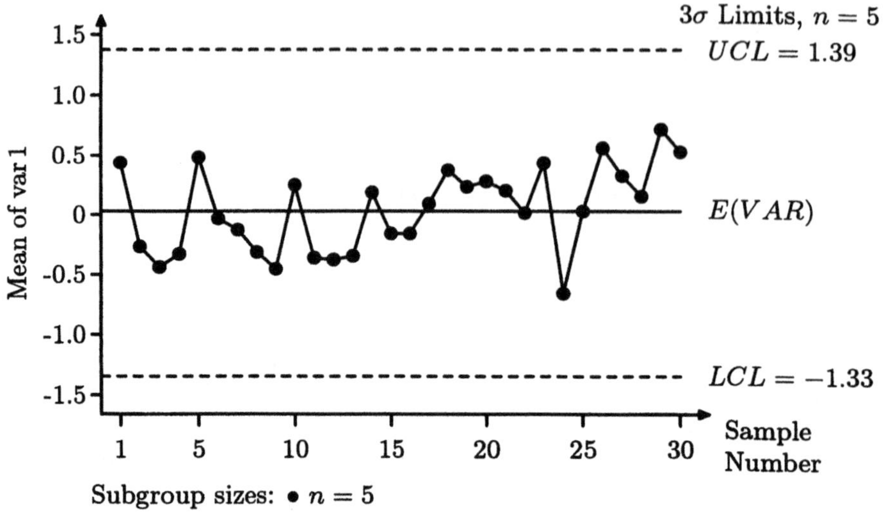

Figure 2.1.

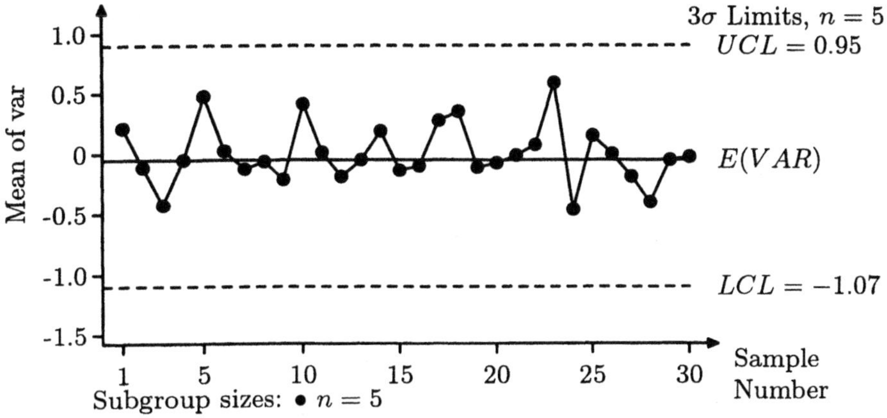

Figure 2.2.

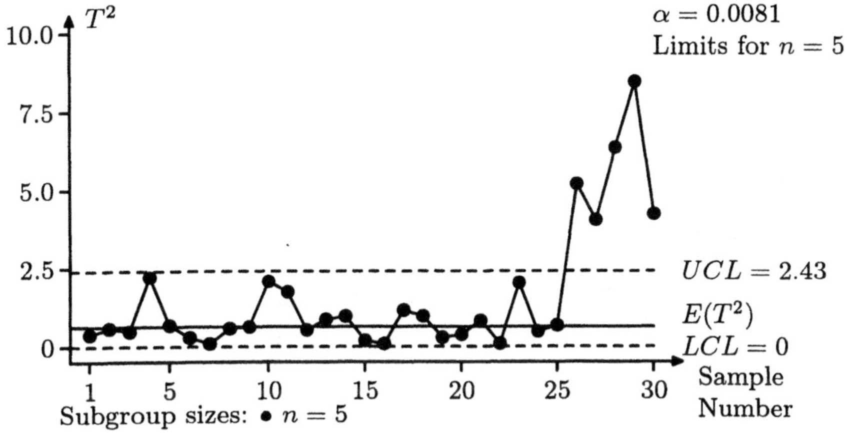

Figure 2.3.

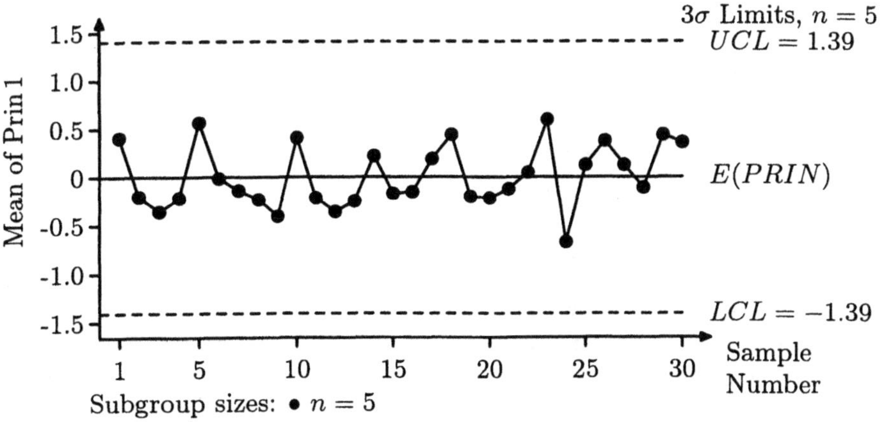

Figure 2.4.

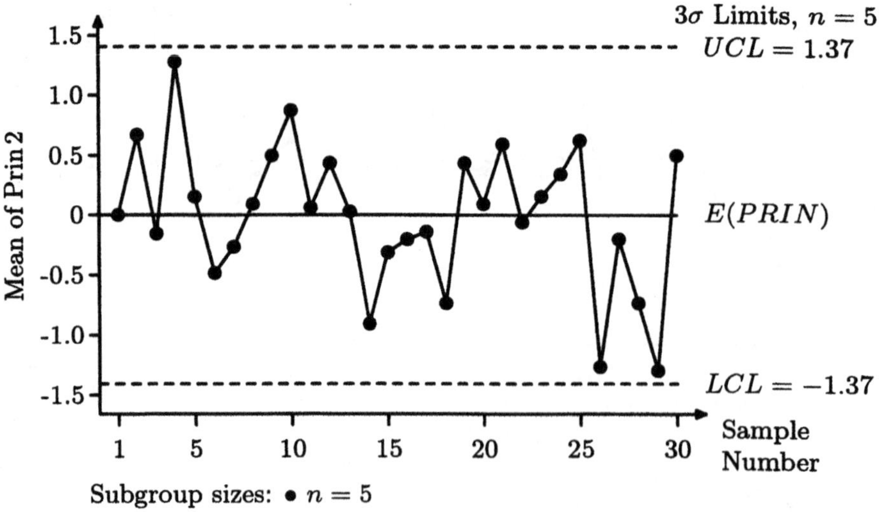

Figure 2.5.

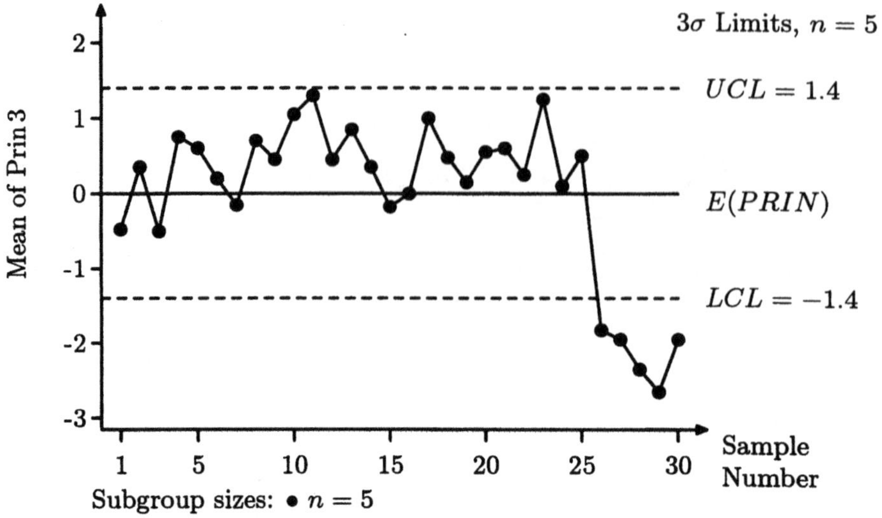

Figure 2.6.

EXAMPLE 2.2. $\boldsymbol{\mu} = (3\sigma_1, 2\sigma_2, 0)'$

This is the case of big shifts respecting the positive correlation between the two first variables. The multivariate chart on Figure 2.7 shows the five out of control points. It is interesting to notice that the univariate chart on the two first initial variables and all the principal component charts (non represented by a figure) detect such big shifts.

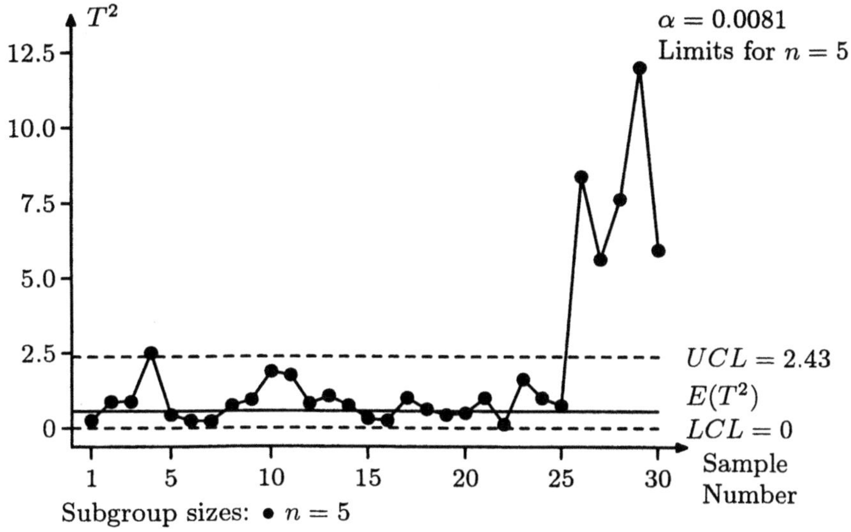

Figure 2.7.

EXAMPLE 2.3. $\mu = (3\sigma_1, -2\sigma_2, 0)'$

We consider here the same values for the mean shifts as in the Example 2.2 but without respecting the correlation between the variables as in Example 2.1. On Figure 2.8 almost all points fall outside the control limits. That leads to exclude too many points in the first stage of Hotelling's chart. This stage is then inapplicable.

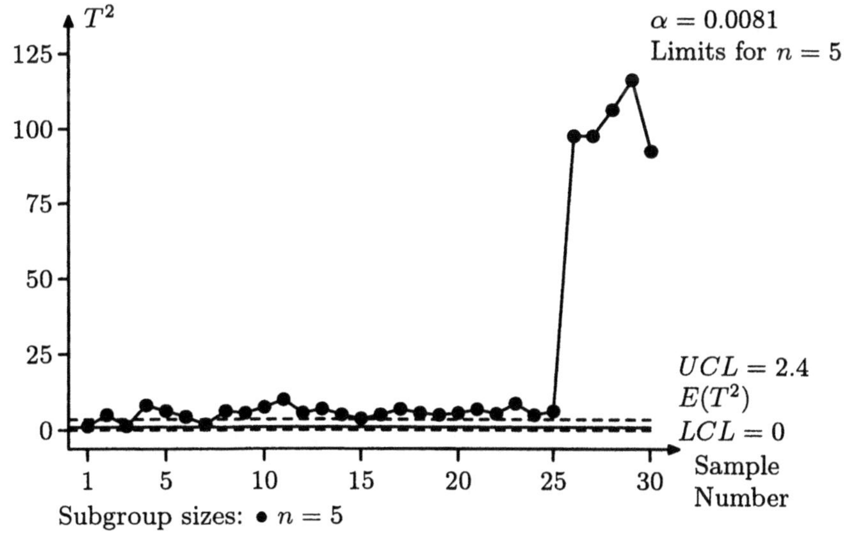

Figure 2.8.

This example shows that the classical estimations for μ and Σ are not optimal in the case of large shifts. Their non-robustness makes the whole analysis polluted by the out of control observations in the case of grouped data.

For the individual measurements case (not illustrated graphically) a kind of compensation is observed and the first stage control chart is inefficient.

Using robust methods allows to remedy to these situations as it will be seen in Section 3 (see Figure 3.1 to Figure 3.3).

3. Robust methods for first stage Shewhart control charts

In this section, robust methods are presented to remove limitations of the first stage Shewhart control charts observed in the numerical examples.

For grouped data the within variance matrix $\bar{S}$ used in the Hotelling statistic is insensitive to the mean shifts usually considered in quality control methods. So, in this case just the process mean needs to be robustly estimated.

In the case of individual measurements, the within variance matrix can no longer be computed. The total variance matrix is used to estimate the process dispersion. This matrix, like the arithmetic mean, is sensitive to the presence of out of control observations. Then, here both process mean and dispersion have to be robustly estimated.

The robust methods proposed later are based on very simple robust estimations of the mean and the variance that are simple versions of M-estimators. This class of robust estimators is briefly presented to facilitate the presentation of the proposed methods. For a complete review of robustness see Huber (1981) or Hampel et al. (1986).

3.1. M-estimators and W-estimators

Several classes of robust estimators have been developed. Hampel et al. (1986), chapter 2, studies the classes of M-, L-, S-, R-,... estimators. Following Huber (1981), we will use here M-estimators which are defined by:

Definition 3.1. *An M-estimator is any estimator T_n solution of*

$$\min_{\theta} \sum_{i=1}^{n} \rho(x_i, \theta) \tag{3.1}$$

or any estimator T_n defined by an implicit equation of the form:

$$\sum_{i=1}^{n} \psi(x_i, T_n) = 0, \tag{3.2}$$

where ρ is an arbitrary function, $\psi(x, \theta) = \frac{\partial}{\partial \theta}\rho(x, \theta)$ (ρ is θ-derivable).

REMARK 3.1. The conditions (3.1) and (3.2) are not always equivalent. The M-estimators are generalizations of maximum likelihood estimators. The maximum likelihood estimator is $T_n = T_n(x_1, x_2, \ldots, x_n)$ that maximizes the likelihood function $L = \prod_{i=1}^{n} f_\theta(x_i)$ or equivalently minimizes

$$-\sum_{i=1}^{n} \ln f_\theta(x_i).$$

Taking $\rho(x, \theta) = -\ln f_\theta(x)$ leads to (3.1).

3.1.1. M-estimator for location

For location parameter the cumulative distribution function F_θ of the sample considered is of the form: $F_\theta = F(x - \theta)$. It is then naturally taken $\psi(x, \theta) = \psi(x - \theta)$.

3.1.2. M-estimator for scale

The cumulative distribution function F_θ of the sample considered is of the form: $F_\theta = F(x/\theta)$. An M-estimator for scale is defined by

$$\sum_{i=1}^{n} \psi(x_i, T_n) = 0 \quad \text{with} \quad \psi(x, \theta) = \psi(x/\theta).$$

3.1.3. W-estimators

They have been introduced by Tukey (1970). They are defined as weighted mean of the observations:

$$T_n(x_1, x_2, \ldots, x_n) = \frac{\sum_{i=1}^{n} w_i x_i}{\sum_{i=1}^{n} w_i}, \tag{3.3}$$

where the weights depend on the observations through the function w:

$$w_i = w(x_i - T_n)$$

Then T_n verifies:

$$T_n = \frac{\sum_{i=1}^{n} w(x_i - T_n) x_i}{\sum_{i=1}^{n} w(x_i - T_n)} \tag{3.4}$$

T_n is obtained by an iterative algorithm; beginning with $T_n^{(0)}$ equal to the median or any other robust estimator the $T_n^{(j)}$ are then computed until convergence, with:

$$T_n^{(j+1)} = \frac{\sum_{i=1}^{n} w\left(x_i - T_n^{(j)}\right) x_i}{\sum_{i=1}^{n} w\left(x_i - T_n^{(j)}\right)} \tag{3.5}$$

Using 3.4 it comes:

$$\frac{\sum_{i=1}^{n} w(x_i - T_n)(x_i - T_n)}{\sum_{i=1}^{n} w(x_i - T_n)} = 0 \tag{3.6}$$

And then T_n verifies:

$$\sum_{i=1}^{n} \psi(x_i - T_n) = 0$$

, where ψ is defined as $\psi(u) = uw(u)$.

The W-estimators are then particular M-estimators and have the same properties (same influence function and asymptotic variance).

Step 1 of the iterative algorithm gives $T_n^{(1)}$, it's a one-step W-estimator noted w-estimator. These w-estimators are more and more used because they are simple to compute, there isn't any problem of convergence and nor of existence and their performances are still good, cf. Hampel et al. (1986), p. 116.

3.1.4. Simultaneous estimation of location and scale

M-estimators presented earlier give separated estimates of location and scale. But sometimes when estimating location it may be necessary to use an estimation of scale, it is the case for example to make the M-estimator scale invariant (Hampel et al. (1986), p. 105). The problem is then to find an estimator of the pair (location, scale), for example $(\boldsymbol{\mu}, \boldsymbol{\Sigma})$. $(\boldsymbol{T}_n, \boldsymbol{S}_n)$ a pair of M-estimators of location and scale is defined as solution of:

$$\begin{cases} \sum_{i=1}^{n} \psi\left(\frac{\boldsymbol{x}_i - \boldsymbol{T}_n}{\boldsymbol{S}_n}\right) = 0 \\ \sum_{i=1}^{n} \chi\left(\frac{\boldsymbol{x}_i - \boldsymbol{T}_n}{\boldsymbol{S}_n}\right) = 0 \end{cases} \tag{3.7}$$

Definition 3.2. *An estimator $(\boldsymbol{T}_n, \boldsymbol{S}_n)$ is affine-equivariant if for any non-singular affine transformation: $\boldsymbol{x} \to \boldsymbol{A}\,\boldsymbol{x} + \boldsymbol{t}$ ($\boldsymbol{t} \in \mathcal{R}_p$, $\boldsymbol{A}$ is $p \times p$ regular matrix) we have:*

$$\begin{aligned} \boldsymbol{T}_n(\boldsymbol{A}\boldsymbol{X}_1 + \boldsymbol{t}, \ldots, \boldsymbol{A}\boldsymbol{X}_1 + \boldsymbol{t}) &= \boldsymbol{A}\,\boldsymbol{T}_n(\boldsymbol{X}_1, \ldots, \boldsymbol{X}_n) + \boldsymbol{t} \\ \boldsymbol{S}_n(\boldsymbol{A}\boldsymbol{X}_1 + \boldsymbol{t}, \ldots, \boldsymbol{A}\boldsymbol{X}_1 + \boldsymbol{t}) &= \boldsymbol{A}\,\boldsymbol{S}_n(\boldsymbol{X}_1, \ldots, \boldsymbol{X}_n)\,\boldsymbol{A}' \end{aligned}$$

REMARK 3.2. The shifts considered in quality control classical applications are affine transformations of this form.

Definition 3.3. *Affine equivariant M-estimators are defined as solution of:*

$$\boldsymbol{\mu}_n = \frac{\sum_{i=1}^{n} w_1\left(\|\boldsymbol{X}_i - \boldsymbol{\mu}_n\|^2_{\boldsymbol{V}_n^{-1}}\right) \boldsymbol{X}_i}{\sum_{i=1}^{n} w_1\left(\|\boldsymbol{X}_i - \boldsymbol{\mu}_n\|^2_{\boldsymbol{V}_n^{-1}}\right)} \tag{3.8}$$

$$\boldsymbol{V}_n = \frac{\sum_{i=1}^{n} w_2\left(\|\boldsymbol{X}_i - \boldsymbol{\mu}_n\|^2_{\boldsymbol{V}_n^{-1}}\right) (\boldsymbol{X}_i - \boldsymbol{\mu}_n)(\boldsymbol{X}_i - \boldsymbol{\mu}_n)'}{\sum_{i=1}^{n} w_3\left(\|\boldsymbol{X}_i - \boldsymbol{\mu}_n\|^2_{\boldsymbol{V}_n^{-1}}\right)}, \tag{3.9}$$

where $(\boldsymbol{X}_1, \boldsymbol{X}_2, \ldots, \boldsymbol{X}_n)$ is a sample of $\mathcal{R}_p$ and w_1, w_2, w_3 are positive functions.

The class of affine equivariant M-estimators is a class of M-estimators of the pair $(\boldsymbol{\mu}, \boldsymbol{\Sigma})$.

3.2. Robust methods for quality control

3.2.1. Robust chart for the process mean in case of grouped data

A simple robust estimation of the process mean

The Hotelling statistic explosion observed in the example in 2.5. is due to the classical estimations for $\boldsymbol{\mu}$ and $\boldsymbol{\Sigma}$ that are not optimal in the case of large shifts. Their non-robustness makes the whole analysis polluted by the out of control observations. This is explained in detail in Niang (1994).

Let $\boldsymbol{X}_{i1}, \ldots, \boldsymbol{X}_{in}$, $i = 1, \ldots, m$, be m samples of n observations of a normal $\mathcal{N}_p(\boldsymbol{\mu}, \boldsymbol{\Sigma})$. Each $\boldsymbol{X}_{ij}$ is a realization of $\boldsymbol{X}$, a $\mathcal{R}_p$-vector: $\boldsymbol{X}' = (X^1, X^2, \ldots, X^p)$; let $\boldsymbol{\mu}' = (\mu^1, \mu^2, \ldots, \mu^p)$.

The mean shifts usually considered in quality control methods are affine translations affecting in the same way all the k out of control points: when a shift occurs the process mean goes from $\boldsymbol{\mu}$ to $\boldsymbol{\mu} + \boldsymbol{\Delta}$ and stays at that value until corrective actions are taken.

The within variance matrix $\bar{\boldsymbol{S}}$ is insensitive to these mean shifts because the variance of each subgroup doesn't change due to its insensitivity to translation. So, in the case of grouped data, only the estimation of the mean is responsible of the explosion of T^2 and it is then what we need to estimate robustly.

The class of M-estimators gives estimators with good properties. But the complexity of implicit expressions and algorithms and the problems of convergence make them difficult to be used in quality control. The w-estimators presented in Section 3.1.3: are easy to compute, there isn't any problem of convergence and their robustness properties are still quite good. We propose a w-estimator for the mean process in the case of grouped data. The determination of this estimator is based on a three-step procedure:

(1) compute $\overline{\overline{\boldsymbol{X}}}_i$, $\bar{\boldsymbol{S}}$ and $T_i^2 = \left(\overline{\boldsymbol{X}}_i - \overline{\overline{\boldsymbol{X}}}\right)' \bar{\boldsymbol{S}}^{-1} \left(\overline{\boldsymbol{X}}_i - \overline{\overline{\boldsymbol{X}}}\right)$

(2) compute p_i: $p_i = w(T_i^2)$ with $w(x) = \exp\left(-\frac{1}{2}\|x\|^2_{\bar{\boldsymbol{S}}^{-1}}\right)$ or any decreasing function.

(3) a robust estimator of the mean is then:

$$\overline{\overline{\boldsymbol{X}}}_R = \frac{\sum_{i=1}^{m} p_i \overline{\boldsymbol{X}}_i}{\sum_{i=1}^{m} p_i} \tag{3.10}$$

$\overline{\overline{\boldsymbol{X}}}_R$ is a kind of M-estimator but without iteration; it's a "one step" M-estimator very simple to compute.

REMARK 3.3. The first estimation of the mean process $\overline{\overline{\boldsymbol{X}}}$ is non-robust and can lead to anormally big values for the estimated Mahalanobis distances and then to too small weights and consequently to underestimate the process mean.

There are several methods to avoid this: truncated mean, winsorized mean etc. It can be more suitable to take a more robust estimation like the median.

Let $\boldsymbol{M} = (\text{med } \overline{\boldsymbol{X}}_1, \ldots, \text{med } \overline{\boldsymbol{X}}_p)'$ the p-vector of the median of the subgroup means, a robust estimation $\overline{\overline{\boldsymbol{X}}}_M$ of the process mean can be obtained as $\overline{\overline{\boldsymbol{X}}}_R$:

(1) with $\boldsymbol{M}$ and $\bar{\boldsymbol{S}}$, compute $T_i^2 = (\boldsymbol{X}_i - \boldsymbol{M})' \bar{\boldsymbol{S}}^{-1} (\boldsymbol{X}_i - \boldsymbol{M})$;

(2) compute p_i: $p_i = w(T_i^2)$ with $w(x) = \exp(-\frac{1}{2}\|x\|^2_{\bar{S}^{-1}})$ or any decreasing function;

(3) $\overline{\boldsymbol{X}}_M = \dfrac{\sum_{i=1}^{m} p_i \overline{\boldsymbol{X}}_i}{\sum_{i=1}^{m} p_i}$.

A simple robust chart:

$$\widehat{\mu} = \overline{\overline{\boldsymbol{X}}}_R = \frac{\sum_{i=1}^{m} p_i \overline{\boldsymbol{X}}_i}{\sum_{i=1}^{m} p_i} \tag{3.11}$$

is used to estimate the unknown mean process and the process matrix dispersion is estimated by the within variance matrix. The control statistic is then:

$$T_R^2 = \left(\overline{\boldsymbol{X}}_i - \overline{\overline{\boldsymbol{X}}}_R\right)' \bar{\boldsymbol{S}}^{-1} \left(\overline{\boldsymbol{X}}_i - \overline{\overline{\boldsymbol{X}}}_R\right). \tag{3.12}$$

Following Alt (1973) we prove that:

$$\frac{k_1^2 n}{k_3} \left(\overline{\boldsymbol{X}}_i - \overline{\overline{\boldsymbol{X}}}_R\right)' \bar{\boldsymbol{S}}^{-1} \left(\overline{\boldsymbol{X}}_i - \overline{\overline{\boldsymbol{X}}}_R\right) \tag{3.13}$$

has a Hotelling distribution which is identifiable to a Fisher:

$$\frac{pm(n-1)}{mn-m-p+1} F_{p,mn-m-p+1}, \tag{3.14}$$

where $k_1 = \sum_{i=1}^m p_i$, $k_2 = \sum_{i=1}^m p_i^2$ and $k_3 = \left(\sum_{i=1}^m p_i\right)^2 + \sum_{i=1}^m p_i^2 = k_1^2 + k_2$.

Firstly the weights p_i are assumed to be constant. Then for all $i = 1, \ldots, m$ $\overline{\boldsymbol{X}}_i$ has a $\mathrm{N}_p(\boldsymbol{\mu}, \boldsymbol{\Sigma}/n)$. And we have:

$$\mathsf{E}(\overline{\overline{\boldsymbol{X}}}_R) = \frac{1}{\sum_{i=1}^{m} p_i} \sum_{i=1}^{m} p_i \, \mathsf{E}(\overline{\boldsymbol{X}}_i) = \mu \tag{3.15}$$

and

$$\mathsf{Var}(\overline{\overline{\boldsymbol{X}}}_R) = \frac{1}{\left(\sum_{i=1}^{m} p_i\right)^2} \sum_{i=1}^{m} p_i^2 \, \mathsf{Var}(\overline{\boldsymbol{X}}_i) = \frac{\sum_{i=1}^{m} p_i^2}{\left(\sum_{i=1}^{m} p_i\right)^2} \frac{1}{n} \boldsymbol{\Sigma} = \frac{k_2}{k_1^2 n} \boldsymbol{\Sigma} \, . \tag{3.16}$$

Then $\overline{\overline{\boldsymbol{X}}}_R$ is then distributed as a normal $\mathrm{N}_p\left(\boldsymbol{\mu}, k_2 \boldsymbol{\Sigma}/k_1^2 n\right)$. Let $\boldsymbol{Y}_i = \overline{\boldsymbol{X}}_i - \overline{\overline{\boldsymbol{X}}}_R$. Without loss of generality, let $i = 1$. We have:

$$\begin{aligned}\boldsymbol{Y}_1 = \overline{\boldsymbol{X}}_1 - \overline{\overline{\boldsymbol{X}}}_R = \overline{\boldsymbol{X}}_1 - \frac{p_1\overline{\boldsymbol{X}}_1}{k_1} - \frac{1}{k_1}\sum_{i=2}^{m} p_i\overline{\boldsymbol{X}}_i \\ = \frac{k_1 - p_1}{k_1}\overline{\boldsymbol{X}}_1 - \frac{\sum_{i=2}^{m} p_i\overline{\boldsymbol{X}}_i}{k_1}.\end{aligned} \tag{3.17}$$

$\frac{k_1-p_1}{k_1}\overline{\boldsymbol{X}}_1$ is distributed as $\mathrm{N}_p\left(\frac{k_1-p_1}{k_1}\boldsymbol{\mu}, \left(\frac{k_1-p_1}{k_1}\right)^2 \frac{1}{n}\boldsymbol{\Sigma}\right)$ and $\frac{1}{k_1}\sum_{i=2}^{m} p_i\overline{\boldsymbol{X}}_i$ is distributed as $\mathrm{N}_p\left(\frac{1}{k_1}\boldsymbol{\mu}\sum_{i=2}^{m} p_i, \frac{1}{k_1^2 n}\boldsymbol{\Sigma}\sum_{i=2}^{m} p_i^2\right)$.
As $\frac{1}{k_1}\sum_{i=2}^{m} p_i = (k_1 - p_1)/k_1$ it comes:

$$\begin{aligned}&\mathsf{E}(\overline{\boldsymbol{X}}_1 - \overline{\overline{\boldsymbol{X}}}_R) = 0, \\ &\mathsf{Var}(\overline{\boldsymbol{X}}_1 - \overline{\overline{\boldsymbol{X}}}_R) = \left(\frac{1}{k_1^2}\Big(\sum_{i=2}^{m} p_i\Big)^2 + \frac{1}{k_1^2}\sum_{i=2}^{m} p_i^2\right)\frac{\boldsymbol{\Sigma}}{n} = \frac{k_3}{k_1^2 n}\boldsymbol{\Sigma}.\end{aligned} \tag{3.18}$$

Then $\overline{\boldsymbol{X}}_i - \overline{\overline{\boldsymbol{X}}}_R$ has a normal distribution with mean $\mathbf{0}$ and variance $k_3\boldsymbol{\Sigma}/k_1^2 n$ and the statistic $k_1\sqrt{n/k_3}(\overline{\boldsymbol{X}}_i - \overline{\overline{\boldsymbol{X}}}_R)$ is distributed as $\mathrm{N}_p(\mathbf{0}, \boldsymbol{\Sigma})$. $\bar{\boldsymbol{S}}$ has a Wishart distribution $\mathrm{W}_p(m(n-1), \boldsymbol{\Sigma})$ independent of $k_1\sqrt{n/k_3}(\overline{\boldsymbol{X}}_i - \overline{\overline{\boldsymbol{X}}}_R)$. So

$$\frac{k_1^2 n}{k_3}\left(\overline{\boldsymbol{X}}_i - \overline{\overline{\boldsymbol{X}}}_R\right)' \bar{\boldsymbol{S}}^{-1}\left(\overline{\boldsymbol{X}}_i - \overline{\overline{\boldsymbol{X}}}_R\right) \tag{3.19}$$

is distributed as Hotelling identifiable to a Fisher:

$$\frac{pm(n-1)}{mn - m - p + 1}F_{p,mn-m-p+1}. \tag{3.20}$$

Now, let $\mathcal{L}(\boldsymbol{Y}|\boldsymbol{X})$ be the distribution of $\boldsymbol{Y}$ conditionally to $\boldsymbol{X}$, then:

$$\begin{aligned}\mathcal{L}\left(\frac{k_1^2 n}{k_3}\left(\overline{\boldsymbol{X}}_i - \overline{\overline{\boldsymbol{X}}}_R\right)' \bar{\boldsymbol{S}}^{-1}\left(\overline{\boldsymbol{X}}_i - \overline{\overline{\boldsymbol{X}}}_R\right)\middle| p_1, p_2, \dots, p_n\right) = \\ = \frac{pm(n-1)}{mn - m - p + 1}F_{p,mn-m-p+1}.\end{aligned} \tag{3.21}$$

Thus the conditional distribution of the control statistic

$$\frac{k_1^2 n}{k_3}\left(\overline{\boldsymbol{X}}_i - \overline{\overline{\boldsymbol{X}}}_R\right)' \bar{\boldsymbol{S}}^{-1}\left(\overline{\boldsymbol{X}}_i - \overline{\overline{\boldsymbol{X}}}_R\right) \tag{3.22}$$

doesn't depend on p_i. So this distribution is also the marginal one. And it can be concluded that control statistic is distributed as Hotelling identifiable to a Fisher:

$$\frac{pm(n-1)}{mn - m - p + 1}F_{p,mn-m-p+1}. \tag{3.23}$$

Then a multivariate control chart for the mean consists in drawing the statistic $(k_1^2 n/k_3)\left(\overline{\boldsymbol{X}}_i - \overline{\overline{\boldsymbol{X}}}_R\right)' \bar{S}^{-1}\left(\overline{\boldsymbol{X}}_i - \overline{\overline{\boldsymbol{X}}}_R\right)$ and compare it to the upper control limit:

$$UCL = \frac{pm(n-1)}{mn - m - p + 1} F_{p,mn-m-p+1,\alpha}. \tag{3.24}$$

The lower control limit is $LCL = 0$.

On numerical Example 2.3, by comparing the values of the control statistic with and without a robust estimate of the mean (Table 3.1). It can be seen that for the proposed robust control chart the explosion of Hotelling's statistic is removed. This is illustrated on Figure 3.3.

It can be also noticed that the last five subgroups have greater values with the robust statistic and then are more proeminent; this is a classical property of robust method. Furthermore the chart has been tested on moderate shifts (Figure 3.1) and on big shifts respecting the correlation (Figure 3.2) and they are still efficient.

Sample	T^2	Robust T^2	Sample	T^2	Robust T^2
1	1.168	0.646	16	2.422	0.043
2	4.096	0.693	17	6.467	1.287
3	1.437	0.604	18	4.524	1.138
4	7.041	2.679	19	3.226	0.278
5	4.784	0.939	20	4.485	0.361
6	3.172	0.292	21	5.097	0.872
7	1.979	0.122	22	3.305	0.096
8	5.085	0.565	23	8.061	2.459
9	4.582	0.640	24	3.339	0.562
10	7.518	2.495	25	4.657	0.845
11	8.038	1.947	26	93.176	158.311
12	4.474	0.549	27	92.992	158.350
13	5.845	0.860	28	101.694	170.855
14	4.185	1.148	29	109.704	182.363
15	2.020	0.158	30	92.681	157.921

Table 3.1.

3.3. Controlling the process mean with individual observations

In the case of individual measurements, the within variance matrix can no longer be computed. The usual method consists in estimating the process dispersion by the total variance matrix. This matrix, like the arithmetic mean, is sensitive to the presence of out of control observations. A multivariate control chart based on robust M-estimators for the mean and dispersion improves consistently the performances of the first stage chart. We propose a "one-step" M-estimator (Ruiz-Gazen and Caussinus (1993)), which is simple to compute and still has good properties of robustness.

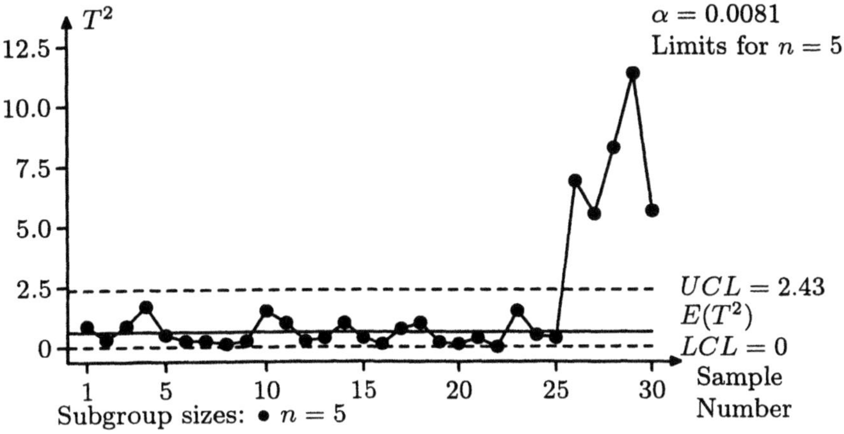

Figure 3.1.

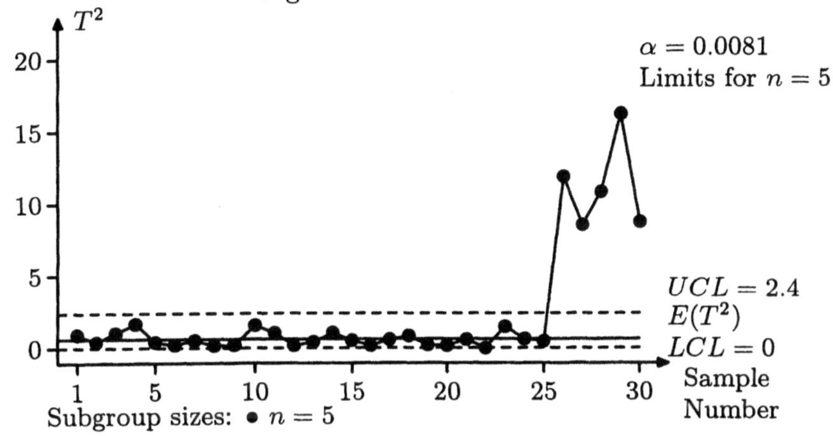

Figure 3.2.

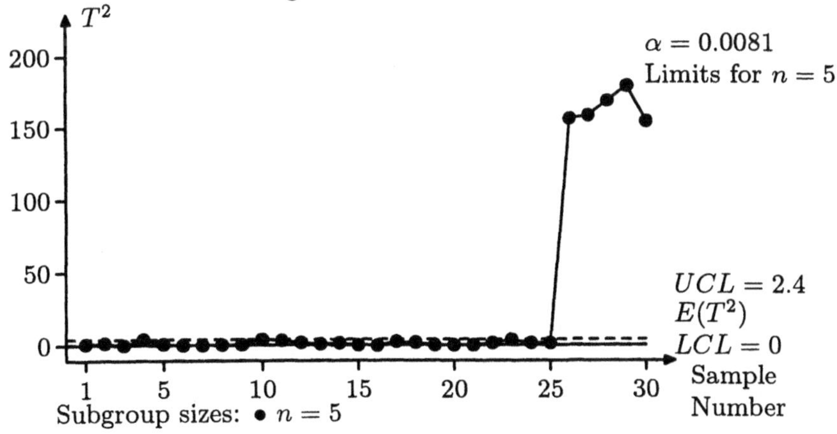

Figure 3.3.

Let

$$V_n = \frac{1}{n}\sum_{i=1}^{n}(X_i - \mu_n)(X_i - \mu_n)' \qquad \text{and} \tag{3.25}$$

$$S_n(\beta) = \frac{\sum_{i=1}^{n} w\left(\beta\|X_i - \overline{X}_n\|^2_{V_n^{-1}}\right)(X_i - \overline{X}_n)(X_i - \overline{X}_n)'}{\sum_{i=1}^{n} w\left(\beta\|X_i - \overline{X}_n\|^2_{V_n^{-1}}\right)}, \tag{3.26}$$

V_n is the classical empirical covariance matrix, μ_n is an estimator for location of the sample $(X_1, X_2, \ldots, X_n)$ and $S_n(\beta)$ is a weighted empirical covariance matrix. The weights are values of a decreasing function of a parameter β and Mahalanobis distances computed with V_n which is sensitive to outliers. $S_n(\beta)$ is then non-robust. For a robust estimation of the dispersion, Ruiz-Gazen and Caussinus (1993) propose to take: $S_R = (S_n^{-1} - \beta V_n^{-1})^{-1}$. The robustness properties of this estimator depend obviously on β and need to be more precisely studied.

A robust estimator for the process mean is obtained as in the case of grouped data by; $\overline{X}_R = (\sum_{i=1}^{m} p_i X_i)/\sum_{i=1}^{m} p_i$ but here the weights are computed using the matrix S_R.

The proposed control chart consists in comparing the statistic $(X_i - \overline{X}_R)' S_R^{-1} (X_i - \overline{X}_R)$ with the limits:

$$\begin{aligned} UCL &= \frac{(m-1)^2}{m}\text{Beta}\left(\frac{\alpha}{2}; \frac{p}{2}, \frac{(m-p-1)}{2}\right) \\ LCL &= \frac{(m-1)^2}{m}\text{Beta}\left(1 - \frac{\alpha}{2}; \frac{p}{2}, \frac{(m-p-1)}{2}\right) \end{aligned} \tag{3.27}$$

The matrix S_R may be also used to realize a robust principal component analysis. Then the process can be controlled by univariate control charts for the uncorrelated principal components. But as noted in Section 2.4., both the first and the last components have to be used. If there is a great number of initial variables, that can lead to too many univariate charts.

Projection pursuit methods more specific to the problem of detecting a structure into multidimensional data can also be used. A special case of projection pursuit adapted from Yenyukov's method more suitable for detecting multivariate outliers can also be used (Niang (1994)). This method consists in a PCA with special metric S_R^{-1}. It leads to an easy detection of out of control points on the first principal components. That reduces the number of univariate charts to construct.

However it must be noted that PCA or projection pursuit are essentially graphical methods. It may append that inferential methods based on the results of this PCA or projection pursuit like control charts for principal components can not detect out of control observations while plots of principal component display them well (Niang (1994)). In fact the two sets of methods are to be used jointly.

Conclusion

We have presented multivariate methods for statistical process control with a focus on the problems of Hotelling charts for the first stage of the control procedure. Robust methods have been proposed to remove limitations of the classical charts for large shifts and to improve the perfomances in general. The described methods have been shown to be efficient on simulated data and some special kind of shift. Studies are in progress for real data, correlated observations and drifts of the mean.

References

Alt F. B. (1985). *Multivariate Quality Control.* Encyclopedia of Statistical Sciences Vol. **6**, Kotz S. et al. eds, 110 – 122, North-Holland, Amsterdam.

Alt F. B. (1973). *Aspects of Multivariate Control Charts.* Master of Science. Georgia Institute of Technology.

Anderson T. W. (1984). *An Introduction to Multivariate Statistical Analysis, 2nd ed.* J. Wiley, New York.

Barnett V. and Lewis T. (1984). *Outliers in Statistical Data, 2nd ed.* J. Wiley, New York.

Crosier R. B. (1988). *Multivariate generalizations of cumulative sum quality control schemes.* Technometrics **30**, 291 – 303.

Flury B. (1988). *Common Principal Components and Related Multivariate Models.* J. Wiley, New York.

Gnanadesikan R. and Kettenring J. R. (1972). *Robust estimation and outlier detection with multiresponse data.* Biometrics **28**, 81 – 124.

Hampel F. R., Ronchetti E. M., Rousseeuw P. J. and Stahel W. A. (1984). *Robust Statistics.* J. Wiley, New York.

Hawkins D. M. (1974). *The detection of errors in multivariate data using principal component.* Journal of the American Statistical Association **69**.

Huber P. J. (1981). *Robust Statistics.* J. Wiley, New York.

Jackson J. E. (1956). *Quality control methods for two related variables.* Industrial Quality Control, 4 – 8.

Jackson J. E. and Morris R. H. (1957). *An application of multivariate quality control to photographic processing.* Journal of the American Statistical Association **52**, 186 – 199.

Jackson J. E. (1959). *Quality control method for several related variables.* Techonometrics **1**.

Jackson J. E. (1991). *A User's Guide to Principal Components.* J. Wiley, New York.

Jolliffe I. T. (1986). *Principal Component Analysis.* Springer-Verlag, Heidelberg.

Lowry C. A., Woodall W. H., Champ C. W. and Rigdon S. E. (1992). *A multivariate exponentially weighted moving average control chart.* Technometrics **34**, 46 – 53.

Montgomery D. C. (1985). *Introduction to Statistical Quality Control.* J. Wiley, New York.

Niang N. (1994). *Méthodes multidimensionnelles pour la Maîtrise Statistique des Procédés.* Doctorat de l'Université Paris Dauphine.

Niang N. and Saporta G. (1995). *Période opérationnelle moyenne de la carte de moyennes mobiles équipondérés pour le contrôle du centrage d'un procédé.* Revue de Statistique Appliquée.

Pignatiello J. J., Jr. and Runger G. C. (1990). *Comparaison of multivariate CUSUM.* Journal of Quality Technology **22**, 173 – 186.

Ruiz-Gazen A. (1993). *Estimation robuste d'une matrice de dispersion et projections révélatrices.* Doctorat de l'Université Paul Sabatier, Toulouse.

Ryan T. P. (1989). *Statistical Methods for Quality Improvement.* J. Wiley, New York.

Tracy N. D., Young J. C. and Mason R. L. (1992). *Multivariate control charts for individual observations.* Journal of Quality Technology **24**.
Tukey J. W. (1970). *Exploratory Data Analysis.* Mimeographed preliminary edition.
Yenyukov I. S. (1988). *Detecting structure by means of projection pursuit.* COMPSTAT'88, Edwards D. ed., 47 – 58, Physica-Verlag, Heidelberg.

PARAMETRIC AND NON PARAMETRIC MULTIVARIATE QUALITY CONTROL CHARTS

Germana Scepi

University of Naples Federico II, Faculty of Economics, Department of Mathematics and Statistics, Via Cintia – Complesso Monte Sant'Angelo, I – 80 126 Naples, Italy; scepi@unina.it

Abstract: This contribution aims at analysing the problem of monitoring products/processes of multivariate quality measurements. Classical and parametric approaches are discussed together with new and non parametric ones. We focus particularly on the non parametric control schemes both for simple data as well as for batch and time dependent data. The advantages and drawbacks of the different approaches are compared by means of some examples and applications. Finally, some new developments on the control of complex and structured data, like symbolic data, are introduced.

1. Introduction

In the following, we present a short discussion about methods for controlling simultaneously more characteristics of a product or a production process. In real life a product is a complex mechanism characterised by different and correlated attributes. Therefore, the most common univariate control charts, such as the *Shewhart* and *CUSUM* charts, may be not adequate for detecting changes in its overall quality. In order to control the overall product variability, it can be necessary to have control charts that can monitor multivariate measurements directly. At this aim, it can be essential to identify a single and common standard for the overall quality of products and, consequently, to determine the random variability around this target. Hence, the problem of building multivariate control schemes consists in computing multivariate control limits.

The assumption of multinormality on the distribution of the variables characterizing the process solves this problem easily. In literature, the well-known multivariate control charts are based on the multinormality assumption (Jackson (1991)) and, as a consequence, have elliptical control boundaries. We refer to schemes both for independent and for sequential samples, such as T^2 control charts, *M-CUSUM* control charts and *EWMA* control

---

2000 *Mathematics Subject Classification.* 62-01, 62G09, 62H25, 62K15, 62P30, 62F40.

Key words and phrases. Bootstrap, three way data analysis, parametric control charts, non parametric schemes, explorative tools, time series data, symbolic data, symplicial depth.

Acknowledgement. The work was partially supported by the MURST grant '98 on *Multivariate Analyses for Total Quality: Methodologies, Computational Aspects and Applications* coordinated by N.C. Lauro.

charts. The same hypothesis holds for control charts based on *Principal Component Analysis (PCA)*.

In Section 2 we discuss and point out the main advantages and drawbacks of some of these parametric schemes.

The hypothesis of multinormality is difficult to verify in real situations, in particular if we deal with qualitative variables. Different authors (Scepi et al. (1993), Liu (1995)) proposed non parametric approaches to building multivariate control schemes for monitoring quality of products (or services, see Lauro et al. (1997)). The non parametric control charts introduced by Scepi et al. (1993) are based on convex and not regular limits.

We discuss the non parametric approach and compare it, see Lauro et al. (1996) for a survey, with the classical one (Section 3). In particular, we underline its usefulness in controlling both single and control schemes for sequential as well as in performing control schemes for batches (Seppala et al. (1995)).

The performance of parametric and non parametric methods must be evaluate on the base of how quickly the schemes detect an actual out of control situation. The sensitivity of multivariate control charts in presence of different shifts from the target is analyzed in Section 4. In particular, the sensitivity of parametric schemes, such as T^2 control charts, is compared with the non parametric one (Scepi & Acconcia (1996), Balbi & Scepi (1996)).

Finally, some new developments are shown. In particular, the problem of controlling peculiar data, like symbolic data, is discussed in Section 5. For these complex and structured data, the approaches proposed for classical data, with some variants, are applied (Giordano & Palumbo (1999), Balbi & Scepi (2000)).

2. Parametric control charts

A control procedure consists generally in two phase. The first (called *pilot phase*) concerns in the construction of the control scheme, in the definition of the control limits and, as a consequence, in the identification of the control region. Samples of observations are randomly selected from the process and, after this, some parameters, measurements of position and/or variability, are calculated for each sample. The aim is to detect the natural tolerance of the process and its natural variability. In this phase, for defining robust limits, it is important to choose statistics (like the median) not sensible to the presence of abnormal values (say *outliers*) in the sampling observations.

When the limits are defined, new measurements are recorded for starting the *on line monitoring* phase. The new sample points are plotted on the chart. A sample point falling in out of control region is interpreted as evidence that the process is out of control. Therefore, for controlling data, it is necessary to use statistics (like the mean) capable of easily detecting some shifts in the process.

As in a statistical test, it is possible that the process is declared out of control when in fact is not. In this case, we say that we have a "false alarm", α. In the pilot phase, it is important to define limits so that the false alarm rate is small.

Finally, fixed α, it is important that limits are chosen both robust to outlier as well as sensible to the presence of real out of control situations (for a reference on the problems of defining limits for different control charts see Does & Schriever (1992)).

In a univariate context, control charts are a sort of graphical confidence intervals. Here, the computation of limits is very simple; since the control parameters (as the mean or the variance) of the process are defined, the limits are easily obtained applying to known distributions. Therefore, the control charts consist of critical values, the *Upper Control Limit* (*UCL*) and the *Lower Control Limit* (*LCL*) for a sample quality measurement. Between the control limits, there is a *Center Line* (*CL*) which represents no deviation from the prescribed distribution. The region above *UCL* or below *LCL* is defined *out of control region.* In univariate control charts clearly evident are shifts in location or in scale, while it is not possible to control the structure of correlation between two (or more) variables. In this sense, it is possible to have two variables that are both in control when separately considered at univariate level but out of control when jointly considered in a multivariate scheme (Wierda (1994)).

In a multivariate approach, the problem of limit computation is more complex. The first action is to define multivariate limits for a multivariate control region. If multinormality is assumed, it is possible to refer to known multivariate distributions, too.

In multinormality hypothesis, the graphical bases of multivariate control charts must be ellipsoids, called *contours*, with shape:

$$(\boldsymbol{x} - \boldsymbol{\mu})'\boldsymbol{\Sigma}^{-1}(\boldsymbol{x} - \boldsymbol{\mu}) = c^2, \tag{2.1}$$

where $\boldsymbol{x}$ has p-dimensional multinormal distribution, $\mathcal{N}_p(\boldsymbol{\mu}, \boldsymbol{\Sigma})$, and c is a constant value.

In practice, it is possible to have two different solutions to the problem of limits computation and representation. The first consists in moving the multivariate scheme at a univariate one, by some transformation on the control statistics; this is the case, for example, of the standard T^2 control chart.

The second deals with confidence regions directly; we refer, for example, to *M-CUSUM* control charts.

In the following, we present the characteristics proper of these two multivariate control charts.

2.1. T^2 control charts

A measure of conformity to the whole system of project standards is given by the following statistics:

$$T^2 = (\boldsymbol{x} - \boldsymbol{x}_0)\boldsymbol{S}^{-1}(\boldsymbol{x} - \boldsymbol{x}_0)', \tag{2.2}$$

where $x(1,p)$ is the vector of the values of variables for each i-th (i=1,...,n) observation, $x_0(1,p)$ is the target vector and S is the sample covariance matrix, unbiased estimator of Σ.

The distribution of T^2 statistics is directly linked to the Snedecor-Fisher distribution as:

$$T^2_{p,n,\alpha} = \frac{p(n-1)}{n-p} F_{p,n-p}(\alpha) \tag{2.3}$$

The theoretical (critical) value, $T^2_{p,n,\alpha}$, is the *UCL* for monitoring each new observation in this multivariate chart. Graphically, the control chart is a simple Cartesian-diagram where the critical value is a parallel line to the abscises axis (Figure 1). The process is considered to be in control if each point is below the central line. An observation that produces a value of T^2 greater than the *UCL* will be out of control on this chart (*A*, *B*, *C*, *D* in Figure 1).

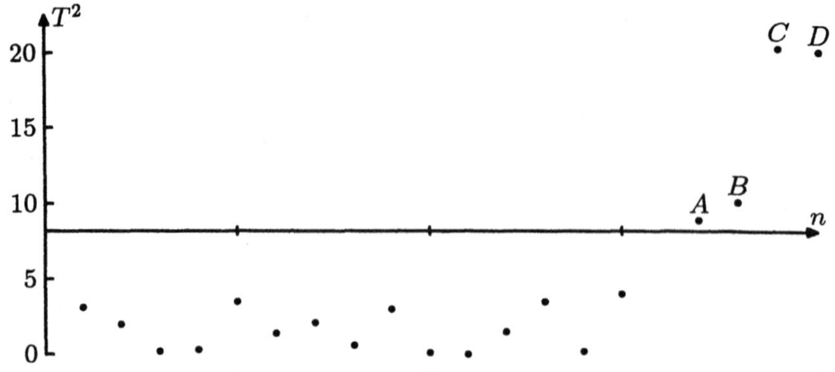

Figure 1. The T^2 control chart.

The use of *PCA* can be extremely useful in quality control applications because it allows to transform a set of correlated variables in a new set of uncorrelated variables that may be easier to monitor with control charts. Principal components resolve the problem of computing the Type I error and, moreover, allow us to minimize the probability of *false alarm*.

T^2 control chart based on *PCA* fulfills three important conditions:

(1) an overall Measure of variability of process is given;
(2) an overall Type I error is specified;
(3) relationships among the variables are taken into account.

The original form of T^2, see (2.2), is here substituted by the following one:

$$T^2 = yy' \tag{2.4}$$

where the vector $\mathbf{y}$ of principal components is computed as:

$$y = W'(x - \bar{x}) \tag{2.5}$$

In (2.5) we indicate with $\bar{\boldsymbol{x}}$ the vector of the mean values of variables and with $\boldsymbol{W} = \boldsymbol{U}\,\boldsymbol{L}^{-1/2}$ the matrix of characteristic vectors $\boldsymbol{U}$ of $\boldsymbol{S}$, scaled to the reciprocal of their characteristic roots (in $\boldsymbol{L}$).

For fixing the theoretical value of T^2 in this case it is necessary to substitute the number of chosen components $q \leq p$ in (2.3). In Figure 1 we show an example of a T^2 control chart (Jackson (1991)) with a control limit of 8.187 for a two-dimensional case. 15 observations can be easily detected in control and 4 observations (A, B, C, D) in out of control.

In a two-dimensional example, the original observation can be displayed in a particular box representing the joint control region of the original control charts. Furthermore, an ellipse can be constructed around the intersection of the means (Figure 2)

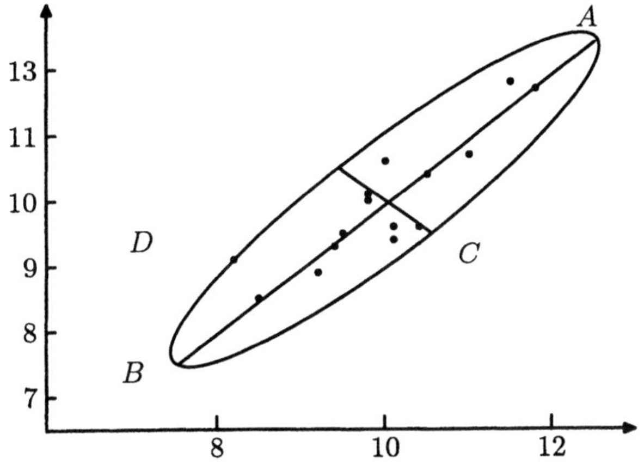

Figure 2. The control ellipses.

Anything that is out of control on the T^2control chart will be outside this ellipse. The length of the ellipses semimajor axis can be easily determined as:

$$\sqrt{\lambda_1 T^2_\alpha}, \tag{2.6}$$

where λ_1 is the first characteristic root; the length of the semiminor axis is:

$$\sqrt{\lambda_2 T^2_\alpha}, \tag{2.7}$$

where λ_2 is the second characteristic root.

The procedure for monitoring a multivariate process using T^2 control chart based on PCA is as follows: for each observation vector, we calculate the coordinates of PCA and, from this, compute T^2. If the value is in control, continue processing. If the value is out of control, it is necessary to examine the Principal Components and, in particular, to build control charts for each single component in order to have some insight into the nature of out of control.

Control Charts for single principal components are built by assuming the normality of the original variables. As a consequence, their standard mean is zero and the $\alpha = 0.05$ control limits are equal to the related values of z-distribution (or t-distribution).

It is possible to subgroup multivariate data aiming at analyzing the different and possible causes of process variability. For each sub-group of size n_k (with $k = 1, \ldots, K$) three different measures of process variability can be obtained, see Hotelling (1947) or Jackson (1991).

The overall variability of each subgroup around the target is measured by:

$$T_0^2 = \sum_{i=1}^{n_k} T_i^2 = tr(\boldsymbol{S}_k \boldsymbol{S}^{-1})/n_k \sim \chi^2_{n,p}, \tag{2.8}$$

where $\boldsymbol{S}_k$ is the covariance matrix of each subgroup.

The dispersion of the mean of each subgroup with respect to the target is measured by:

$$T_M^2 = n_k(\bar{\boldsymbol{x}}_k - \boldsymbol{x}_0)\boldsymbol{S}^{-1}(\bar{\boldsymbol{x}}_k - \boldsymbol{x}_0)' \sim \chi^2_p, \tag{2.9}$$

where $\bar{\boldsymbol{x}}_k$ is the mean of each subgroup.

Finally, the dispersion of each subgroup with respect to its own mean value is measured by:

$$T_D^2 = (\boldsymbol{x} - \bar{\boldsymbol{x}}_k)\boldsymbol{S}^{-1}(\boldsymbol{x} - \bar{\boldsymbol{x}}_k)' = T_0^2 - T_M^2 \sim \chi^2_{(n-1)p}. \tag{2.10}$$

It is a very simple matter to adapt the previous formulas for use with principal components and to build corresponding T^2 control charts.

The drawbacks of T^2 control charts can be synthesized in the following points:

(1) the assumption of multinormality is difficult to verify or to justify in reality;
(2) T^2 statistics is not so sensitive to specific mean changes of one or more variables;
(3) T^2 statistics is a quadratic form so that it does not distinguish between changes in mean and changes in variance;
(4) the variables are treated simultaneously so that the diagnosis of causes of out of control is uneasy;
(5) T^2 statistics is used as a repeated test so that each point is considered a single and separate sample.

For taking into account some of these points we discuss in the following subsection the characteristics of *M-CUSUM* control charts.

2.2. *M-CUSUM* control charts

The T^2 control charts, described in Section 2.1., have the feature that, although points are plotted in time, any testing done on them does not take previous values into account. The peculiarity of *M-CUSUM* control charts is that the control procedure attempts to include information contained in the past in order to comment on current performance (Page (1954)).

The *M-CUSUM* control charts are based on sequential probability ratio tests (Healy (1987)). Let $X_1, X_2, \dots, X_j, \dots$ be a sequence of iid random variables. If the behaviour of X_j is in control, the variables have a good (in-control) density function f_0. At an unknown point in time ($t = t_1, \dots$) it is possible that the distribution function shifts and becomes out-of control (f_1). The problem is to determine *if* and *when* the shift occurs.

The *M-CUSUM* control scheme signals an out of control at time t_{max} if for the variables $X_{t_{max}-t+1}, X_{t_{max}-t+2}, \dots, X_{t_{max}}$:

$$S_{t_{\max}} = \sum_{j=t_{\max}-t-1}^{t_{\max}} \log \frac{f_1(X_j)}{f_0(X_j)} > h, \tag{2.11}$$

where h is a constant that determines the operating characteristics of the procedure. $S_{t_{\max}}$ can be calculated recursively in the following way:

$$S_0 = 0, \quad S_{t_{\max}} = \max\left\{S_{t_{\max}-1} + \log\left[f_1(X_{t_{\max}})/f_0(X_{t_{\max}})\right], 0\right\} \tag{2.12}$$

The initial value for S_0 is some number between 0 and h. A careful choice of S_0 gives a CUSUM having improved properties in terms of robustness (Lucas & Crosier (1982a)).

In order to detect a shift in either direction in the sequence of parameters it is possible to use a two-sided *M-CUSUM* procedure in which two one-sided procedures are performed simultaneously (Lucas & Crosier (1982b), Crosier (1986)). Implementation of a two-sided *M-CUSUM* scheme requires two cumulative sums:

$$S_{H,t_{\max}} = \max\left\{S_{H,(t_{\max}-1)} + (X_{t_{\max}} - r), 0\right\} \tag{2.13}$$

and

$$S_{L,t_{\max}} = \min\left\{S_{L(t_{\max}-1)} + (X_{t_{\max}} + r), 0\right\}, \tag{2.14}$$

where $r \geq 0$ serves as the reference value of the scheme.

If both $S_{H,t_{\max}}$ and $S_{L,t_{\max}}$ becomes greater than the decision interval value h, it is considered to be out of statistical control.

The most common application of this scheme is to the mean of a multivariate distribution. In this case, it is necessary to assume that the X_j comes from a multivariate normal distribution. Therefore, in the case of two variables, the control boundary of *M-CUSUM* is an ellipse, like in T^2 control chart.

The *M-CUSUM*, like the T^2 control chart, can be used jointly with *PCA* for solving the problem of multicollinearity. The main advantage of *M-CUSUM* control scheme with respect to the T^2 one is that here it is easy to see whether the out of control is due to a specific variable being below/above the target, or to the two variables jointly deviating from the target. Hence the scatterplot (Figure 3) indicates which variables deviate from the target and the direction of any location shift from the target. If the process is in control, the coordinates should follow a two-dimensional random walk within the elliptical control boundary. If the consecutive coordinates show a trend towards the boundary of the elliptical control limit, then this indicates a location shift form the target. The direction in which the points move towards the boundary indicates which variables are going out-of control, and it also signals whether the shift in location is higher or lower than the respective elements of the target values.

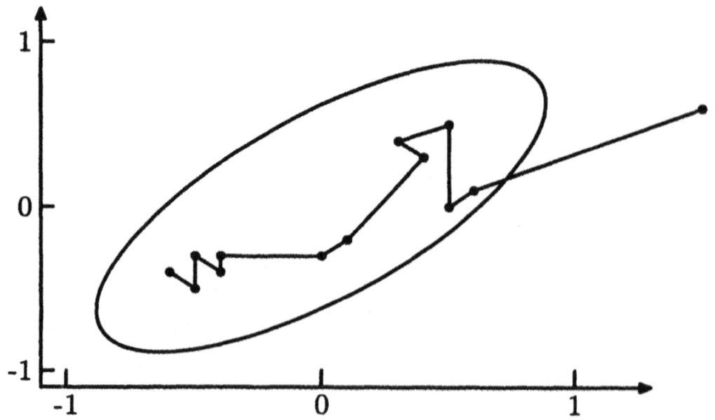

Figure 3. The *M-CUSUM* control chart.

The *M-CUSUM* scheme gives faster detection of small shifts in the mean vector than the T^2 one and, moreover, it is less sensitive to the presence of *outliers* (Crosier (1988)).

Both the *M-CUSUM* control chart and the T^2 control chart, like all known methods for constructing multivariate control charts, see (Alt & Smith (1988)) for a survey, rely heavily on the assumption of multinormality, difficult to justify or verify in real applications. For this reason, in the following, we show different and original approaches to build multivariate control charts without imposing distributional assumptions on the variables.

3. Non parametric control charts

In this paragraph we present and discuss different methods proposed in a non parametric context for constructing multivariate control charts.

It is interesting to underline a difference between the considered approaches. The main idea behind the control charts introduced by (Liu & Singh (1993) and Liu (1995)) is to reduce each multivariate measurement to a univariate index — namely, its relative center-outward ranking induced by a *data depth*. The original observations are reduced in ranks and a new observation is controlled by comparing its depth to the depth of the observations supposed in control. For these control charts it is not required to know the sampling distribution of the original distribution, on the contrary of T^2 statistic.

The non parametric control charts introduced by other authors (Scepi et al. (1993), Scepi (1994), Scepi & Acconcia (1996)) are derived by jointly using non parametric resampling algorithms and multidimensional factorial data techniques.

The idea is to replace the equally concentrated ellipsoids, coming from multinormal distribution, by empirical confidence regions. These regions are determined through the use of the bootstrap technique and are represented graphically in suitable factorial planes.

Hence the first non parametric approach is based on the control of the values of a particular statistic with respect to empirical control limits, derived from the empirical distribution of the statistic. Differently, the second non parametric approach aims at controlling the values of factorial coordinates with respect to the limits of an empirical confidence region, built on a factorial plane by resampling the factorial coordinates of a training data set.

3.1. The *r-chart*

The *r-chart* (Liu (1995)) is based on statistics derived from *data depth.* There exist different notions of *data depth* (Tukey (1975), Liu & Singh (1993)). For any point $\boldsymbol{x} \in \mathcal{R}_p$ (where p are the characteristics of each product used to determine its quality or the number of the control variables chosen) the simplician depth of $\boldsymbol{x}$ with respect to the p-dimensional distribution f is defined to be:

$$SD_f(\boldsymbol{x}) = P_f\Big\{\boldsymbol{x} \in s\big[\boldsymbol{x}_1, \dots, \boldsymbol{x}_{p+1}\big]\Big\}, \tag{3.1}$$

where $s[\boldsymbol{x}_1, \dots, \boldsymbol{x}_{p+1}]$ is the open simplex whose vertices $\boldsymbol{x}_1, \dots, \boldsymbol{x}_{p+1}$ are $(p+1)$ random observations from f.

The value of SD_f is the measure of "how" deep, or "how" central $\boldsymbol{x}$ is with respect to f. When f is unknown and only a sample $\boldsymbol{x}_1, \dots, \boldsymbol{x}_n$ is given, the simplician depth of $\boldsymbol{x}$ is defined as:

$$SD_{g_p}(\boldsymbol{x}) = \binom{n}{p+1}^{-1} \sum I(\boldsymbol{x} \in s\left[\boldsymbol{x}_{i_1}, \dots, \boldsymbol{x}_{i_{p+1}}\right]) \tag{3.2}$$

where $I(.)$ denotes an indicator function and $g(.)$ the empirical distribution.

The (3.2) measures how deep $\boldsymbol{x}$ is within the data cloud $\boldsymbol{x}_1, \dots, \boldsymbol{x}_n$. It is possible to obtain a similar information by using a Mahalanobis distance. In this case the depth is measured by how small the quadratic distance of $\boldsymbol{x}$ is to the mean.

A data depth induces a center-outward ordering of the sample points if depth values for all points are computed and compared. After arranging all $SD_g(\boldsymbol{x}_i)$'s in an ascending order, it is possible to identify the smallest depth value and to compute the most central point. The smaller the order (the rank) of a point, the more outlying that point with respect to the underlying distribution.

Control charts based on data depth are similar to univariate control charts. We must choose an UCL and an LCL so that the false alarm rate α is small (Section 2).

Let us assume to have $x_1, \ldots, x_n$ observations in control and to observe $x_1^+, x_2^+, \ldots$ new observations, the *r-chart* is constructed by computing, for each new observation, the following statistic:

$$r_{g_p}(x_i^+) = \frac{\#\left\{x_i \mid SD_{g_p}(x_i) \leq SD_{g_p}(x_i^+)\right\}}{p}, \tag{3.3}$$

where $\#\{\mathcal{A}\}$ denotes cardinality of $\{\mathcal{A}\}$.

Therefore, the *r-chart* is the plot of $r_{g_p}(x_i^+)$'s against time i, with central limit=0.5 and control limit α. The process is declared out of control if $r_{g_p}(\cdot)$ falls below α. In fact, the *r-chart* has only the $LCL = \alpha$ but no UCL.

The expression in (3.3) furnishes a measure of how outlying x_i^+ is with respect to the original and in control data cloud. Since $r_{g_p}(\cdot)$ is defined according to the data depth, the possible deviation can be a shift in "center" and/or an increase in scale. Thus the *r-chart* with $LCL = \alpha$ corresponds to an α-level test of the following hypotheses:

$H_0 : f = g$

H_1 : there is a location shift and/or a scale increase from g to f

The *r-chart* has different advantages with respect to the standard control charts. It does not require the assumption of multinormality, which is needed in standard approaches. Furthermore, it allows detecting simultaneously location shift and scale increase of the process, unlike the methods based on T^2 statistic. Finally, it is possible to compute a culpability index vector that indicates which variable can be considered mainly culpable for an out of control situation.

Recall that Liu (1995) suggested other interesting non parametric control charts also based on data depth, the *Q-chart* which has the same aim of the univariate $\bar{X}$ chart and the *S-chart* similar to the *CUSUM* control chart.

3.2. The IS-CO control charts

The aim of the approach proposed by Scepi et al. (1993) is to supply a strategy in building multivariate control charts by exploiting the tools derived from the evaluation of stability for factorial techniques in multidimensional data analysis (Balbi & Scepi (1996)). The features of these control schemes are the absence of distributional assumptions as well as the significant role assigned to the geometrical point of view.

The non parametric control strategy applies resampling techniques together with three-way methods of data analysis. In particular, starting from an observed data matrix $\boldsymbol{X}(n,p)$, the bootstrap technique (Efron & Tibshirani (1993)) is here used for obtaining B matrices $\boldsymbol{X}_b^*$, $b = 1, \ldots, B$, by resampling the rows of $\boldsymbol{X}$ with replacement.

Following Greenacre (1984), both the original and the replaced rows are projected on common factorial planes. Suitable planes (Holmes (1989)) are derived from three-way data methods, like Statis (Escoufier (1987)) or Principal Matrices Analyses (Rizzi (1989)). On these planes, convex and not regular empirical confidence regions are defined in order to provide non parametric control charts.

Let $\boldsymbol{X}$ be a sample data matrix based on n measures of p characteristics of a product/process, the general control strategy for building non parametric charts consists in the following steps:

(1) *Resampling*: B matrices $\boldsymbol{X}_b^*$, $b = 1, \ldots, B$, are obtained by resampling the rows of $\boldsymbol{X}$ with replacement. These B matrices play the role of the original samples in the pilot phase.

(2) *Analysis*: According to Statis, an interstructure-compromise-intrastructure approach is applied to the three-way data structure of the $\boldsymbol{X}_b^*$ replicated matrices. This approach furnishes two basic matrices.
a) The *interstructure* matrix $\boldsymbol{IS}$ showing the relationships between different matrices. The general element of the $\boldsymbol{IS}$ is the Escoufier's vector correlation coefficient:

$$Rv_{bb'} = tr(\Theta_b \Theta_{b'}) / \sqrt{tr(\Theta_b)^2 tr(\Theta_{b'})^2}, \tag{3.4}$$

where the covariance matrix $\Theta_b = \frac{1}{n} \boldsymbol{X}_b^{*\prime} \boldsymbol{X}_b^*$ is assumed to be the characteristic element associated to each matrix.
b) the *compromise* matrix $\boldsymbol{CO}$ which is defined as linear combination of the Θ_b's:

$$\boldsymbol{CO} = \sum_{b=1}^{B} u_b \Theta_b, \tag{3.5}$$

where u_b is the b-th element of the eigenvector associated to the highest eigenvalue obtained by diagonalizing $\boldsymbol{IS}$.

(3) *Construction of the* $\boldsymbol{IS}$ *Control Chart.* In the pilot phase, by diagonalizing the matrix $\boldsymbol{IS}$, it is possible to obtain a factorial plane (the first principal plane) on which to project jointly the original data matrix and, as supplementary, the B replicated ones. The cloud of the overall points provides an empirical confidence region. The $\boldsymbol{IS}$-Control Chart (Figure 4) is determined by *peeling* subsequent *convex hulls* (Green (1981)); the confidence level of the control chart is here computed by eliminating a fixed percentage α of more external points. In the monitoring phase, this control chart allows obtaining an overall evaluation of the behaviour of a new sample $\boldsymbol{X}^+(n, p)$. If the projection of its characteristic element Θ_b^+ as supplementary is outside the control region, we have an out of control signal.

(4) *Construction of the* $\boldsymbol{CO}$ *Control Chart*: the $\boldsymbol{CO}$ Control Chart answers to the following question: if the $\boldsymbol{IS}$ Control Chart signals an out of control, which variable is mainly culpable? In the pilot phase, the original variables and the variables associated to each replication are projected on the factorial plane determined by diagonalizing the $\boldsymbol{CO}$ matrix. Subsequently, an empirical confidence region for each variable is built by peeling the related *convex hull*. The width of each *convex hull* represents a signal of the variability of each variable

and so it shows the risk of out-of control for each variable. In the monitoring phase, if the projection of a variable of the new sample $\boldsymbol{X}^{+}(n,p)$ falls out of its relative region, we have an out of control signal for this variable.

It is important to underline that if the aim is controlling single units (the rows of $\boldsymbol{X}$) rather than variables, it is possible to build non parametric control charts for units with the same procedure but considering as general element of $\boldsymbol{IS}$: $\boldsymbol{\Theta}_b = \frac{1}{n}\boldsymbol{X}_b^*\boldsymbol{X}_b^{*\prime}$.

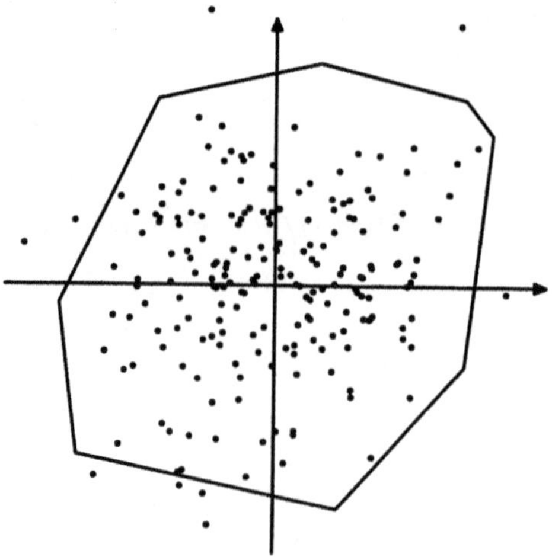

Figure 4. The $\boldsymbol{IS}$-control chart, $B = 200$.

The proposed general strategy furnishes two non parametric control charts, the first gives a global evaluation of the performances reached by a sample and the second gives an indication of the causes of a possible out-of control. This procedure has been applied for obtaining non parametric control charts for sequential data in Section 3.3. and for controlling data in batches in Section 3.4..

3.3. The IS and CO-chart for sequential data

The non parametric approach, above introduced, can be applied also dealing with time series (Scepi & Acconcia (1996)).

Let $\boldsymbol{X}(T,p)$ be a data sample based on T measures of p continuous variables. Note that the rows of matrix, previously indicated with n, because indicators of units are, in this paragraph, indicated with T because of being indicators of times.

Suppose that each column vector of $\boldsymbol{X}(T,p)$ is the empirical realization of a strictly stationary and weakly dependent sequence of random variables. Considering the p series as being observed from a process in control, the first assumption is quite reasonable. The second assumption allows us to extend the applicability of these non parametric control charts to a sequence of variables, which are not independently distributed.

To extend the strategy suggested in Section 3.2. to this kind of data it is necessary to consider a particular resampling method that allows retaining the stationarity property of the original series. Therefore, conditionally on the original data, we generate B matrices $\boldsymbol{X}_b^*$, $b = 1, \ldots, B$, of (pseudo) time series by using the stationary bootstrap (Politis & Romano (1994)). The B matrices can be thought as sample observations of the controlled process in different days, while the T rows $(t_1, t_2, \ldots, T)$ of each matrix represent sample observations taken at different times in the day.

The generic matrix $\boldsymbol{X}_b^*$ consisting of p columns of pseudo time series, is obtained by blocks of row vectors $\boldsymbol{x}$ generated by the following algorithm:

(1) Fix q in [0,1].
(2) Generate a sample sequence $L_1, L_2, \ldots$ from a geometric distribution with parameter q.
(3) Independently of the original $\boldsymbol{X}$ and L_i, generate a sample sequence $I_1, I_2, \ldots$ from a discrete uniform distribution on $(1, \ldots, T)$.
(4) Obtain the i-th block of row vectors of $\boldsymbol{X}_b^*$ by the L_i row vectors of the original matrix starting from the I_i row: $\boldsymbol{x}_{I_i}, \ldots, \boldsymbol{x}_{I_i+j}, \ldots, \boldsymbol{x}_{I_i+L_i-1}$. In particular, if I_{i+j} is greater than T, the algorithm assumes that $\boldsymbol{x}_1$ "follows" $\boldsymbol{x}_T$ (in other words, the *stationary bootstrap* method wraps the data around in a circle), this contributes to the stationarity of the re-sampled time series.
(5) Stop the process once T row vectors in the $\boldsymbol{X}_b^*$ have been generated. Of course, repeating the steps from (2) to (5) it is possible to generate a new matrix.

The general strategy of Section 3.2. can be applied on the so generated three-way data structure. In choosing the characteristic element of the inter-structure matrix, we must take in mind that the aim is here controlling single observations, or times. In the monitoring phase, the $\boldsymbol{IS}$-control chart allows obtaining an overall evaluation of the behaviour of a new sample $\boldsymbol{X}^+(T, p)$. If the projection of its characteristic element $\boldsymbol{\Theta}_b^+$ as supplementary is outside the control region, we have an overall out of control signal.

The $\boldsymbol{CO}$-control chart is here constituted by an empirical confidence region for each time. The aim is to control the influence of the time T in the overall out of control signaled by the $\boldsymbol{IS}$-control chart. In the pilot phase, the original T observations and the observations associated to each replication are projected on the factorial plane determined by diagonalizing the $\boldsymbol{CO}$ matrix. In the monitoring phase, if the projection of a new t_i observation is outside of its related region, we have an out-of-control signal for this observation.

3.4. The IS and CO-chart for batches

This approach aims at controlling subgroups of homogeneous observations defined by structural and controllable characteristics of a product/process (Scepi (1994)).

In particular, we suppose having two groups of variables examined on the same units: the *a-priori information* and the *interest variables*. The first group of variables can be constituted, for example, by technical and structural characteristics concerning the product/process while the second one is composed by characteristics of product used to control its quality.

A different and non symmetric role to the two groups of variables is assigned. The *a-priori* information is supposed explicative with respect to the *interest variables*. In particular, the explicative variables are considered the factors of an experimental design allowing to cluster observations in experimental and homogeneous conditions (batches) with respect to controlling the values of the interest variables.

Let us suppose the sample data matrix $\boldsymbol{X}(n,p)$ partitioned into two submatrices, the matrix $\boldsymbol{E}(n,p_1)$ of the explicative variables and the matrix $\boldsymbol{C}(n,p)$ of the interest or control variables. The interest variables are quantitative, while the explicative variables can be categorical or categorised.

The variables in $\boldsymbol{E}$ are here considered the factors of a design matrix $\boldsymbol{D}(K,M)$. In particular, the columns of $\boldsymbol{D}$ are the M categories (levels) of the p_1 factors of $\boldsymbol{E}$, while the rows are the K factor-level combinations (experimental conditions), where $M = \sum_{j=1}^{p_1} m_j$ and $K \leq \prod_{j=1}^{p_1} m_j$, with m_j the number of levels of the j-th factor.

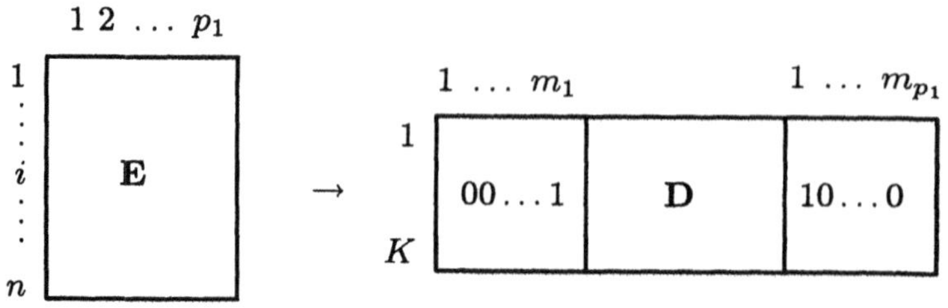

In order to build multivariate control charts, we apply to the observed data matrix $\boldsymbol{X}$ the step 1 of the general procedure of Section 3.2.. Therefore, $\boldsymbol{X}_b^*$, $b = 1, \ldots, B$ matrices are obtained by resampling the rows of $\boldsymbol{X}$ with replacement. Consequently, B $\boldsymbol{E}_b^*(n,p_1)$ and B $\boldsymbol{C}_b^*(n,p_2)$ matrices are defined. The experimental matrix $\boldsymbol{D}(K,M)$ is considered common to the different replications. Taking into account $\boldsymbol{D}$, the n observations of each matrix $\boldsymbol{C}_b^*$ are rearranged according to their own experimental pattern.

Therefore, from now onwards we deal with the matrices $\boldsymbol{M}_b^*$:

$$\boldsymbol{M}_b^* = \boldsymbol{W}_b^{-1}\boldsymbol{H}_b\boldsymbol{C}_b^*, \tag{3.6}$$

where $\boldsymbol{H}_b$ is an indicator matrix of general term:

$$h_{ikb} = \begin{cases} 1: & \text{if the } i^{th} \text{ observation} \in \text{ to the } k^{th} \text{ condition} \\ 0: & \text{otherwise} \end{cases}$$

and $\boldsymbol{W}_b = diag(\boldsymbol{H}'_b \mathbf{1})$.

The so obtained $\boldsymbol{M}^*_b(K, p_2)$ matrices have K rows, corresponding to the experimental conditions. The general element of each matrix is a mean of the values assumed by the p_2 variables of interest, for the observations belonging to the generic k experimental condition ($k = 1, \ldots, K$) in each b replication, $b = 1, \ldots, B$.

The three-way data structure constituted by the B matrices $\boldsymbol{M}^*_b$ can be analyzed by a three-way data method for identifying a suitable factorial plane and applying the step (2), (3) and (4) of the general non parametric control strategy described in Section 3.2..

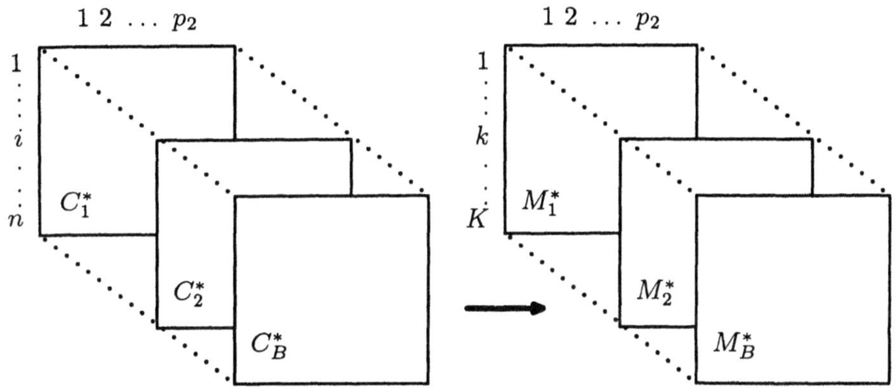

In order to take into account the different role of the variables, a non symmetric three-way method called *Principal Matrices Analysis into Experimental Design* (PMAD), (Balbi et al. (1994), Lauro et al. (1995)) is applied. PMAD deals with data structured as in *Principal Component Analysis onto a Reference subspace* (PCAR) (D'Ambra & Lauro (1982)) and with this technique shares the aim of analyzing non symmetrical relations between variables. The idea consists in reading the relationships directly on the factorial plane, by projecting the interest variables on the factorial plane spanned by the explicative ones.

Here, the first step of PMAD consists in projecting the $\boldsymbol{M}^*_b$ matrices onto the reference subspace spanned by the K experimental conditions by means of an orthogonal projector:

$$\boldsymbol{A}^*_b = \boldsymbol{P}_b \boldsymbol{M}^*_b, \tag{3.7}$$

where $\boldsymbol{P}_b = \boldsymbol{D}(\boldsymbol{D}'\boldsymbol{W}_b\boldsymbol{D})^{-1}\boldsymbol{D}'\boldsymbol{W}_b$.

The so obtained $\boldsymbol{A}^*_b$ are analyzed by a *Principal Matrices Analysis* (PMA), (Rizzi (1989)). PMA follows an interstructure-compromise-intrastructure approach like STATIS, but with respect to STATIS has the advantage of considering as characteristic element associated to each matrix, the matrix itself.

Therefore the $\boldsymbol{IS}$ matrix (3.4) has here the following general element:

$$Rv_{bb'} = tr(\boldsymbol{A}_b^* \boldsymbol{A}_{b'}^*)/\sqrt{tr(\boldsymbol{A}_b^*)^2 tr(\boldsymbol{A}_{b'}^*)^2}. \tag{3.8}$$

Thus, it is possible to define $(R \leq B)$ compromise matrices as linear combination of the $\boldsymbol{A}_b^*$'s, i.e.

$$\boldsymbol{CO}_r = \sum_{b=1}^{B} \boldsymbol{u}_{r_b} \boldsymbol{A}_b^*, \quad r = 1, \ldots, R, \tag{3.9}$$

where $\boldsymbol{u}_{r_b}$ is the b-th element of the eigenvector associated to the highest eigenvalue obtained by diagonalising $\boldsymbol{IS}$.

Starting from the $\boldsymbol{IS}$ (3.4) and the $\boldsymbol{CO}_1$ (see (3.9) for $r = 1$) matrix, non parametric control charts are built. In particular, an empirical confidence region for each experimental condition (batch) is visualized on the first principal plane of the most informative $\boldsymbol{CO}$.

The principal advantage of the non parametric control charts for experimental conditions is in the opportunity of projecting the interest variables as supplementary on the $\boldsymbol{CO}$-chart. It allows to characterize the factorial axes and so to explain the control behaviour of different experimental conditions.

The interpretation of these control schemes is clearer in presence of well definite (low internal variability) and well separate (high external variability) *convex hulls*. In fact, the width of each *convex hull* represents a signal of the internal variability of each experimental condition, while it is possible to control the variability of the product/process by measuring the influence of the interest variables on the product/process quality.

The suggested approach (Scepi (1994)) has been applied on a sample of 76 beams (Crowder et al. (1991)). Each beam is characterized by the values of 3 structural (supposed *explicative*) variables (width, length, and weight) and 3 control (supposed of *interest*) variables (breaking load, failure time, and sag of the beam at the first crack).

	76 BEAMS			
	6 VARIABLES			
3 FACTORS	**LEVELS**			
Width	2	1-thin	2-wide	
Length	2	1-short	2-long	
Weight	3	1-light	2-medium	3-heavy
3 RESPONSES				
Breaking load (BL)				
Failure time (FT)				
Sag of the beam at the first crack (SB)				

The theoretical experimental conditions are 12 (2*2*3) but the observed ones in the sample are 7. Therefore the experimental matrix $\boldsymbol{D}$ is the following:

$$\begin{pmatrix} 1 & 111 & 1 & 0 & 1 & 0 & 1 & 0 & 0 \\ 2 & 112 & 1 & 0 & 1 & 0 & 0 & 1 & 0 \\ 3 & 121 & 1 & 0 & 0 & 1 & 1 & 0 & 0 \\ 4 & 212 & 0 & 1 & 1 & 0 & 0 & 1 & 0 \\ 5 & 213 & 0 & 1 & 1 & 0 & 0 & 0 & 1 \\ 6 & 222 & 0 & 1 & 0 & 1 & 0 & 1 & 0 \\ 7 & 223 & 0 & 1 & 0 & 1 & 0 & 0 & 1 \end{pmatrix}$$

For obtaining non parametric control charts, the original matrix is resampled $B = 200$ times. By considering $\boldsymbol{D}$, the observations are rearranged with respect to the 7 experimental conditions and, successively, the variables of interest are projected on the subspace spanned by the experimental conditions.

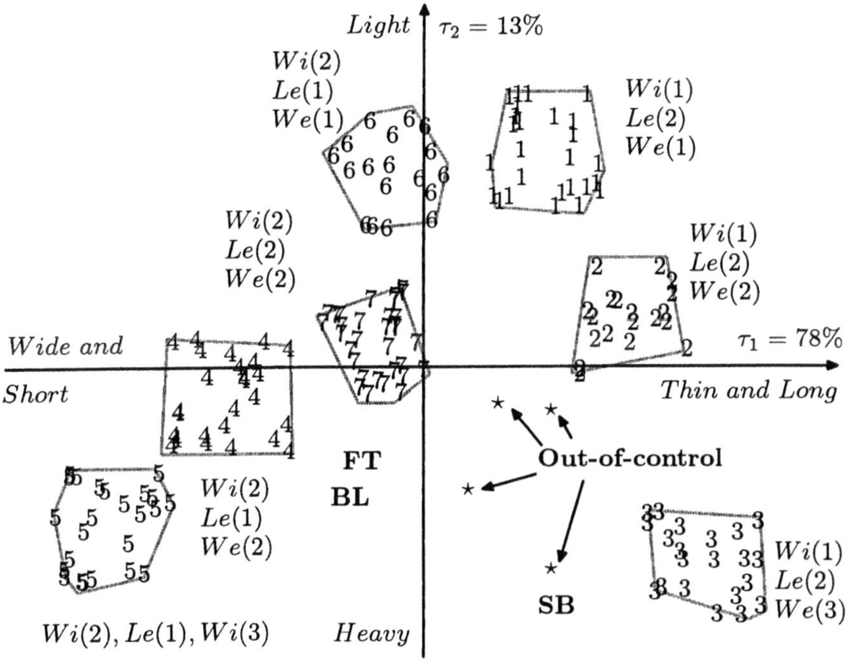

Figure 5. ***CO***-control chart for Buffon's data.

Therefore $B = 200$ $\boldsymbol{A}_b^*$'s are obtained. In Figure 5 the first factorial plane obtained by diagonalizing the first ***CO*** matrix (that explains the 85 % of the overall variability) is shown. In particular, an empirical confidence region for each experimental condition is defined. The variables of interest are projected as supplementary in order to have more control information.

Let us note that the variability is explained for the 91% by the first two factorial axes. For the interpretation, it is interesting to underline the role of the interest variables and their position on the graph. The variable *Sag* of the *Beam* is positively correlated with the first axis while the variables *Failure Time* and *Breaking Load* show a negative correlation. The location of the beams with respect to the first axis can be considered as a signal of their *breaking resistance.* Thin and not strong beams (1st, 2nd and 3rd experimental condition) are located on the right side while short and wide beams (4th and 5th experimental condition) on the left.

The second axis displays the *type of breaking.* Therefore, it is possible to note the opposition between more regular beams (1st and 2nd experimental condition) and peculiar beams (3rd, 4th and 5th experimental condition); the first group with superficial sags, the second one with deep sags. The beams belonging to the third experimental condition are the more unstable. Further information can be obtained by the width of *convex hull* that indicates a larger within variability in the 3rd, 4th and 5th condition.

4. Sensitivity of multivariate control charts

When comparing multivariate control schemes (Mason et al. (1997)) there is a performance aspect that should be discussed. This aspect concerns the question of how quickly the scheme generates a signal when an actual change in the process has occurred. Clearly, the quicker a scheme responds to a real change, the better. A control scheme that can quickly detect real changes, while not being overly sensitive to *false alarm*, is desired. In particular, it is possible to identify two different situations: the control chart signals an out of control but the process is in control (*false alarm*); the chart signals an out of control and the process is actually in out of control.

The number of samples required to detect a real change in the process is measured by the *Run Length* or, if we refer to the expected value, by the *Average Run Length (ARL)*. Therefore, a good performance of a control scheme is obtained if the *ARL* is low in out of control situations (where each variable is distributed with density function f_1) and high in control ones (where each variable is distributed with density function f_0).

The calculation of the *ARL* for T^2 control charts is relatively straightforward (Crosier (1988)):

$$ARL = 1/P \quad where \quad P = Pr(T^2 \geq UCL) \tag{4.1}$$

It is interesting to note that the *UCL* must be increased as the number of variables increases in order to obtain the same on target *ARL*.

A practical problem with T^2-control charts is that they are very sensitive to *false alarms.*

The *ARL* of *M-CUSUM* control charts was based on the Markov-Chain approach (Brook & Evans (1972)) and is generally computed by simulations. In control situation, the *ARL* of these schemes is lower than the T^2 one. Furthermore, the *M-CUSUM* control charts give faster detection of small shifts than T^2 control charts. However, the *ARL* of *M-CUSUM* depends on the direction of the shift.

A lot of interesting simulation examples, for computing both the *ARL* of T^2 control charts and the *ARL* of *M-CUSUM* control charts, are in Crosier (1988).

The sensitivity of the T^2 control charts with respect to the non parametric ones, has been compared (Balbi & Scepi (1996)). In particular, the behaviour of the two types of control schemes in presence of a shift of the mean and distribution contamination has been observed.

From a multinormal distribution $\mathcal{N}(\mathbf{0}, \boldsymbol{I})$, 1 000 samples $\boldsymbol{X}(20,4) \sim \mathcal{N}(\delta\mathbf{1}, \boldsymbol{I})$ have been generated. Initially, a shift in the mean equal to +2 is supposed, as following:

$$\begin{cases} \delta = 0 & for \quad i = 1, \ldots, 10 \\ \delta = 2 & for \quad i = 11, \ldots, 20 \end{cases}$$

A comparison among the different distributions of the out of control signals given by the T^2 control chart is shown in Figure 6. In particular, different control charts are built by considering 1 (SCL=5.12), 2 (SCL=10.03), 3 (SCL=16.7) and 4 (SCL=27) principal components. Nearly the 70 % of the generated out of control is detected when the T^2-control chart is based on 4 components; only the 30 % in the T^2-control chart based on 2 components.

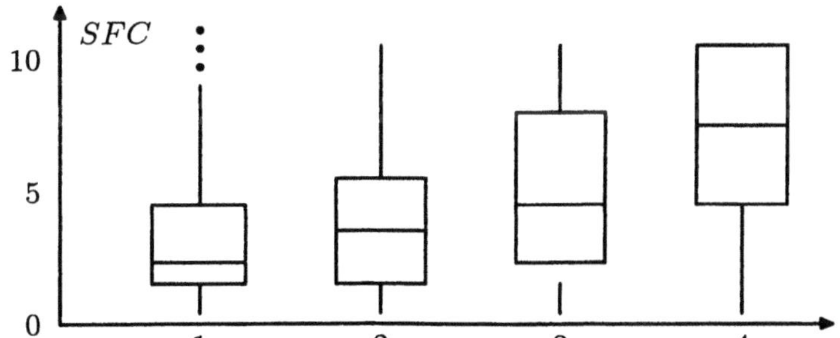

Figure 6. SFC Boxplot – shift in the mean.

For evaluating the sensitivity of the T^2 scheme to distribution contaminations, a change in the distribution of the variables is supposed, as following:

$$\boldsymbol{X}_{ij} \sim \begin{cases} N(0,1), & i = 1, \ldots, 10, \\ N(0,1) + U(2,3), & i = 11, \ldots, 20. \end{cases}$$

The control chart based on the overall components detects the 80 % of the out of control, while in the case of two components, only the 50 % is detected

(Figure 7). For small shifts, the T^2 control chart does not detect the out of control whatever number of components may be chosen.

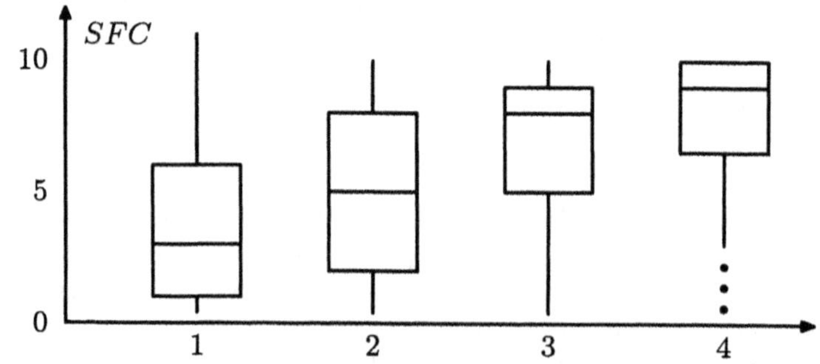

Figure 7. SFC Boxplot – distribution contamination.

The sensitivity of the non parametric control chart $\boldsymbol{IS}$ is evaluated and is compared with the T^2 one. The general non parametric control strategy (Section 3.2.) is applied to the matrix $\boldsymbol{X}(20,4) \sim \mathcal{N}(\delta\mathbf{1}, \boldsymbol{I})$ by supposing δ=0. Starting form $B = 200$ replication samples, the empirical confidence region is obtained.

The 1 000 simulated samples $\boldsymbol{X} \sim \mathcal{N}(\delta\mathbf{1}, \boldsymbol{I})$ with $\delta = 2$ are projected as supplementary on the $\boldsymbol{IS}$-control chart. The 100 % of the out of control is out of the control region while simulated control samples lie inside the control region, see Figure 8. The 100 % of the out of control generated by distribution contamination is also detected by the $\boldsymbol{IS}$.

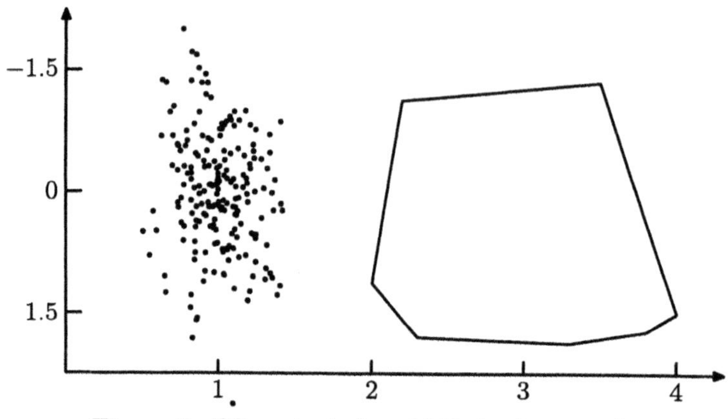

Figure 8. $\boldsymbol{IS}$-control chart(shift in the mean).

An empirical way to calculate the ARL of non parametric control charts for sequential data has been proposed by Scepi & Acconcia (1996). In the case of non parametric control charts, the function f_0 is unknown and is replaced by an empirical density function $\hat{f}_0$. The latter function is simulated by the stationary bootstrap that assigns the probability $\frac{1/p}{T}$ to each row of the data matrix $\boldsymbol{X}(T, Q)$. In the same way, the out-of-control function $\hat{f}_1$ can be derived starting from an out-of-control matrix.

The ARL function, in the sequential analysis, plays a similar role as that of the power function in the hypotheses testing. The empirical confidence region (EC_0), furnished by the general control procedure, can be seen as a multivariate region under $\hat{f}_0$, at the same time, multivariate empirical confidence region (EC_1) could be obtained under $\hat{f}_1$. Therefore, the ARL function is given by:

$$ARL(\cdot) = 1/\Phi, \tag{4.2}$$

where Φ is the function $\hat{f}_1$ defined on the complementary space to (EC_0).

For evaluating the sensitivity of these control charts, some examples are shown in Scepi & Acconcia (1996). Let us consider a sample matrix $\boldsymbol{X}(70, 2)$ of empirical data consisting of a pair of time series, realizationes of a hypothetical bivariate stationary stochastic process. The original data represent the input gas feed rate and the carbon dioxide concentration from a gas furnace read off at equispaced times of nine seconds (Box et al. (1994)); in particular, first differences of original data are used.

Initially $B = 200$ pseudo bivariate time series (Figure 9 and Figure 10) have been generated by means of the stationary bootstrap (with $p = 0.05$) and consequently the $\boldsymbol{IS}$-control chart has been obtained.

Furthermore, 100 sample matrices $\boldsymbol{X}_s^+$ of the size $(70, 2)$ have been simulated as follows:

$$\boldsymbol{X}_s^+ = \boldsymbol{X} + \boldsymbol{Y}_s, \quad s = 1, \ldots, 100.$$

In the first example, non stationarity is simulated. Each $\boldsymbol{Y}_s$ has, as a general row, $\boldsymbol{y}_{t_i} = \boldsymbol{y}_{t_i - 1} + \boldsymbol{\varepsilon}_t$ with $\boldsymbol{\varepsilon}_t \sim \mathcal{N}(0, \Sigma)$ and $\boldsymbol{\Sigma} = diag(0.35, 0.67)$.

In a second example, each $\boldsymbol{X}_s^+$ matrix, $s = 1, \ldots, 100$, consists of the first 100 bootstrap matrices $\boldsymbol{X}_b$ except for a location shift γ in some observations (the 5th, 10th, 15th, 30th, 50th ones). In particular, the data are simulated by varying shifts, so that $\gamma = 1.0, 1.5, 2.0, 2.5, 3$.

A third example is based on $\boldsymbol{Y}_s$ matrices with rows drawn from a normal distribution with zero mean vector and $\boldsymbol{\Sigma} = diag(0.9, 1)$.

The results show that the $\boldsymbol{IS}$-Control Chart detects the 100 % of out of control in the first case as well as in the second case and the 97 % in the third case.

With the aim of evaluating the behaviour of single observations, the $\boldsymbol{CO}-$Control Chart has been built and so an empirical confidence region for each observation has been obtained. In order to control the observations associated with each kind of out-of-control matrix previously examined, the rows of $\boldsymbol{X}_s^+$ are individually projected as supplementary onto the plane of their empirical confidence region. On the whole, results show a good sensibility of the $\boldsymbol{CO}-$Control Chart to detect out of control situations. The minor differences among the three cases are connected to the characteristics of the different generated out-of-control.

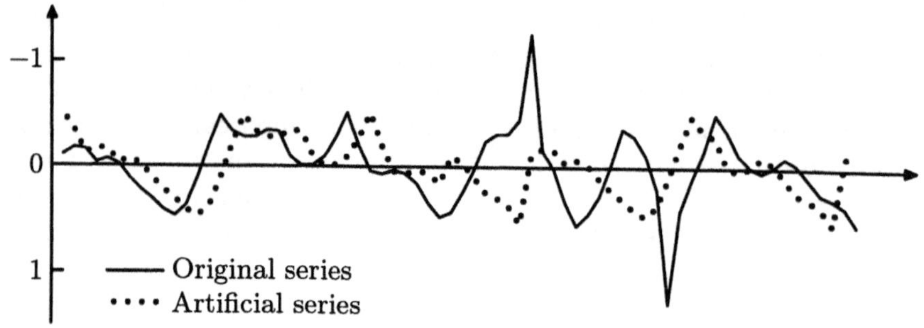

Figure 9. First series: the original and the replicated one.

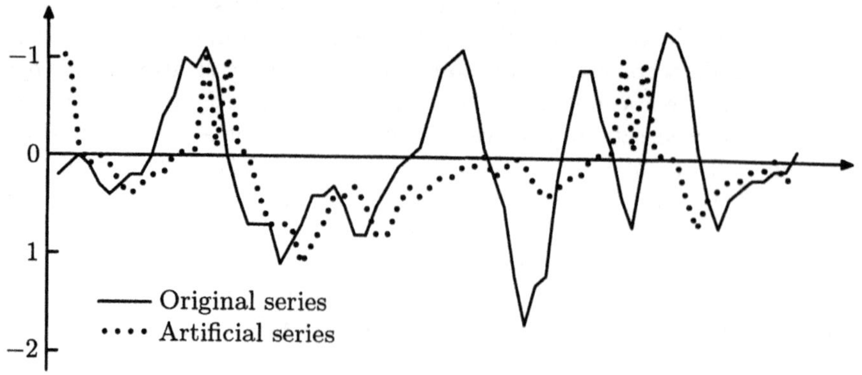

Figure 10. Second series: the original and the replicated one.

The chart signals 100% of out-of-control for the last 30 observations of each matrix, in the first example. At the same time, a very good percentage of out-of-control for the first 40 observations is retained as indicated by the value of *ARL* that is 1.8. In the second example, the 80 % of the generated out-of-control are detected when the shift is 1. The percentage increases for larger shifts (100% when the shift is 3). The values of *ARL* for all shifts examined in the second example are recorded in Table 1. The estimate of *ARL* (value: 4.8), in the third example, shows that the control chart easily detects real changes in the variability of the process.

γ	1.0	1.5	2.0	2.5	3.0
shift	6.5	6.5	5.5	5.0	4.5

Table 1. Values of ARL.

5. New developments: multivariate control charts for symbolic data

Let us now consider dealing with complex and structured data, coming from large databases. In literature, this particular kind of data is called *symbolic object* (Bock & Diday (2000)).

Fitting methods for analyzing *symbolic data* are suggested starting from traditional techniques. In particular, some authors have proposed suitable control charts, parametric and non parametric, for controlling the behaviour of processes/products well represented by such so complex data. In the following, we introduce briefly some factorial techniques for analyzing *symbolic data* in order to show successively the procedure for constructing non parametric control schemes.

5.1. Principal component analysis for symbolic objects

Let us denote with $\boldsymbol{O}_i$, $i = 1, \ldots, n$ *Symbolic Objects* (SO) that are described by p variables, or descriptors, X_j, assuming values within the domains $D_j, j = 1, \ldots, p$. Each $\boldsymbol{O}_i$ is a complex statistical unit defined by the logical conjunction of some events:

$$[X_1 \in d_1] \cap \cdots \cap [X_j \in d_j] \cap \cdots \cap [X_p \in d_p], \tag{5.1}$$

where $d_j \subseteq D_j$, $j = 1, \ldots, p$, can be, according to the characteristic of X_j, an interval value or a set of modalities of X_j.

In particular, the $\boldsymbol{O}_i$ can be described by interval variables, qualitative variables and probability distributions. In the following, in according to the control aim, we focus on interval variables. Therefore, the sample matrix $\boldsymbol{X}$ has dimensions $(n, 2p)$; each object is re-coded combining two values (the *min* and the *max* value) for each descriptor. If all possible combinations between the *min* and the *max* value are considered, the vertices data matrix $\boldsymbol{Z}(n2^p, p)$ is obtained.

A particular transformation of $\boldsymbol{X}$ is suggested by Lauro & Palumbo (2000). In order to take into account the symbolic objects structural elements (size, shape), it is possible to use a *range transformation.* The matrix $\boldsymbol{Z}^{\#}(n, p)$, having as generic term the range of the interval for each descriptor, is so computed.

For analyzing SO's and visualizing them onto a suitable subspace, different PCA have been proposed. *Vertices Principal Component Analysis (V-PCA)* is the extension of *PCA* to interval data proposed by Chouakria et al. (1998). *V-PCA* consists in performing a classical *PCA* on the matrix $\boldsymbol{Z}$.

A variant to this approach is the *Symbolic Object Principal Component Analysis* introduced by Lauro & Palumbo (2000). The main idea consists in maximizing the between SO's variance, instead of the total vertices variance, in order to take into account the necessary vertices cohesion constraint.

A new approach, named *Principal Components Analysis on the Range Transformations (RT-PCA)* has been proposed by Lauro & Palumbo (2000). *PCA* is here carried on the range matrix $\boldsymbol{Z}^{\#}(n, p)$. The range transformation implies a translation of each object; from a geometric point of view, the vertices associated to the *min* values are reported in the origin. This simple representation allows interpreting the relationships among the variables and consequently identifying SO's characterized by the same descriptors.

Different versions of *PCA* are used in building multivariate control charts in parametric (Giordano & Palumbo (1999)) and a non parametric context (Balbi & Scepi (2000)).

5.2. Non parametric control charts for interval data

The parametric approach of Giordano & Palumbo (1999) aims at separately monitoring the SO's position and their size/shape on the basis of two control charts. The basic assumption relies on the independence between the SO's position and their size/shape.

The control charts referring to size and shape are built like classical T^2 control charts but are based on *Principal Components Analysis on the Range Transformations.* With respect to the SO's position, the target limits are fixed by representing, on the plane derived from a *Symbolic Object Principal Component Analysis*, the maximum covering area rectangle as a generalization of all SO's representations. The control procedure can be viewed as a particular sequential test of Wald (1947) on the proportion of defects.

The starting point of a non parametric approach (Balbi & Scepi (2000)) is the difficulty to carry out control techniques based on multinormality hypothesis, particularly for complex and structured data. The authors suggest to apply the non parametric schemes proposed for classical data with some variants. In particular, non parametric control charts for sequential objects (Section 3.3.), as well as for objects as batches (Section 3.4.), are defined.

Sequential objects can be considered as T measurements in T consecutive times, for example hours, of some characteristics of a process, inspected in interval times, for example minutes.

The sample matrix $\boldsymbol{X}(T, 2p)$ is transformed in the range matrix $\boldsymbol{Z}^{\#}(T, p)$; on this matrix the control procedure for sequential data is carried out. Therefore, B $\boldsymbol{Z}^{\#}(T, p)$ are obtained by stationary bootstrap and $\boldsymbol{IS}$, as well as $\boldsymbol{CO}$ control charts, are built for each time.

It is important to underline that the *PCA* used for analyzing the data is here the *Principal Components Analysis on the Range Transformations.* In this way, these schemes are suitable for controlling simultaneously the size and shape of the SO's that are represented by the first and the second principal component. A previous control of the position is supposed. Furthermore, as each object is geometrically represented by the maximum vertex, the *peeling* of *convex hulls* is carried out by eliminating a fixed percentage of points with extreme coordinates of this vertex.

The $\boldsymbol{CO}$ control scheme signals an out of control if an object, projected as supplementary on the chart, has coordinates of its maximum vertex out of the associated *convex hull.* For evaluating the different role of size and shape in causing the out of control, empirical box plots of the maximum coordinates of each object, in different replications, can be constructed.

A symbolic object can be regarded as a set of individuals (batch) that satisfy conditions defined by the proper definition of the object. According to this idea, it is possible to assume that variables with different role can be associated to the same object. In particular, it is possible to distinguish the descriptors of the object from the variables that we are interested to control. If, for example, we assume the variables associated to Buffon's data (Section 3.4.) to be interval variables, we have the problem of controlling symbolic batches.

The non parametric procedure for building non parametric control charts for classical batches (Section 3.4.) is here applied with some variants. In particular, the matrix $\boldsymbol{X}(n, 2p)$, supposed partitioned into two matrices, the matrix $\boldsymbol{E}(n, 2p_1)$ of descriptors and the matrix $\boldsymbol{C}(n, 2p_2)$ of control variables, is transformed into the matrix $\boldsymbol{Z}^{\#}(n, p)$ partitioned into $\boldsymbol{E}^{\#}(n, p_1)$ and $\boldsymbol{C}^{\#}(n, p_2)$.

The $\boldsymbol{E}^{\#}$'s are here considered as design matrices and so the p_2 control variables are directly projected on the subspace defined by the p_1 descriptors. The advantage is in considering the role of the different interval domains of descriptors in determining the possible out of control. The width of each convex hull is a signal of the internal variability of each batch.

The problem of controlling complex and structured data is actually studied and a lot of interesting points can be developed. In particular, operations proper of symbolic objects, like generalization or specialization, can be used for evaluating causes of out of control. Furthermore, proper symbolic methods can be employed for building suitable symbolic control charts.

References

Alt F. B. and Smith N. D. (1988). *Multivariate process control.* In: Handbook of Statistics **7**, P. R. Krishnaiah and C.R. Rao (eds.), Amsterdam, 333 – 351.

Balbi S. and Scepi G. (1996). *Sensibilita' delle carte di controllo multivariato basate sulle coordinate fattoriali.* In: Atti della XXXIIII Riunione Scientifica della SIS, A. Maggioli ed., Rimini, 431 – 438.

Balbi S. and Scepi G. (2000). *Carte di controllo non parametriche per oggetti simbolici.* XL Riunione Scientifica SIS, Aprile, Firenze.

Balbi S., Lauro N. C. and Scepi G. (1994). *A multiway data analysis technique for comparing surveys.* Methodologica **3**, 79 – 90.

Bock H. H. and Diday E. (eds.) (2000). *Analysis of Symbolic Data.* Springer Verlag, Heidelberg.

Box G. E., Jenkins G. M. and Reinsel G. C. (1994). *Time Series Analysis.* P. Hall, New York.

Brook D. and Evans D. (1972). *An approach to the Probability Distribution of CUSUM Run Length.* Biometrika **59**, 539 – 549.

Chouakria A., Diday E. and Cazes P. (1998). *An Improved Factorial Representation of Symbolic Objects.* KESDA '98, Luxembourg.

Crowder M. J., Kimber A. C., Smith R. L. and Sweeting T. J. (1991). *Statistical Analysis of Reliability Data.* Chapman and Hall, London.

Crosier R. B. (1986). *A new two side cumulative sum quality-control scheme.* Technometrics **3**, 187 – 194.

Crosier R. B. (1988). *Multivariate generalizations of cumulative sum quality-control schemes.* Technometrics **3**, 291 – 303.

D'Ambra L. and Lauro N. C. (1982). *Analisi in componenti principali in rapporto ad un sottospazio di riferimento.* Rivista di Statistica Applicata **1**, 51 – 67.

Does R. and Schriever K. (1992). *Variables control charts limits and test for special causes.* Statistica Neerlandica **46**, 229 – 245.

Efron B. and Tibshirani R. J. (1993). *An Introduction to the Bootstrap.* Chapman and Hall, London.

Escoufier Y. (1987). *Three-mode data analysis: the STATIS method, Methods for multidimensional data analysis.* In: ECAS, Fichet B. and Lauro N. C. (eds.), 259 – 272.

Giordano G., Palumbo F. (1999). *A New Statistical Quality Control Tool Based on PCA of Interval Data.* In: Proceedings of CLADAG '99, Roma, 197 – 200.

Green P. J. (1981). *Peeling bivariate data.* In: Interpreting Multivariate Data, Barnett V. et al. (eds.), J. Wiley, New York.

Greenacre M. J. (1984). *Theory and Applications of Correspondence Analysis.* Academic Press, New York.

Healy J. D. (1987). *A note on multivariate CUSUM procedures.* Technometrics **29**, 409 – 412.

Holmes S. (1989). *Using the bootstrap and the Rv coefficient in the multivariate context.* In: Data Analysis Learning Symbolic and Numeric Knowledge, E. Diday (ed.), Nova Science Publishers, New York.

Hotelling H. (1947). *Multivariate quality-control, illustrated by the air testing of sample bombsites.* In: Selected Techniques of Statistical Analysis, C. Eisenhart et al. (eds.), McGraw-Hill, New York, 111 – 184.

Jackson J. E. (1991). *A User's Guide to Principal Components.* J. Wiley, New York.

Lauro N. C. and Palumbo F. (2000). *Principal Component Analysis of Interval Data: a Symbolic Data Analysis Approach.* Computational Statistics **15**, 73 – 87.

Lauro N. C., Balbi S., and Scepi G. (1995). *The Analysis of Repeated Surveys on Italian Manufacturing Enterprises: A Multidimensional Approach.* In: Proceedings of the First International Workshop on Les Techniques de Panel d'Enterprises, Luxembourg, 344 – 356.

Lauro N. C., Scepi G. and Balbi S. (1996). *Differenti approcci nella costruzione di carte di controllo multivariato.* In: Scritti in onore di G. Landenna, Giuffre (ed.), 261 – 291.

Lauro N. C., Balbi S., and Scepi G. (1997). *L'analisi multidimensionale dei dati per la misurazione della customer satisfaction nei servizi* . In: Atti del Convegno SIS La Statistica per le Imprese, Tirrenia (ed.), 567 – 583.

Liu R. Y. (1995). *Control Charts for multivariate processes.* Journal of American Statistical Association **42**, 1380 – 1387.

Liu R. Y. and Singh K. (1993). *A quality index based on data depth and multivariate rank tests.* Journal of American Statistical Association **88**, 252 – 260.

Lucas J. M. and Crosier R. B. (1982a). *Fast initial response for CUSUM quality-control schemes: Give your CUSUM a head start.* Technometrics **24**, 199 – 204.

Lucas J. M. and Crosier R. B. (1982b). *Robust CUSUM: A robustness study for CUSUM quality-control schemes.* Comm.Stat.-Theor.Meth **11**, 2669 – 2687.

Mason et al. (1997). *Assessment of multivariate process control.* Journal Quality Technology **29**, 140 – 143.

Page E. S. (1954). *Continous inspection schemes.* Biometrika **41**, 110 – 114.

Politis D. N. and Romano J. P. (1994). *The Stationary Bootstrap.* Journal of American Statistical Association **89**, 1303 – 1313.

Rizzi A. (1989). *On the synthesis of three-way data matrices.* In: Multiway Data Analysis, R.Coppi and S. Bolasco (eds.), North Holland, 143 – 154.

Scepi G. (1994). *Approcci parametrici e non parametrici alla progettazione e al controllo multivariato della qualita'*. Ph.D.Thesis, University Federico II, Naples.

Scepi G., Acconcia A. (1996). *Parametric Control Charts for Sequential Process*. In: Proceedings in Computational Statistics, Physica Verlag, Barcellona, 447 – 452.

Scepi G., Lauro N. C. and Balbi S. (1993). *Empirical confidence regions for multidimensional control charts*. In: Proceedings of the 49-th ISI Session, Contributed Papers **2**, Firenze, 379 – 380.

Seppala et al. (1995). *Statistical Process Control via Subgroup Bootstrap*. Journal Quality Technology **27**, 139 – 153.

Tukey J. W. (1975). *Mathematics and Picturing Data*. In: Proceedings of the 1975 International Congress of Mathematics **2**, 523 – 531.

Wald A. (1947). *Sequential Analysis*. J. Wiley, New York.

Wierda A. (1994). *Multivariate statistical process control: Recent results and directions for future research*. Statistica Neerlandica **38**, 147 – 168.

NON–SYMMETRICAL EXPLORATORY COMPARATIVE ANALYSIS

Vincenzo Esposito Vinzi

University of Naples Federico II, Faculty of Economics, Department of Mathematics and Statistics, Via Cintia – Complesso Monte Sant'Angelo, I – 80 126 Naples, Italy; binci@unina.it

Abstract: The first part of this paper intends to show the fundamental ideas behind the methodological achievements of *Non–Symmetrical Data Analysis* (*NSDA*) from a geometrical point of view. We then focus on some of the most recent extensions of *NSDA* which aim at studying multiple tables with comparative purposes by stressing their graphical aspects and giving insights on their interpretation tools. The methods presented in this part are then applied on a couple of data sets from the field of sensory analysis and water quality control. Finally, we outline some of the most promising directions of further research in this field. The author has developed S-Plus® codes that implement the four non–symmetrical *comparative analyses* presented in the paper. Interested readers may directly contact the author to get the codes and relative instructions so as to apply the methods and practice with them on their own datasets.

1 Introduction

In the highly complex world of industrial production, the statistical quality control (*SQC*) procedures are assuming an always more relevant role. Since the past decade, in fact, *SQC* has evolved as it has not been limited just to the production technical aspects. In the perspective of advantages over the competitors, it is no longer enough to express quality in terms of absence of defects in the product but rather in terms of customer satisfaction as shown by Lauro, Balbi & Scepi (1997). Therefore, it becomes fashionable and much more impressive to speak of *Total Quality* which actually involves all components of an enterprise, not only the production department. Statistics may and must play a prominent role in this scenario as being capable of providing suitable tools for *Total Quality Measurement* (*TQM*).

TQM is a key-point. It constitutes the basis for *Total Quality Design and Assurance* that may be all meant as part of the so-called *Total Quality Management.* By referring to the well-known Deming's cycle (see Fig. 1), it

---

1991 *Mathematics Subject Classification.* 62-01, 62H20, 62H25, 62HH99, 62P30.

Key words and phrases. Dependence analysis; principal component analysis onto a reference subspace; Procrustean rotations; co–inertia analysis; graphical representations; multi-occasion data; S-Plus®.

Acknowledgement. The work was partially supported by the MURST (Italian Ministry for University and Scientific and Technological Research) 1999 – 2000 grant on *Multivariate Analyses for Total Quality: Methodologies, Computational Aspects and Applications* (resp.: Prof. Carlo Lauro).

is important to enhance that the role of Statistics is actually very prominent within the steps of *Planning* and *Checking* by means of, respectively, *Quality Design* and *Quality Control* techniques.

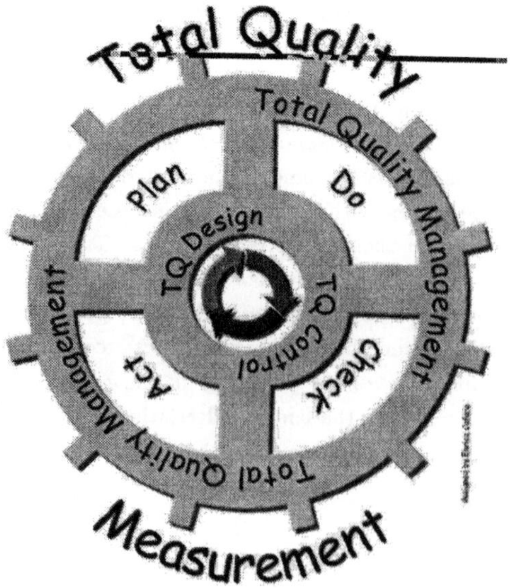

Figure 1. The Deming's cycle in the *TQM* perspective.

The steps of *Doing* and *Acting* are instead of main interest for the *Total Quality Management.* Throughout the paper, we will focus on *TQM* with the objective of visualising relationships within quality characteristics as well as between these ones and the process variables for monitoring the industrial process itself. For the sake of brevity, we will confine our considerations to *TQM* in the case of products regardless of any quality aspects related to services for which numerous contributions have been developed also in the context of *Conjoint Analysis* (see Lauro, Balbi & Scepi (1997) and Lauro, Giordano & Verde (1998).)

In most industry applications, the economic relevance of moving from Quality Control to Quality Design has demanded the quality measure to be predicted from the process variables thus implying the identification of a relationship between them. Inferential estimation on multiple linear regressions usually accomplishes this task but, in *TQM*, it fails as it is not capable to provide the user with a graphical interpretation of the results and to retain the directional information due to the interactions between variables.

There are usually several applications related to the real-world problems where, in describing the dependence structure between a set of response variables and a set of explanatory ones, almost no information is available to define an a priori mathematical model. In such a context, geometrically-based techniques have been developed aiming at exploring and describing the data structure at hand.

Most of the well-known and widely used techniques of multidimensional data analysis originated from *Canonical Correlation Analysis* (*CCA*) proposed by Hotelling (1936), e.g. *Discriminant Analysis* and *Correspondence Analysis*. However, the application context of *CCA* is restricted to symmetrical relationships between two sets of quantitative variables or properly coded qualitative ones so that any a-priori information on the different roles played by the observed variables is neglected and trivial results may be reached. In order to encompass this limit, some techniques have been developed that actually constitute the framework of *Non Symmetrical Data Analysis* (*NSDA*).

Classical references are represented by the *Principal Component Analysis with Instrumental Variables* (*PCAIV*) proposed by Rao (1964) which is clearly set in a multivariate regression-like approach and *Redundancy Analysis* (*RA*) proposed by Van den Wollenberg (1977) as an alternative to *CCA*.

In a geometrical context, D'Ambra & Lauro (1982), and Lauro & D'Ambra (1984) developed new methods, such as *Principal Component Analysis onto a Reference Subspace* (*PCAR*) for quantitative variables and *Non Symmetrical Correspondence Analysis* (*NSCA*) for qualitative variables (see also Lauro & Siciliano (2000)). These methods enriched the interpretation, extended the analysis to the study of more sets of variables and put new insight in *NSDA*. Since then, particular attention was paid to inferential problems in terms of suitable models introduced in Lauro & Siciliano (1989), Siciliano, Mooijaart & Van der Heijden (1993) and Esposito (1998), as well as to stability and validation of results (see Balbi (1992)).

The most recent developments of *PCAR* extend the treatment to mixed variables (quantitative, ordinal and categorical data) and multiway data. This allows to improve the practice in various applications related to Quality Control. In this paper, we refer to some theoretical relationships and interpretation tools which enable to graphically compare multiway data by means of suitable strategies. These strategies combine *PCAR* with *Procrustean rotations* and *co-inertia analyses* which help in detecting, interpreting and displaying interesting structures in the data.

2 The geometrical approach to NSDA

Principal Component Analysis, as originally developed by Pearson (1901) (see Fig. 2), aims at fitting a line or a plane to a cloud of n points in a p-dimensional space spanned by numerical variables all playing the same role in the analysis, being $\boldsymbol{X}_{n,p}$ the data matrix under study.

In the framework of numerical variables, Rao (1964) extends *PCA* so as to take into account also a set of q instrumental (explanatory) numerical variables collected in the matrix $\boldsymbol{Y}_{n,q}$. However, his approach lacks with a

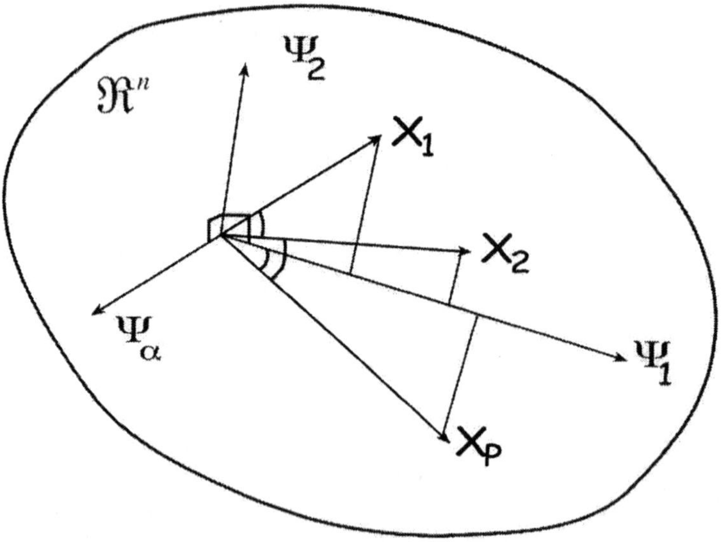

Figure 2. Looking for principal components of $\boldsymbol{X}_{n,p}$.

suitable geometrical environment where to visualise, on factorial planes, the dependence structure among the variables.

Some years later, Van den Wollenberg (1977) proposes the Redundancy Analysis as an alternative to *CCA*. In fact, *CCA* (see Fig. 3) searches for the linear combinations of $\boldsymbol{X}(\xi_1)$ and $\boldsymbol{Y}(\eta_1)$ that are most highly correlated among them (i.e. the angle θ being as little as possible), thus studying the interdependence between the $\boldsymbol{X}$- and $\boldsymbol{Y}$-variables.

CCA actually suffers from the following major advantages: it forces correlations, which might not actually exist, between the two sets of variables; it usually yields a pair of strongly linked factors, but uninteresting with respect to their weak relationship with the original variables, and a pair of more interesting factors which, however, are weakly related between themselves.

On the other hand, *RA* searches for the linear combinations of the variables in $\boldsymbol{Y}$ that maximise the explained variance of the variables in $\boldsymbol{X}$ through optimising the redundancy index Stewart & Love (1968). The solution is the same as for *PCAIV*.

Finally, *PCAR* is developed in order to study, in a proper geometrical framework, the structure of the relationships between $\boldsymbol{X}$ and $\boldsymbol{Y}$. Without loss of generality, the variables in $\boldsymbol{X}$ and $\boldsymbol{Y}$ are both centered with respect to their means and divided by the scalar $\sqrt{n}$ so that their cross-products directly define variance-covariance $\boldsymbol{V}_{XX}$ and inter-covariance matrices ($\boldsymbol{V}_{YX}$ and $\boldsymbol{V}_{XY}$).

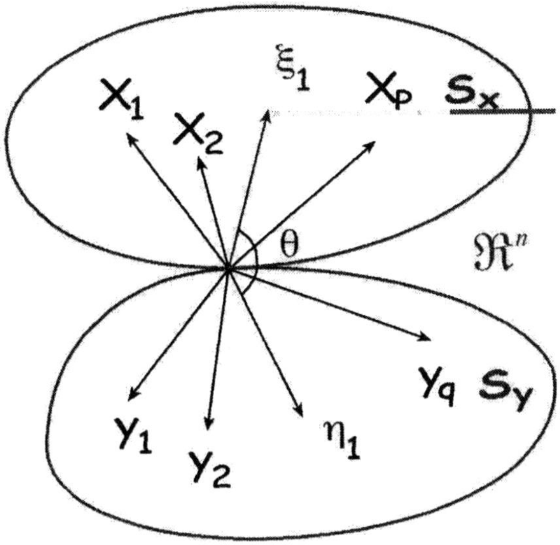

Figure 3. A geometrical insight on *CCA*.

Namely, *PCAR* (see Fig. 4) searches for the principal components of the image, obtained by means of orthogonal projections, of the dependent variables, in $\mathcal{S}_X$, on the subspace $\mathcal{S}_Y$ spanned by the explanatory variables.

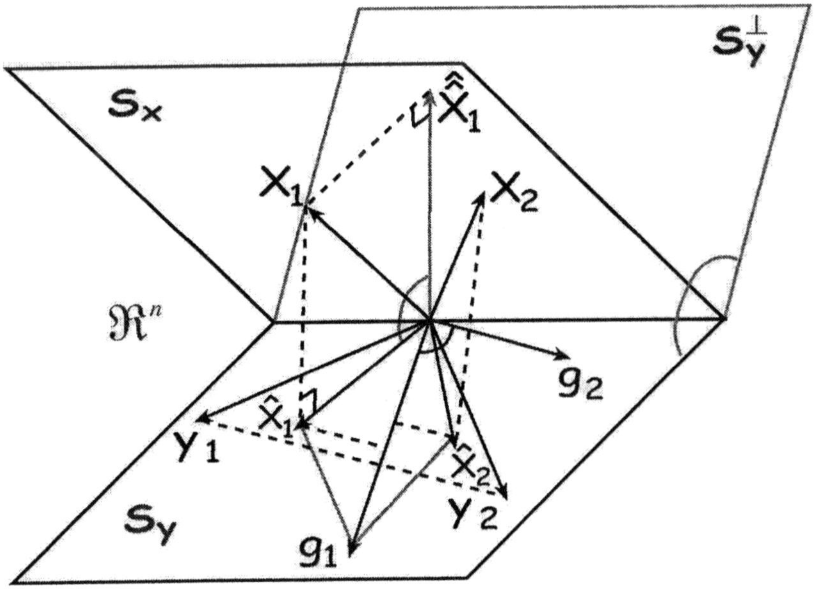

Figure 4. A geometrical insight on *PCAR*.

The solution is given by a *PCA* performed on the matrix $\boldsymbol{P_Y X}$, i.e. by the eigenvectors $\boldsymbol{v}_k$ associated with the highest eigenvalues λ_k of:

$$\boldsymbol{X}'\boldsymbol{P_Y}\boldsymbol{X}\boldsymbol{v}_k = \lambda_k \boldsymbol{v}_k \tag{2.1}$$

with $k = 1, \ldots, min\left[rank(\boldsymbol{X}), rank(\boldsymbol{Y})\right]$ and $\boldsymbol{P_Y} = \boldsymbol{Y}\left(\boldsymbol{Y}'\boldsymbol{W}_n\boldsymbol{Y}\right)^{-1}\boldsymbol{Y}'\boldsymbol{W}_n$ is the $\boldsymbol{W}_n$-orthogonal projector on the subspace spanned by the variables in $\boldsymbol{Y}$, with $\boldsymbol{W}_n$ the weight system assigned to the statistical units in the variables space $\mathcal{R}^n$.

It is worth noticing that with respect to classical *PCA*, the preliminary geometrical transformation by projection in *PCAR* (see Fig. 4) may be meant as an optimal quantification (or coding) of the variables in according with the explanatory objective of the analysis. In this case, in fact, the predictive efficiency measure is given by the portion of explained variance

$$\frac{trace\left(\boldsymbol{X}'\boldsymbol{P_Y}\boldsymbol{X}\right)}{trace\left(\boldsymbol{X}'\boldsymbol{X}\right)} \tag{2.2}$$

and this portion is exactly the one decomposed by *PCAR*, i.e. the sum of the multiple correlation coefficients between each dependent variable and the set of q explanatory variables.

Very often, in real applications, it is of great interest to analyse the relationships among dependent variables after taking away the influence of the explanatory ones. Thus, a *partial analysis* is needed. *PCAR* can easily deal with this need by referring to the projection operator $\left(\boldsymbol{I}_n - \boldsymbol{P_Y}\right)$ that, geometrically speaking, allows performing the analysis on the ortho–complementary subspace ($S_Y^{\perp}$ in Fig. 4).

With respect to the *PCAIV* solution, *PCAR* has the same non-trivial eigenvalues, simple relations among the eigenvectors and the rank depending upon $\boldsymbol{X}'\boldsymbol{Y}$.

Throughout the paper, we will be focusing on the geometrical aspects of *NSDA* leaving apart any probability distribution assumptions. In fact, our aim is to improve the quality of data description performed by the analysis according with the complex data structure at hand, rather than to improve the quality of interpretation of results.

2.1 PCAR biplot representations. The graphical aspects are of main interest for the interpretation of *PCAR* and represent its major advantages with respect to other techniques dealing with the same data structure. As it concerns the building of representation planes for the units and the variables, it is helpful to enhance that, because the matrix in (2.1) is symmetric, the following relations hold:

$$\boldsymbol{v}_k'\boldsymbol{v}_k = \lambda_k \quad \text{and} \quad \boldsymbol{v}_k'\boldsymbol{v}_{k'} = 0, \tag{2.3}$$

from which derives the orthogonality and normalisation to λ_k of the principal components:

$$g_k = \boldsymbol{P_Y X v_k}. \tag{2.4}$$

However, for interpretation scopes, it is very helpful to set both statistical units and variables in the same geometry. Therefore, we ensure that the graphical display in the reduced space is a *biplot* (i.e. a plot where two sets of objects may be *simultaneously* represented) by dealing with components normalised to 1 and defined as:

$$g_k^* = \boldsymbol{P_Y X v_k}/\sqrt{\lambda_k}. \tag{2.5}$$

The number k of retained components for yielding graphical displays is chosen by means of the usual variance–based criteria. However, by referring to (2.2), the percentage of retained variance is computed with respect to the variance accounted for by the explanatory variables rather than to the total variance of the response ones.

In *PCAR*, the co-ordinates of dependent variables are obtained as the correlation between the variables themselves and the principal components, i.e., taking into account the expression ((2.1)):

$$\boldsymbol{X}'g_k^* = \boldsymbol{X}'\boldsymbol{P_Y X v_k}/\sqrt{\lambda_k} = \boldsymbol{v}_k\sqrt{\lambda_k}. \tag{2.6}$$

The same applies to the explanatory variables as, after simple algebra, we have:

$$\boldsymbol{Y}'g_k^* = \boldsymbol{Y}'\boldsymbol{X v_k}/\sqrt{\lambda_k}. \tag{2.7}$$

The equation (2.7) enhances how *PCAR* is able to reconstruct the image of the correlations between the two sets $\boldsymbol{X}$ and $\boldsymbol{Y}$ directly on the principal axes. Therefore, differently from *CCA*, we are enabled to read on a unique factorial plane both the correlations *within* the same set of variables as well as the correlations *between* the two sets in terms of proximities between the relative represented points.

It is finally worth pointing out that the k-th eigenvalue of *PCAR* is the sum of squared correlations between the variables in $\boldsymbol{X}$ and the principal components of the reference subspace. Therefore, it represents the explanatory power of the principal component, associated with the k-th eigenvector, with respect to the dependent variables.

In order to visualise the statistical units, we refer to the dual analysis in $\mathcal{R}^{p+q}$ whose characteristic equation is:

$$\boldsymbol{P_Y X X' W_n P_Y X v_k} = \lambda_k \boldsymbol{P_Y X v_k} \tag{2.8}$$

so that the principal vectors $\boldsymbol{P_Y X v_k}$ in $\mathcal{R}^{p+q}$ actually represent the principal components, relative to the space $\mathcal{R}^n$. They can be also expressed as linear combinations of the variables $\boldsymbol{Y}$:

$$g_k = \boldsymbol{P_Y X v_k} = \boldsymbol{Y}\left[\boldsymbol{Y}\left(\boldsymbol{Y}'\boldsymbol{W}_n\boldsymbol{Y}\right)^{-1}\boldsymbol{Y}'\boldsymbol{W}_n\right] = \boldsymbol{Y}\beta_k. \tag{2.9}$$

3 Non-Symmetrical comparative analyses

We now refer to a powerful set of comparative analyses for the treatment of multiple tables which are especially recurrent and important in the field of quality control. The particular, but highly frequent, data structure we are dealing with, is made up of: C sets $\boldsymbol{X}_c$'s ($c = 1, \ldots, C$) of p dependent variables repeatedly observed on the same statistical units under C observational conditions (totally paired matrices); a set $\boldsymbol{Y}$ of q condition-invariant explanatory variables. As a general purpose, we aim at comparing and visualising the behaviours of the paired units and variables in the different matrices with respect to the dependence structure between the two sets of variables.

In the following, we refer to the quality control case where we are interested in comparing judgements expressed by different evaluators on the quality characteristic of a product in terms of its structural features yielded by the process variables.

In this perspective, several approaches have been recently proposed. We will firstly show some results referred to the joint use of multidimensional analyses, such as *CCA* and *PCAR*, with *Procrustean rotations* in Section 3.1 and then some analyses based on the criterion of *co-inertia* in Section 3.2.

3.1 Analyses based on Procrustean rotations. In the perspective of comparing and visualising the elements of different matrices, Procrustean rotations play a fundamental role as they classically aim at finding the so-called *maximal agreement* plane by adapting one matrix to another chosen as a reference.

Before going deep into the different combined strategies between *PCAR* and Procrustean rotations, a reminder on the algebra and the meaning of this latter method is provided.

3.1.1 Orthogonal Procrustean analysis. *Procrustean analysis* is a classical technique commonly used for comparing different sets of judgements. In fact, starting from two different configurations with the same dimensions, Procrustean analysis allows to detect how much they differ, or at which extent they are similar, by taking one configuration as a target. In interpreting different sets of judgements expressed on the same objects, several sources of variation need to be taken into account as they affect the sensory data and originate from the different groups of judges. Namely, the judges may differ in :

- the general level of scores assigned to the objects;
- the different terms, or combinations of them, they use for describing the same stimulus;
- the scale interval they refer to;
- the different behaviours they follow in replicated judging sessions.

By means of scaling transformations, such as translation, rotation and ranking, Procrustean analysis yields a maximal agreement configuration starting from two different initial configurations.

The translation phase eliminates the *level effect* (due to the fact that the judges vary the attributes by using different scale levels) by moving the attributes towards a common centre. The rotation phase eliminates the *interpretation effect* rising from the different attributes used for describing the objects. The last phase of a Procrustean analysis eliminates the *interval effect* (due to the use of different intervals of the response scales by the judges) by shrinking or extending the axes of each configuration.

From the methodological point of view, the aim of *orthogonal Procrustean analysis* (Green (1952), Gower (1971)) is to compare two sets of coordinates by optimizing some goodness–of–fit criterion, i.e. by rotating, reflecting, and translating one configuration in order to minimize its *distance* from the other one.

Given two configurations of n points $\boldsymbol{X}_1$ and $\boldsymbol{X}_2$ in a p-dimensional space, for convenience shifted to the origin, the *best rotation* (in a least squares sense) of $\boldsymbol{X}_2$ relative to $\boldsymbol{X}_1$ is $\boldsymbol{X}_2\boldsymbol{R}$, where $\boldsymbol{R}$ is given by

$$\boldsymbol{R} = \boldsymbol{W}\boldsymbol{U}' = \sum_{j=1}^{r} \boldsymbol{w}_j \boldsymbol{u}_j', \tag{3.1}$$

$\boldsymbol{W}$ and $\boldsymbol{U}$ are square matrices of order p having in columns the left singular vectors $\boldsymbol{w}_j$ and the right ones $\boldsymbol{u}_j$ normalized to 1 $\left(\boldsymbol{u}_j'\boldsymbol{u}_j = \boldsymbol{w}_j'\boldsymbol{w}_j = 1\right)$ of $\boldsymbol{V}_{X_2X_1}$, whose rank is $r \leq p$. $\boldsymbol{R}$ is said the orthogonal polar factor of $\boldsymbol{V}_{X_2X_1}$ (Golub & Van Loan (1989)) and it is constrained to be $\boldsymbol{R}'\boldsymbol{R} = \boldsymbol{I}_p$. Additionally, note that, after some elementary algebra, $\boldsymbol{R}$ can be rewritten in terms of cross-covariance matrices so that

$$\boldsymbol{R} = \boldsymbol{V}_{X_2X_1}\left(\boldsymbol{V}_{X_1X_2}\boldsymbol{V}_{X_2X_1}\right)^{-\frac{1}{2}}. \tag{3.2}$$

This basic approach to orthogonal rotations has been developed in several directions (e.g. different normalization constraint, as in Lafosse (1986)), and mainly for the case in which more than two configurations of points are involved as in generalized Procrustean analysis, see Gower (1975)).

Procrustean analysis has often been utilized in the geometric framework of data analysis: it is generally performed *after* a classical multidimensional analysis in order to compare factorial configurations obtained analyzing different data sets or, less frequently, obtained by different analyzes performed on the same data set (Mardia, Kent & Bibby (1979)).

Furthermore, the idea of performing a Procrustean rotation *before* analysing two (or more) data configurations of paired points is strictly connected with a problem frequently faced in a multidimensional data analysis approach. We refer to the need of finding a space in which to compare the similarity of two, or more, (factorial) configurations.

In a symmetrical framework, Lafosse (1989) proposes a *joint use of Procrustean rotations* and *CCA* in order to analyze similarities in two totally paired tables (obtained by measuring twice the same variables on the same cases). The first technique aims at finding the *maximal agreement*, by adapting one matrix to the other, chosen as a reference. The second one aims at finding the main structure of similarity, enabling to represent how the two configurations differ.

In the following, it is of some consequences to note that, when we perform a *CCA* after a Procrustean rotation, we work with the matrix

$$\boldsymbol{V}_{X_1X_2}\boldsymbol{R} = \left(\boldsymbol{V}_{X_1X_2}\boldsymbol{V}_{X_2X_1}\right)^{\frac{1}{2}}. \tag{3.3}$$

This matrix is symmetric and its spectral decomposition is given by:

$$\boldsymbol{V}_{X_1X_2}\boldsymbol{R} = \sum_{j=1}^{p} \zeta_j \boldsymbol{u}_j \boldsymbol{w}'_j \sum_{j=1}^{p} \boldsymbol{w}_j \boldsymbol{u}'_j = \sum_{j=1}^{p} \zeta_j \boldsymbol{u}_j \boldsymbol{u}'_j, \tag{3.4}$$

where ζ_j is the j-th singular value of both $\boldsymbol{V}_{X_1X_2}$ and $\boldsymbol{V}_{X_2X_1}$. The result in (3.4) implies that the linear combinations relative to $\boldsymbol{X}_1$ and $\boldsymbol{X}_2$ are associated with the same vectors $\boldsymbol{u}_j$. Conversely, they would be associated to the same vectors $\boldsymbol{w}_j$ if $\boldsymbol{X}_2$, instead of $\boldsymbol{X}_1$, was chosen as the rotation reference.

This is a very appealing feature of this approach as it solves out a classical problem of *CCA*: having two sets of different linear combinations to choose from in order to build the graphical displays with dramatic consequences on interpretation.

3.1.2 Rotated canonical analysis onto a reference subspace. Starting from the just mentioned result, but extending it in order to actually allow for external information provided by the set $\boldsymbol{Y}$ of explanatory variables, Balbi & Esposito (2000) propose a rotated canonical analysis onto a reference subspace (*RCAR*) and demonstrate several mathematical properties and interpretation facilities enjoyed by this method.

In the field of *Customer Satisfaction*, we generally aim at measuring the gap between perceived and expected quality by the customers of a product/service with respect to a pre-defined set of scenarios (Lauro, Balbi & Scepi (1997)). This objective is achieved by representing the principal structure of similarity between conditions (perceived/expected), once the variability of each set of dependent variables has been decomposed into a portion explained by the variables (scenarios) in $\boldsymbol{Y}$ and a residual one. In this direction, taking $C = 2$, we first project both $\boldsymbol{X}_1$ (expected quality) and $\boldsymbol{X}_2$ (perceived quality) onto the subspace spanned by the columns of $\boldsymbol{Y}$, then, after an orthogonal Procrustean rotation of $\boldsymbol{P_Y X}_2$ towards $\boldsymbol{P_Y X}_1$ taken as a reference, we perform a *CCA* (see Fig. 5).

The rotation step is performed by means of the operator:

$$\boldsymbol{R}_{\boldsymbol{P_Y}} = \boldsymbol{X}'_2\boldsymbol{P_Y}\boldsymbol{X}_1\left(\boldsymbol{X}'_1\boldsymbol{P_Y}\boldsymbol{X}_2\boldsymbol{X}'_2\boldsymbol{P_Y}\boldsymbol{X}_1\right)^{-\frac{1}{2}}, \tag{3.5}$$

thus defining the transformation $\boldsymbol{P_Y X}_2\boldsymbol{R}_{\boldsymbol{P_Y}}$.

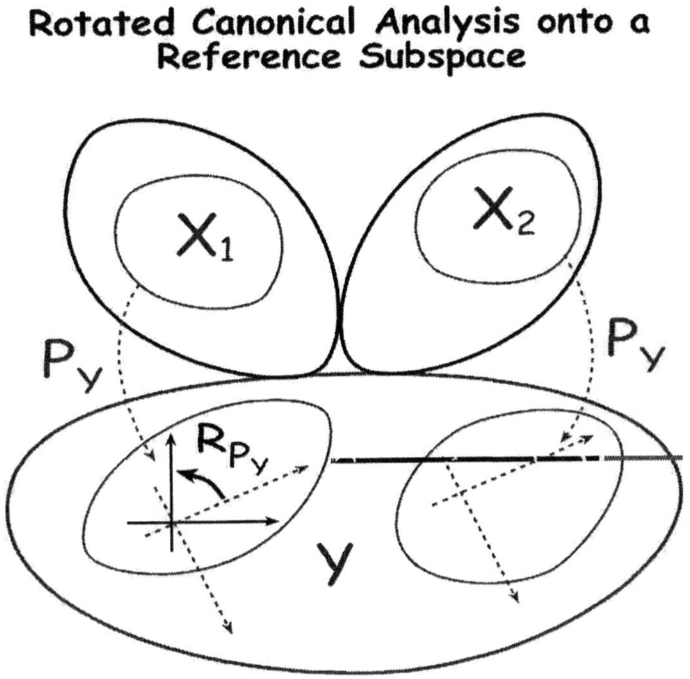

Figure 5. A geometrical insight on *RCAR*.

The core of the analysis consists in the spectral decomposition of the inter-covariance matrix between $\boldsymbol{P_Y X_1}$ and $\boldsymbol{P_Y X_2 R_{P_Y}}$ that, after simple algebra, becomes :

$$\left(\boldsymbol{V}_{X_1Y}\boldsymbol{V}_{YY}^{-1}\boldsymbol{V}_{YX_2}\boldsymbol{V}_{X_2Y}\boldsymbol{V}_{YY}{}^{-1}\boldsymbol{V}_{YX_1}\right)^{\frac{1}{2}}. \tag{3.6}$$

From (3.6) it is clear that, though we actually perform a symmetrical analysis in the sense of *CCA*, the variables in $\boldsymbol{Y}$ act as conditioning variables of the comparison between $\boldsymbol{X}_1$ and $\boldsymbol{X}_2$. The decomposition in (3.6) may be also expressed in terms of singular values and singular vectors as :

$$\sum_{j=1}^{p}\mu_j\boldsymbol{b}_j\boldsymbol{a}_j'\sum_{j=1}^{p}\boldsymbol{a}_j\boldsymbol{b}_j' = \sum_{j=1}^{p}\mu_j\boldsymbol{b}_j\boldsymbol{b}_j' \tag{3.7}$$

with μ_j the j-th singular value of $\boldsymbol{X}_2'\boldsymbol{P_Y}\boldsymbol{X}_1$ and $\boldsymbol{b}_j$ and $\boldsymbol{a}_j$ the respective right and left singular vectors.

Furthermore, among other properties, it is demonstrated that, differently from *CCA*, the two sets of canonical variates are here associated with the same eigenvectors so that we have a unique factorial representation with nice consequences on the interpretation of the results.

By following the mentioned strategy, it is possible to monitor on the same factorial plane both the inter-conditions and intra-conditions variability. In fact, the canonical variates are interpreted as a measure of the likeness between projected clouds of points. This likeness is evaluated in terms of residual distances between matched points. In particular, the length of the residual distance gives information on the size of the difference while its direction with respect to the axes' orientation allows explaining the eventual difference in terms of the original variables.

With respect to the graphical displays, it is worth mentioning that *RCAR* inherits from *PCAR* the property of directly representing the image of correlations between the response variables and the explanatory ones as the co-ordinates of the latter are:

$$\boldsymbol{Y}'\boldsymbol{P}_{\boldsymbol{Y}}\boldsymbol{X}_1\boldsymbol{b}_j = \boldsymbol{Y}'\boldsymbol{X}_1\boldsymbol{b}_j. \tag{3.8}$$

When dealing with Procrustean rotations, the choice of a proper scaling of the variables is of great relevance. In fact, depending on that, the cross-product matrices assume different meanings and so does their spectral decomposition. In our case, the usual choice of scaling all sets of variables to unit variance may be misleading as the interpretation of distances on canonical planes would be severely damaged. Moreover, as we aim at identifying structures of similarities, we do not want to lose any differences in the variability of the phenomenon in the different occasions. For these reasons, we impose the same scale for $\boldsymbol{X}_1$ and $\boldsymbol{X}_2$ by post-multiplying the centred variables in the two matrices by:

$$\left(\sqrt{n}\boldsymbol{\sigma}_{X_1}\boldsymbol{I}_q\right)^{-1}, \tag{3.9}$$

where $\boldsymbol{\sigma}_{X_1}$ is the vector of the standard deviations of the response variables in the first condition. These transformations lead the cross-product matrix $\boldsymbol{V}_{X_2X_1}$ to contain simple regression coefficients of $\boldsymbol{X}_2$ with respect to $\boldsymbol{X}_1$. This result even more stresses the relevance of the choice of the rotation reference between the observational conditions. In fact, these conditions clearly play a non-symmetrical role in the analysis.

From the point of view of the spectral decomposition, this means that we are decomposing the dependence structure so to grasp the main features of the regression coefficients matrix by representing its reduced rank approximation (Ter Braak (1990), Lauro, Giordano & Verde (1998)). Therefore, the factorial planes will represent conditional effects, rather than marginal ones, of one condition on the other and these effects may be interpreted with respect to the explanatory variables.

3.1.3 Simultaneous PCA onto a reference subspace. *RCAR*, as it has been presented, is interesting for representing the principal structure of similarity once the variability of each dependent set is decomposed in terms of an explained part plus a residual one. This is, in any case, just one way of looking at the problem. In fact, it is usually of great relevance, for the same data structure as in Section 3.1.2, also to detect the differences in the overall structure of dependent variables and then to explain these differences in terms of the explanatory variables.

For this purpose, Esposito & Balbi (1999) have developed a *simultaneous PCAR* (*SPCAR*). This strategy (see Fig. 6) represents a direct extension of *PCAR* for multiple sets of dependent variables (D'Ambra & Lauro (1982)). Here, we firstly aim at detecting the differences in the overall structure of dependent variables and then at explaining these differences in terms of $\boldsymbol{Y}$.

This is a general problem in multivariate quality control when one wants to compare the really observed quality characteristics with the *in-control* situation and explain the eventual differences with respect to the process variables. In fact, once an atypical matching is detected, the relative causes need to be identified in one or some of the process variables.

In this direction, a Procrustean rotation of $\boldsymbol{X}_2$ towards $\boldsymbol{X}_1$ is performed, as a preliminary transformation, by means of the operator in (3.2) which, differently from the one defined in (3.5), is of a symmetrical nature as it does not depend upon $\boldsymbol{Y}$.

The core of the analysis, in a multiple factor analysis fashion, consists of a *PCA* on:

$$\tilde{\boldsymbol{X}} = \boldsymbol{P_Y}\,[\boldsymbol{X}_1 \,|\, \boldsymbol{X}_2\boldsymbol{R}]\,. \tag{3.10}$$

In fact, $\tilde{\boldsymbol{X}}$ contains the structure of variability due to the variables in $\boldsymbol{Y}$ and its *PCA* helps in identifying the main components of such a variability, as a linear combination of variables in both $\boldsymbol{X}_1$ and $\boldsymbol{X}_2\boldsymbol{R}$.

As to the graphical outputs, the matched representation of variable-points on the previously defined common plane allows to read similarities and, most importantly, to explain differences by interpreting residual distances between matched points. Additionally, it is remarkable that, being the $\boldsymbol{X}_c$'s matrices totally paired, the statistical units belong to the same multidimensional space. Therefore, gaps between conditions can be individually measured and represented on the factorial planes.

Three kinds of plots are yielded. Namely, one for the gaps between the structures in $\boldsymbol{X}_1$ and $\boldsymbol{X}_2$; two for the observed units. The representation of both sets of dependent variables in the space of the variables in $\boldsymbol{Y}$ is enriched by projecting the explanatory variables, as supplementary, so as to explain the eventual gaps in terms of the explanatory structures.

In the ortho-complementary space, an analogous representation reveals residual gaps, after deducting the effects of the variables in $\boldsymbol{Y}$.

As to the individuals, in one plot we represent each unit, with co-ordinate on the k-th axis:

$$\boldsymbol{P_Y}\big[\boldsymbol{X}_1 \,|\, \boldsymbol{X}_2\boldsymbol{R}\big]\xi_k/\sqrt{\upsilon_k} \quad \text{with} \quad k = 1, \ldots, rank\big(\boldsymbol{P_Y}\big[\boldsymbol{X}_1 \,|\, \boldsymbol{X}_2\boldsymbol{R}\big]\big), \tag{3.11}$$

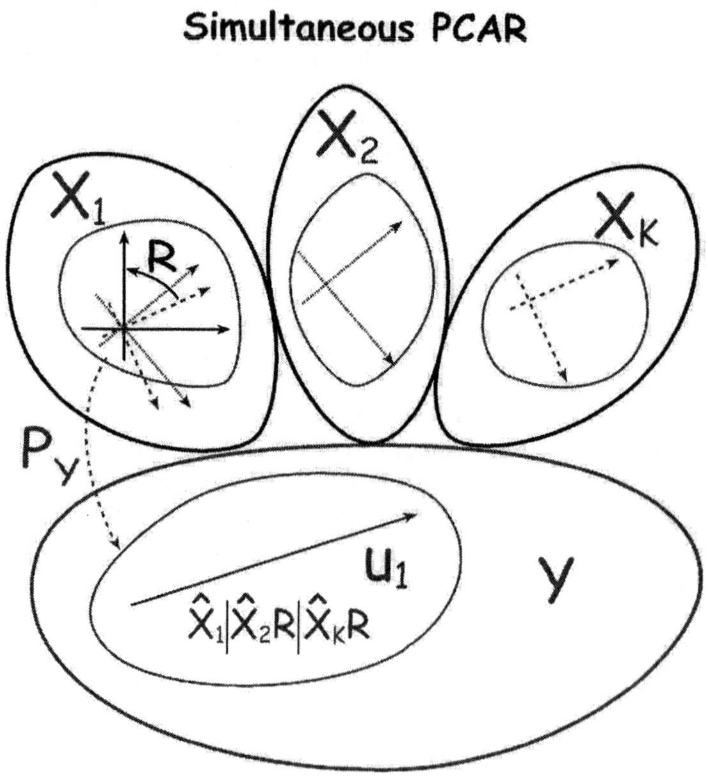

Figure 6. A geometrical insight on *SPCAR*.

where ξ_k is the eigenvector of the analysis associated to the eigenvalue υ_k.

The co-ordinates defined in (3.11) allow to represent each individual as a single point which is a synthesis of its behaviour in the different observational occasions.

On a different plot, it is possible to represent gaps, by simultaneously displaying the configurations relative to $\boldsymbol{X}_1$ and $\boldsymbol{X}_2$, with properly normalised co-ordinates on the k-th axis:

$$\frac{\boldsymbol{P_Y X}_1\xi_{1k}}{\|\boldsymbol{P_Y X}_1\xi_{1k}\|} \quad \text{and} \quad \frac{\boldsymbol{P_Y X}_2\boldsymbol{R}\xi_{2k}}{\|\boldsymbol{P_Y X}_2\boldsymbol{R}\xi_{2k}\|}, \tag{3.12}$$

where ξ_{1k} and ξ_{2k} are, respectively, the k-th eigenvectors of $\boldsymbol{X}_1'\boldsymbol{P_Y X}_1$ and $\boldsymbol{X}_2'\boldsymbol{P_Y X}_2$.

Moreover, both variables and statistical units may be represented on a common principal axes system defined by the so-called *auxiliary variables.*

Finally, the property of visualising the correlation structure between the $\boldsymbol{X}_c$'s and $\boldsymbol{Y}$ holds. This has a remarkable advantage in controlling multivariate processes where distortion causes affect not only the process mean and variability but also the correlation structure both among the quality characteristics and between these ones and the process variables.

This procedure is easily extendable to $C > 2$ through successively juxtaposing the different rotated conditions in (3.10).

3.2 Analyses based on co-inertia criteria. The above mentioned analyses based on Procrustean rotations refer, in any case, to *correlations-based criteria.* In this section, we discuss some developments in the framework of co–inertia analysis where covariances between the obtained latent variables are maximised rather than correlations. At the same time, some relations are shown that link these methods to the rotations-based ones. The general reference scheme of classical co–inertia analysis is the same as of *CCA* so that a symmetrical relationship is assumed between the variables. Therefore, the data matrices are meant as being *partially paired* as they describe the same set of statistical units by means of two distinct sets of variables. This implies that the two configurations of statistical units belong to different spaces.

Before discussing the non–symmetrical comparative analyses based on the co–inertia criterion, it is worth giving some hints on the links between *PCAR* and the most famous co–inertia based analysis: *PLS* regression.

3.2.1 Comparing PCAR with PLS regression. Nowadays, *PLS* (*Partial Least Squares* or *Projection to Latent Structures*) is a very fashionable algorithmic approach to the analysis of dependence where missing data and multicollinearity problems among explanatory variables exist.

The criterion that is maximized in *PCAR* may be also expressed as the sum of the squared correlations between the response variables and the newly defined linear combinations of the explanatory ones. Actually, by referring just to the k-th component, this may be written as:

$$cor^2(\boldsymbol{X}, \boldsymbol{Y}\boldsymbol{v}_k) = \frac{cov^2(\boldsymbol{X}, \boldsymbol{Y}\boldsymbol{v}_k)}{var(\boldsymbol{X}) \cdot var(\boldsymbol{Y}\boldsymbol{v}_k)}, \tag{3.13}$$

where *cor*, *cov* and *var* stand, respectively, for the correlation, covariance and variance operators. The latter expression enhances how *PCAR* actually aims at maximising the covariance while minimising the variances.

Therefore, it may yield principal components which are highly correlated with the response variables but little explicative of the explanatory ones. This is, however, a problem common to all correlations-based analyses. In order to encompass this limit, co-inertia analysis (Chessel & Mercier (1993)), for symmetrical relationships, and *PLS regression* (Tenenhaus (1998)), for non symmetrical ones, in the previous expression aim at maximising covariances rather than correlations.

In the first step, *PLS regression* performs an inter-battery factor analysis (Tucker (1958)). After that, it works on the residuals of the regressions of $\boldsymbol{X}$ and $\boldsymbol{Y}$ on linear combinations of $\boldsymbol{Y}$, whose coefficients are given by the elements of eigenvectors computed in the first step. Therefore, both *PCAR* and *PLS* yield orthogonal linear combinations of the explanatory variables oriented to the explanation of the response variables structure. However, while *PLS* is based on the criterion of co-inertia, *PCAR* refers to the explained inertia (Scippacercola, Amenta & D'Ambra (1995)) by achieving a structural decomposition of the variability shown by the response variables.

Moreover, *PLS* basically originated from an algorithmic approach to the problem of multicollinearity, while *PCAR* has a more geometrically-oriented foundation which justifies its appeal for interpretation purposes especially on the side of end-users working in several application fields. Both approaches may, in any case, be interpreted in terms of forecasting equations in a multivariate regression-like fashion. Several relationships may be demonstrated (Burnham, Viveros & MacGregor (1996)) between *PCAR*, *PLS*, *CCA* and *Tucker's analysis* in terms of generalisations of metrics and projection operators. However, we will focus our attention on the feasibility of the developed techniques to peculiar structures of data and on the interpretation of the graphical outputs rather than on algebraic methodological extensions.

A PCAR–PLS integrated approach may help in coping with related general questions (e.g. ill-conditioned data, non-linearity, missing data, predictions, decomposition of regression coefficients, double-sided dependence) as well as in giving further insights on graphical interpretation tools (biplots, triplots).

Moreover, the interpretation of PLS as a soft–modeling technique (Tenenhaus (1999)), together with its established links with the latent variable path modeling, may be exploited in terms of both validation of results and a confirmatory approach to non–symmetrical analysis in order to have a broader understanding and a deeper inspection of the data.

3.2.2 Non-symmetrical co-inertia analysis. With respect to the data structure that we are dealing with in this paper, we know that the $\boldsymbol{X}_c$'s are totally paired matrices so that the statistical units all belong to the same space. From the analytical viewpoint, this means that we may search (see Fig. 7) for single axes $\boldsymbol{z}_j, j = 1, \ldots, q$, rather than couples, maximising the covariance of the projections co–ordinates on them.

The criterion to be maximised in Non Symmetrical Co-Inertia Analysis (*NSCOI*, Esposito (1997)) is thus:

$$cov^2(\boldsymbol{P_Y X}_1 \boldsymbol{z}_1, \boldsymbol{P_Y X}_2 \boldsymbol{z}_1) \tag{3.14}$$

which is a non symmetrical version of the criterion proposed by Torre & Chessel (1995) in their co–structure analysis for paired tables.

Namely, in this case, the solution is given by the singular value decomposition of the matrix:

$$\frac{\boldsymbol{X}_2' \boldsymbol{P_Y X}_1 + \boldsymbol{X}_1' \boldsymbol{P_Y X}_2}{2}. \tag{3.15}$$

By considering also the successive orthogonal eigenvectors of the latter matrix, ordered in a decreasing manner with respect to the module of the

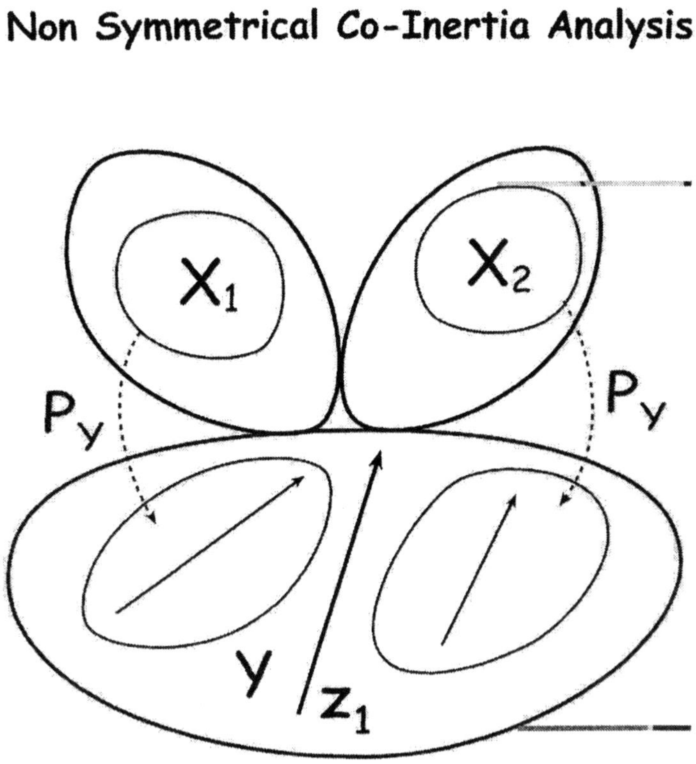

Figure 7. A geometrical insight on *NSCOI*.

associated eigenvalues, it is possible to build factorial planes. Since this result maximises a covariance criterion rather than a correlation one, the analysis is actually a compromise between performing two separate *PCAR*'s for each condition and a global *CCA* on projected variables.

This analysis may be profitable applied in the field of *Quality Sensory Data Analysis* where the prime interest is to compare, on the basis of a common structure, the judgements expressed by different groups of tasters with respect to the chemical-physical features of a product (wine, oil, foods, etc.).

From a statistical point of view, the objective is then to identify a structure common to the different observational occasions and to compare the different statistical units configurations with respect to this structure.

The graphical representations yielded by *NSCOI*, due to the maximisation criterion in (3.14), may not be interpreted in terms of explained variance. Therefore, they are not optimal from the point of view of *PCAR* in expressing the dependence relationship between the response variables and the explanatory ones. However, several empirical trials have shown that the z_j's usually explain a percentage of inertia of the two clouds of points which is always less but not significantly different from the case of performing two separate *PCAR*'s. For this reason, the compromise between explaining less variance for each matrix and gaining information on a *common structure* between them comes out to be advantageous. Moreover, the interpretation of *NSCOI* may be enriched by referring to its links with other analyses as the *trace* of (3.15) is equal to:

$$\frac{In_{BETWEEN} + In_{WITHIN} - In_{DIFFERENCE}}{2}, \tag{3.16}$$

where $In_{BETWEEN}$, In_{WITHIN} and $In_{DIFFERENCE}$ are, respectively, the inertias maximised by the non-symmetrical analyses between the conditions, within each condition and on the difference between the conditions themselves.

Finally, there exists a relationship between *NSCOI* and *RCAR*. In fact, if we performed a *NSCOI* on the matrices $\boldsymbol{P_Y X}_1$ and $\boldsymbol{P_Y X}_2\boldsymbol{R}_{P_Y}$, we would get the same results as performing a *RCAR* on the raw data. Therefore, *NSCOI* is actually a particular case of *RCAR* where the rotation matrix is taken to be the identity matrix. Namely, the Procrustean rotation, which is invariant for any traslation and orthogonal transformation of the $\boldsymbol{X}_c$'s, introduces a linear modelling of data in *RCAR* by imposing the juxtaposition on the same plane of the clouds of points relative to the different observational conditions. For this reason, the comparison is carried out by evaluating eventual differences after rotating. On the other hand, without any modelling, *NSCOI* allows studying a structure which is common to the different conditions and by means of which to compare the relative configurations of statistical units.

As to the graphical representations, a paired representation of the statistical units for the two observational conditions is yielded where eventual gaps between conditions may be detected by means of straight lines. With respect to the variables, only one representation is yielded. However, this latter one is extremely important because, accordingly to the aims of the analysis, it shows the relationships common to both conditions.

3.2.3 Non-symmetrical generalised co-structure analysis. The feasibility of *NSCOI* as well as of *RCAR* is limited to the case where both $C = 2$ and totally paired matrices occurr. Actually, these techniques may be somehow extended to the case of $C > 2$ but not in a straightforward way.

More generally, it may happen that the data structure may reveal C tables $\boldsymbol{X}_c$'s that are just row-wise paired. This is the case when aiming at the building of *non-parametric multivariate control charts* where the whole set of quality characteristics may be split into differently sized groups whose contents are defined according to a specified expert's criterion.

An extension of *NSCOI* to the case of multiple row–wise paired tables is represented by *Non–Symmetrical Generalised Co–Structure Analysis* (*NSGCI*, Esposito & Scepi (1999)). The data structure addressed by *NSGCI* is similar to that of *Generalised Canonical Correlation Analysis* (*GCCA*,

Carroll (1968)) so that it represents an extension of *NSCOI* in the sense of *GCCA* so that its geometrical interpretation is the same as in Fig. 7 but for the extension to more than 2 matrices.

In fact, with respect to the response variables, it comprises C sets of p_c ($c = 1, \ldots, C$) numerical variables observed on the same n statistical units, represented by the C matrices $\boldsymbol{X}_c$ of dimensions (n, p_c). In the following, *NSGCI* is developed in the direction of Multiple Co–Inertia Analysis but in a non–symmetrical context so that it will be shown to be set in the geometrical framework of *PCAR*.

In the core of the analysis, we first search for C block–normalised vectors $\boldsymbol{v}_c^1$'s, one for each subspace $\mathcal{R}^{p_c}$ spanned by the variables in $\boldsymbol{P_Y X}_c$, as well as for one normalised auxiliary variable $\boldsymbol{z}^1$ in the statistical units subspace $\mathcal{R}^n$. After that, vectors $\boldsymbol{v}_c^2$'s and $\boldsymbol{z}^2$ are searched to be orthogonal with, respectively, $\boldsymbol{v}_c^1$'s and $\boldsymbol{z}^1$. The choice of these vectors is based on the maximisation of:

$$\sum_{c=1}^{C} \pi_c \left(\boldsymbol{P_Y X}_c \boldsymbol{v}_c \mid \boldsymbol{z}\right)^2 = \sum_{c=1}^{C} \pi_c var\left(\boldsymbol{P_Y X}_c \boldsymbol{v}_c\right) cor^2\left(\boldsymbol{P_Y X}_c \boldsymbol{v}_c, \boldsymbol{z}\right) \qquad (3.17)$$

where each π_c represents the weight assigned to each $\boldsymbol{P_Y X}_c$. Such a weighting system is necessary in order to take into account the different number of response variables in each $\boldsymbol{X}_c$, i.e. $\pi_c = p_c \,/ \sum_c p_c$, or the different variability of each response variables set, i.e. $\pi_c = \mathrm{var}\, X_c \,/ \sum_c \mathrm{var}\, \boldsymbol{X}_c$.

The equivalence in (3.17) points out that *NSGCI* jointly maximises the criteria of, respectively, *PCAR* (the term $var\left(\boldsymbol{P_Y X}_c \boldsymbol{v}_c\right)$) and *GCCA* (the term $cor^2\left(\boldsymbol{P_Y X}_c \boldsymbol{v}_c, \boldsymbol{z}\right)$). Consequently, it succeeds in performing C separate analyses together with a global one of the projected response variables. As indicated in Chessel & Hanafi (1996), by referring to the Cauchy-Schwartz inequality, it can be easily shown that the majoring quantity for the maximisation problem in (3.17) is defined by:

$$\sum_{c=1}^{C} \pi_c \left\| \boldsymbol{P_Y X}_c' \boldsymbol{z} \right\|^2 = \boldsymbol{z}' \left(\sum_{c=1}^{C} \pi_c \boldsymbol{P_Y X}_c \boldsymbol{X}_c' \boldsymbol{P_Y} \right) \boldsymbol{z}. \qquad (3.18)$$

The matrices $\boldsymbol{X}_c$'s are by rows so that the first order solutions $\boldsymbol{v}_c^1$'s and $\boldsymbol{z}^1$ are derived from a *PCA* performed on the pooled projected matrix $\boldsymbol{P_Y}\left[\boldsymbol{X}_1 | \boldsymbol{X}_2 | \ldots | \ldots | \boldsymbol{X}_C\right]$, that is from a multiple *PCAR*.

Namely, the axes $\boldsymbol{v}_c^1$'s are the p_c-dimensional vectors of the principal axis associated to the highest eigenvalue of a *PCAR* on the pooled C sets $\boldsymbol{X}_c$'s with respect to $\boldsymbol{Y}$, while the variable $\boldsymbol{z}^1$ is given by the relative principal component.

This is a very appealing result because it shows how, starting from a co–inertia criterion, we actually end up with performing a classical *PCAR*.

With respect to the second order solution $\boldsymbol{v}_c^2$'s and $\boldsymbol{z}^2$, we work on the residual matrices $\boldsymbol{E}_c = \boldsymbol{P_Y}\boldsymbol{X}_c - \boldsymbol{P_Y}\boldsymbol{X}_c\boldsymbol{P}^1_{v_c}$ where $\boldsymbol{P}^1_{v_c}$ is the orthogonal projection operator on the subspace spanned by the vector $\boldsymbol{v}_c^1$.

By juxtaposing all $\boldsymbol{E}_c$'s, we define a pooled matrix $\left[\boldsymbol{E}_1 \,|\, \boldsymbol{E}_2 \,|\, \ldots \,|\, \boldsymbol{E}_C\right]$. Finally, the first order solutions of this matrix represent the second order solutions of the problem. Similarly, by iterating the procedure, the generic r-th order solution may be found.

NSGCI provides one system of co–inertia axes where to project the C clouds of statistical units relative to each set of projected response variables. These representations, according with the chosen maximisation criterion, allow a global comparison of C configurations of the statistical units, so that peculiar configurations/behaviours of the units may be inspected.

On the same system, we can display the axes of C separate *PCAR*'s with the objective of representing the inertia structure of each table itself.

More importantly, after a proper normalisation, aiming at enhancing a typology of the units, we may simultaneously project, on a common display, both the C representations of each unit and the relative components of the auxiliary variables so as to visualise both specific and average behaviours.

Thereafter, *star–plots* are drawn which display the C behaviours of each unit with respect to a synthesis of theirs so as to represent the inertia structure of each table itself and comparing it to the global co–inertia structure.

Finally, both explanatory and response variables are represented on the plane spanned by the auxiliary variables in order to show the links between the C sets $\boldsymbol{X}_c$'s as well as the influence of the variables in $\boldsymbol{Y}$ on each of these sets.

4 Applications on real data

In the following we show two applications of the methods presented in this paper. For the sake of brevity and simplicity, the core of the results will be shown. Please refer to the original papers Balbi & Esposito (2000), Esposito & Balbi (1999) and Esposito (1998), Esposito & Scepi (1999) for further details.

4.1 An application from quality sensory data analysis. The data analysed in this first application refer to a typical quality problem in sensory analysis: how to detect differences in judgements on sensory variables (quality characteristics) expressed, on the same objects (foods, very commonly), by different groups of individuals and how to explain these differences in terms of the objects' chemical-physical features (process variables).

In our case, we deal with a North Italian white wine, the *Tocai Friuliano* and refer to 22 different wineries producing this wine (hereinafter labelled as WP). Organoleptic characteristics (i.e. sight, taste, smell, and aftertaste) have been evaluated on a 5 level (0-4) ordinal scale by a panel of wine–tasters (hereinafter labelled as sight1, taste1, smell1, and aft1) and a group of ordi-

Data for the first example

Judgements expressed by experts — Judgements expressed by consumers

WP	*sight1*	*smell1*	*taste1*	*aft1*	WP	*sight2*	*smell2*	*taste2*	*aft2*
e1	3	2	1.5	1	*c1*	2	2	2	2
e2	3	2	2	2	*c2*	2	2	2	3
e3	2.5	1.5	2	1.5	*c3*	2	2	2	1
e4	3	2	2	2	*c4*	2	2	2	1
e5	3	1.5	2	2	*c5*	2	2	2	2
e6	3	2	2	2	*c6*	2	2	2	2
e7	3	2	2	2	*c7*	2	1	1	2
e8	3	2	2	2	*c8*	2	2	2	2
e9	3	2	2	1	*c9*	2	2	1	2
e10	3	2	2	2	*c10*	2	1	3	2
e11	3	2	2	2	*c11*	2	2	2	2
e12	3	2	2	2	*c12*	2	2	2	1
e13	3	2	2	2	*c13*	2	2	2	2
e14	3	2	1	1	*c14*	2	2	3	2
e15	3	2	2	2	*c15*	3	2	2	2
e16	2	1	1	1	*c16*	2	2	2	2
e17	3	1	1	1	*c17*	2	3	2	2
e18	3	2	2	1	*c18*	2	1	1	2
e19	3	2	1	1	*c19*	2	1	0	1
e20	3	1	1	1	*c20*	3	3	2	2
e21	3	1.5	2	1	*c21*	2	1	2	2
e22	3	2	1.5	2	*c22*	2	2	3	1

Chemical–physical features of the wines

WP	*alcol*	*pH*	*sugar*	*metan*	*SOfree*	*absor*	*etace*
1	12.63	3.25	34	5.65	32	2.28	64
2	12.26	3.39	23	4.79	39	10.15	32
3	12.30	3.33	34	4.65	27	10.28	8
4	12.46	3.37	24	4.85	33	2.01	16
5	13.11	3.35	24	6.75	30	2.12	20
6	13.33	3.31	16	5.97	30	1.12	23
7	12.39	3.44	37	5.54	44	2.30	18
8	12.30	3.31	44	5.49	40	1.43	27
9	12.78	3.48	17	7.10	30	3.74	36
10	13.13	3.32	16	4.88	45	3.93	32
11	12.77	3.37	12	5.68	30	3.87	17
12	13.08	3.25	15	6.40	44	4.21	35
13	13.17	3.18	28	6.33	33	2.49	28
14	12.63	3.39	22	5.62	26	8.23	29
15	12.58	3.22	22	6.17	24	1.86	23
16	12.87	3.43	34	6.52	29	1.89	31
17	12.14	3.01	21	3.55	32	2.03	18
18	12.53	3.17	36	4.83	29	4.69	18
19	11.37	2.99	15	5.17	43	5.80	33
20	11.73	3.32	24	6.43	18	1.91	10
21	12.10	3.21	39	6.03	19	6.42	44
22	12.61	3.18	18	4.58	23	1.91	43

nary consumers during a specialised exhibition of wines (hereinafter labelled as sight2, taste2, smell2, and aft2), collected, respectively, in the matrices $\boldsymbol{X}_1$ and $\boldsymbol{X}_2$.

The 22 wines are described by 7 physical-chemical features: alcohol, pH degree, sugar, methanol, free sulphur dioxide, optical absorbency, ethyl acetate (hereinafter labelled as alcol, pH, sugar, metan, SOfree, absor, etace), collected in the matrix $\boldsymbol{Y}$. By taking the experts as our rotation reference, we firstly apply *RCAR*. The first two axes of the analysis explain 87.7% of the total variability. Therefore, the representation on the plane built by the first two axes (see Fig. 8) reveals the main structure of similarity between the experts and ordinary people judgements.

Because of the performed rotation, we are allowed to link the judgements relative to the same sensory variables with straight lines and interpret their directions with respect to the explanatory variables.

The first axis mainly enhances the differences in judging aftertaste. Actually, the evaluation of aftertaste is quite sophisticated and is related to very specific structural characteristics. In fact, with respect to the explanatory variables, the first axis opposes the edible and fragrant alcohol's (ethyl acetate and ethanol) against the bad one (methanol).

The second axis refers to differences in judging smell and taste. With respect to the explanatory variables, it is characterised by the opposition between sweetness (sugar) and acidity (pH). Evidently, consumers appreciate sugar in wine better than experts do. As to the sight, an almost complete matching between judgements occurs as it represents the feature of wine most difficult to differentiate and is highly related to the optical absorbency.

On the same data set, we also perform an *SPCAR* in order to have a more complete understanding of the dependence structure under study. Differently from *RCAR*, *SPCAR* aims at studying the differences in the overall preference structure explained separately by the physical-chemical characteristics of wines and by the remaining ones.

Actually, we first detect, by means of a rotation, the differences between $\boldsymbol{X}_1$ and $\boldsymbol{X}_2$ and then explain these differences in terms of the variables in $\boldsymbol{Y}$. This different way of approaching the data structure is clearly reflected on the different meaning that the axes of the analysis assume as well as in the different interpretation yielded by the patterns of variables on the related factorial planes.

As a matter of fact, the first axis of *SPCAR* (see Fig. 9) is mainly characterised by the differences in judging the sight, which is the easiest feature to be judged.

On the other side, the second axis is mainly characterised by the difference in judging smell and taste, which are more difficult to be perceived and evaluated.

With respect to the explanatory variables, the difference in judging sight is still naturally mostly explained by the optical absorbency of the wines. The

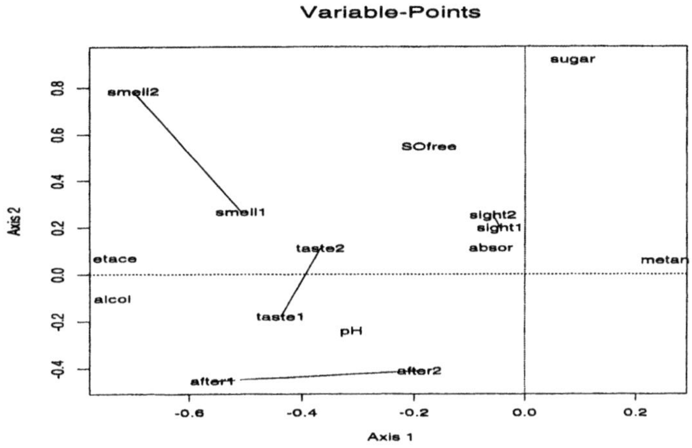

Figure 8. Variable points on the *RCAR* factorial plane.

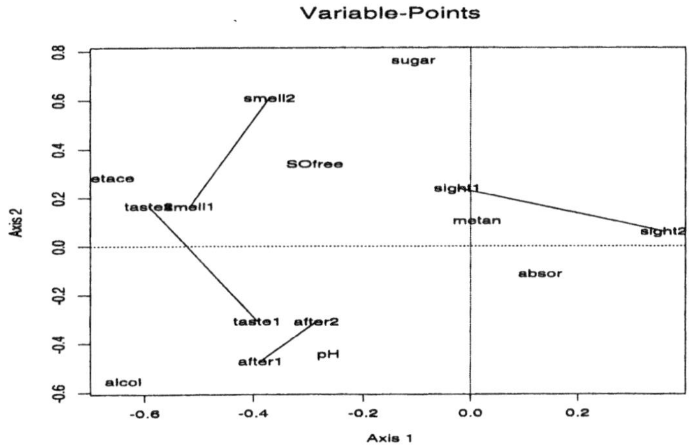

Figure 9. Variable points on the *SPCAR* factorial plane.

remaining chemical and physical characteristics contribute at different levels to the explanation of the differences in judging smell, taste and aftertaste. In particular, the degree of acidity (pH) seems to characterise the evaluation in terms of aftertaste similarly in the two groups of judges.

On the contrary, the presence of ethanol, the edible alcohol, is better perceived and correctly evaluated by experts in terms of smell and taste, while it seems that consumers are not able to consider this index of quality when formulating their judgements.

4.2 An application from water quality control. In this second application, we show the feasibility of *NSGCI* for visualising quality control multivariate problems, see Lauro, Balbi & Scepi (1997) for a review.

In particular, *NSGCI* helps in identifying both a control typology and the causes of eventual out of controls. The latter is a very important topic in the analysis of a multivariate control process in which it is very difficult to detect the variables actually determining out–of–control situations.

In the following example, we refer to the well-known data in Verneaux (1973) relative to a study in hydrobiology. The aim is to investigate the water quality of the French River Doubs taking into account both its ichtyologic fauna and its chemical, physical and biological components.

Therefore, two data matrices are available. The first one crosses 35 sites observed along the river with 11 variables (explanatory ones) referring to morphological features of the river (distance from the source, altitude, slope, minimum mean flow) and to water quality indicators (pH, total hardness, phosphates, nitrates, ammoniacal azote, oxygen, Biological Request of Oxygen). The second one crosses the same sites with an index of presence of 27 species (response variables) of fishes in each site.

These data have been analysed by Chessel, Lebreton & Yoccoz (1987) in the framework of *Canonical Correspondence Analysis.* They result in partitioning the 27 species into 8 groups. In our analysis, we take into account their partition. However, as there are two groups where each is formed by just one species, we first aggregate these groups to the nearest groups on the factorial plane therein obtained. We thus finally consider a partition in 6 groups.

By applying our technique to the same data, our added value consists in identifying a common structure of the sites so as to detect *anomalous situations.*

For the sake of brevity, the results hereinafter shown relate just to group 3 which is the one with the highest overall discrimination power.

Fig. 10 refers to the representation of sites in the species space in group 3. The axes displayed as full lines are the ones obtained by performing a single *PCAR* on group 3 and, very importantly, they almost match with the principal axes of *GCCA* (dotted lines).

The first axis (the horizontal one in Fig. 10) discriminates between sites nearest to the source (on the left with low numbers) and those far from the source itself (on the right with high numbers). In order to understand the meaning of the axes, we must interpret the display in Fig. 11 that represents both explanatory variables and the 9 species (long words in lower-case) in Group 3.

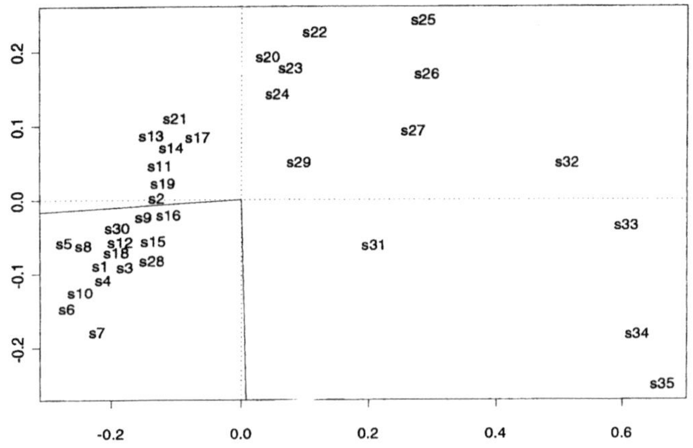

Figure 10. Sites representation using PCAR and GCCA.

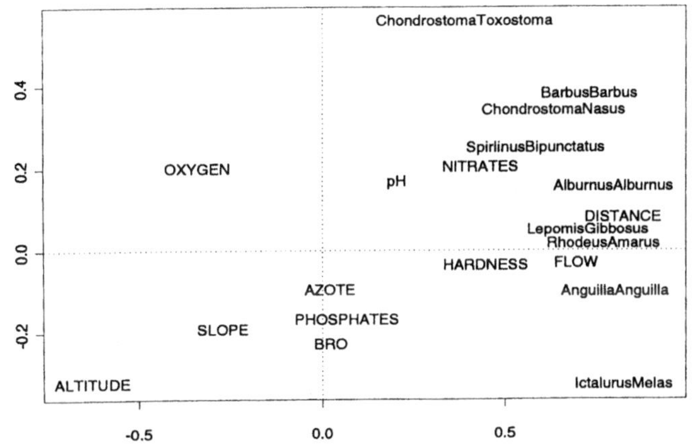

Figure 11. Variables representation.

The first axis (see Fig. 11) opposes altitude to distance, as they are logically inversely related, and discriminates among the other morphological features (slope, hardness, and flow) relative to the position of the observed sites. The grouping of the species on the right hand side of this axis implies that their presence is very much influenced by the distance from the source. They are actually species usually living very far from the source.

The second axis, instead, discriminates among the chemical components of the water thus representing a synthesis of its quality. In particular, species suited to live in waters rich of azote, phosphates and BRO oppose to those who prefer oxygen, pH and nitrates.

In Fig. 12, the star-plots of the behaviour of just the *anomalous sites* in the different groups are represented.

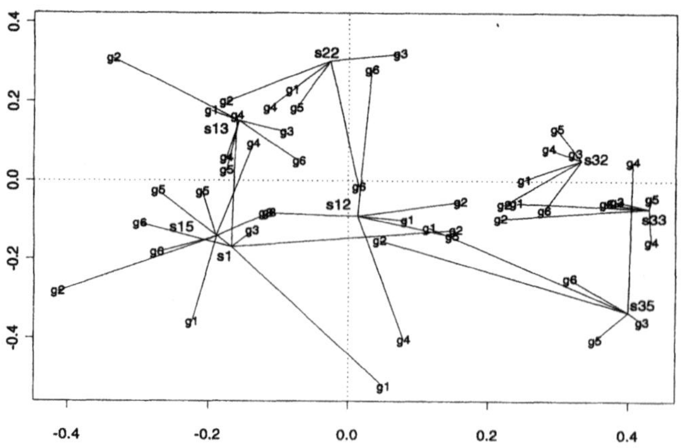

Figure 12. Synthesis representation of sites.

The centre of each star relates to the synthesis representation while its edges relate to the behaviours in the different groups. Moreover, by 'anomalous' sites we mean those who do not strictly conform themselves to a common behaviour around the origin as most of the sites do.

The sites very far from the source have synthesis configurations that, with respect to the shape, are very similar to each other and altogether form a configuration substantially different from the other sites.

However, each of them has a very different variability, e.g. $s32$ has a low variability since its representations are very close to each other, while $s35$ has a very high variability due to its different behaviours in groups 1, 2 and 4.

On the other side of the first axis, it is worth noting, for example, that $s1$ has the highest variability among all sites, and its synthesis may be considered quite anomalous due to its behaviour in group 1.

5 Concluding remarks and perspectives of research

In the paper we have shown that a sort of combination between *PCAR* and *Procrustean rotations* helps in identifying, on the variables side, structural changes in the dependence structure and, on the statistical units side, anomalous situations with respect to a common behaviour.

Because of their easily interpretable graphical properties, it is natural to aim at dynamically extending the mentioned strategies to monitoring observations collected sequentially through time. Changes in the pattern of points in the statistical units space as well as in the lengths and directions of the variables vectors would provide with rich information as to the evolutions of variables means, variances and correlation structures.

Graphically, performing the analysis and updating the plots each time a new observation comes in may help but it is essential to keep track of the previous configuration of paired variables so to be able to dynamically interpret the results of rotations.

Furthermore, it must be enhanced that the dynamic approach would also let the explanatory variables vary for each new observation. This is essential for detecting which variable is responsible of a strange behaviour.

The time dimension has always represented a challenge for data analysts. In fact, though easy to be considered from a technical point of view, it often lacks of a proper interpretation.

Geometrically–based techniques usually consider this dimension only implicitly, thus interpreting the ordinal feature of time a posteriori on the graphical displays regardless of its being a real variable that should be taken into account in the core of the analysis. This is a matter that still needs much discussion and work.

After putting so much emphasis on a geometrical approach, we feel like stressing that an inferential approach needs, however, not to be neglected and may still represent a fertile research field in *TQM*.

Two different approaches to *TQM* have been pointed out in the present paper and the other ones included in this book. The first approach relates to classical *hypothesis testing* where all data are used for testing but strong assumptions are needed and graphical interpretation is missing. The second one relates to *dimension reduction* where relationships among variables are displayed on factorial planes, no distributional assumptions are needed but some information is lost due to lower-dimensional approximations.

We truly believe in the need of integrating both approaches in order to have a broader understanding and a deeper inspection of the quality data structure at hand. *Dynamic biplot* (*DB*, Sparks, Adolphson & Phatak (1997)) is an attempt in the direction of integrating the two approaches. It is a tool, for representing both variables and statistical units, based on the so-called GH–factorisation of the interest variables (the quality characteristics in a *TQM* environment). The dynamic peculiarity of this technique consists in continuously re–estimating the GH–factors and consequently re–plotting the graphical displays each time a new observation comes in. Moreover, it refers to multivariate tests for detecting the changes from the in–control situation (such as shifts in the mean both global and local, changes in the covariance structure, changes in the fraction of explained variation), to univariate tests for determining the quality variables responsible for the alert signals and, finally, to ad–hoc developed graphical tools for visualising the results. However, in its present form, dynamic biplot is actually unaware of any interaction between the quality characteristics and the process variables or any external information that may help in interpreting the out–of–control signals and, even more importantly, to identify any needed correcting action. For this reason, a dynamic extension of *PCAR* in the sense of *DB* may be capable of coping with many of the problems mentioned throughout the paper thus providing the end–users with easily interpretable graphical displays and problem–oriented analytical tools.

When the observed variables are related to the performance of a product/process in time, it may be too restrictive and a waste of information to summarise these temporal observations by a mean or any other synthetic index. In this direction, *Symbolic Data Analysis* (*SDA*) may be a suitable approach (Bock & Diday (2000), Lauro & Palumbo (2000), Lauro, Verde & Palumbo (2000)).

SDA is actually based on the idea that each cell of a data matrix should contain more complex data than a mono-valued one. A range, a multi–valued function or a random variable represented by a probability distribution should better describe the behaviour over time of the observed statistical units in relation to specific attributes.

The newly–defined statistical units are known as symbolic objects and the analysis is carried out on these complex statistical units. This may be a suitable approach to typical problems within *TQM* where the variables related to the performance of a product/process in time may be better represented by a probability distribution than by a mean. In this respect, symbolic data analysis may play an important role.

Taking into account *off–line* information, discovered in a *retrospective* phase of the analysis or given a priori, is essential when applying *on–line* control procedures. In fact, such a knowledge leads to quicker and more effective decisions in a total quality perspective.

References

Balbi S. (1992). *On Stability in non–symmetrical correspondence analysis using bootstrap.* Statistica Applicata **4**, 543 – 552.

Balbi S. and Esposito V. (2000). *Rotated Canonical Analysis onto a Reference Subspace.* Computational Statistics and Data Analysis **32**, 395 – 410.

Bock H. H. and Diday E. (2000). *Analysis of Symbolic Data.* Springer–Verlag, Berlin.

Burnham A. J., Viveros R. and MacGregor J. F. (1996). *Frameworks for latent variable multivariate regression.* Journal of Chemometrics **10**, 31 – 45.

Carrol J. D. (1968). *A generalization of canonical correlation analysis to three or more sets of variables.* In: Proceedings of the 76th Convention of the American Psychological Association **3**, 227 – 228.

Chessel D. and Hanafi M. (1996). *Analyse de la co–inertie de K nuages de points.* Revue de Statistique Appliquée **44**, 35 – 60.

Chessel D., Lebreton J. D. and Yoccoz N. (1987). *Propriétés de l'analyse canonique des correspondances: une illustration en hydrobiologie.* Revue de Statistique Appliquée **35**, 55 – 72.

Chessel D. and Mercier P. (1993). *Couplage de triplets statistiques et liaisons espèces-environment.* In: Biométrie et Environment, Lebreton J. D. and Asselain B. (eds.), Masson, Paris, 15 – 44.

D'Ambra L. and Lauro C. (1982). *Analisi in Componenti Principali in rapporto ad un sottospazio di Riferimento.* Rivista di Statistica Applicata **15**, 51 – 67.

Esposito V. (1997). *A non-symmetrical comparative analysis with stratified observations (in Italian).* In: Proceedings of the Italian Statistical Society Conference on Statistics for Enterprises, Tirrenia Stampatori, Torino, **2**, 343 – 350.

Esposito V. (1998). *Deterministic and Probabilistic Models for Symmetrical and Non-Symmetrical Principal Component Analysis.* Metron, International Journal of Statistics **56**, 139 – 154.

Esposito V. and Balbi S. (1999). *Representing Gaps in Sensory Perceptions by a Simultaneous Principal Component Analysis onto a Reference Subspace.* Applied Stochastic Models in Business and Industry, John Wiley & Sons, Chichester, **15**, 301 – 309.

Esposito V. and Scepi G. (1999). *A Non-Symmetrical Generalised Co-Structure Analysis for Inspecting Quality Control Data.* In: Classification and Data Analysis. Theory and Application, Vichi M. and Opitz O. (eds.), Springer-Verlag, 179 – 186.

Golub G. H. and Van Loan C. F. (1989). *Matrix Computations.* The Johns Hopkins University Press, Baltimore.

Gower J. C. (1971). *Statistical methods of comparing different multivariate analyses of the same data.* In: Mathematics in the Archaeological and Historical Sciences, Hudson F. R., Kendall D. G. and Tăutu P. (eds.), University Press, Edinburgh, 138 – 149.

Gower J. C. (1975). *Generalized Procrustean Analysis.* Psychometrika **40** 33 – 51.

Green B. F. (1952). *The orthogonal approximation of an oblique structure in factor analysis.* Psychometrika **17** 429 – 440.

Hotelling H. (1936). *Relations between two sets of variables.* Biometrika **28** 321 – 377.

Lafosse R. (1986). *Metriques et Analyse Factorielle Comparative de Deux Tableaux ou Plus.* Statistique et Analyse des Données **11** 51 – 75.

Lafosse R. (1989). *Ressemblance et difference entre deux tableaux totalement appariés.* Statistique et Analyse des Données **14** 1 – 24.

Lauro C., Balbi S. and Scepi G. (1997). *Multidimensional Data Analysis for Customer Satisfaction in Services.* (in Italian) In: Proceedings of the Italian Statistical Society Conference on Statistics for Enterprises, Tirrenia Stampatori, Torino, **1**, 567 – 581.

Lauro C. and D'Ambra L. (1984). *L'analyse non symetrique des correspondances.* In: Data Analysis and Informatics, Diday E. et al. (eds.), North-Holland, Amsterdam, **3**, 433 – 446.

Lauro C., Giordano G. and Verde R. (1998). *A multidimensional approach to Conjoint Analysis.* Applied Stochastic Models and Data Analysis, John Wiley & Sons, Chichester, **14**, 265 – 274.

Lauro C. and Palumbo F. (2000). *Principal Component Analysis of interval data: a symbolic data analysis approach.* Computational Statistics **15**, 73 – 87.

Lauro C. and Siciliano R. (1989). *Exploratory methods and modelling for contingency tables analysis: an integrated approach.* Statistica Applicata **1**, 5 – 32.

Lauro C. and Siciliano R. (2000). *Analyse non symétrique des correspondances pour des tables de contingences.* In: L'Analyse des correspondances et les techniques connexes, Moreau J. et al. (eds.), Springer-Verlag, 183 – 210.

Lauro C., Verde R. and Palumbo F. (2000). *Factorial Discriminant Analysis on Symbolic Objects.* In: Analysis of Symbolic Data, Bock H.H. et al. (eds.), Springer-Verlag, 212 – 233.

Mardia K. V., Kent J. T. and Bibby J. M. (1979). *Multivariate Analysis.* Academic Press, London.

Pearson K. (1901). *On line and planes of closest fit to systems of points in space.* Phil.Mag. **2**, 559 – 572.

Rao C. R. (1964). *The use and interpretation of principal component analysis in applied research.* Sankhya **26** 329 – 358.

Scippacercola S., Amenta P. and D'Ambra L. (1995). *Il Partial Least Squares e l'Analisi in Componenti Principali rispetto ad un sottospazio di Riferimento. Un'applicazione per la rilevazione di inquinamento di ambienti fluviali.* Boll. Soc. Natur. Napoli, **43**, 7 – 63.

Siciliano R., Mooijaart A. and Van der Heijden P. G. M. (1993). *A probabilistic model for non–symmetric correspondence analysis and prediction in contingency tables.* Journal of Italian Statistical Society **1**, 85 – 106.

Sparks R.S., Adolphson A. and Phatak A. (1997). *Multivariate Process Monitoring Using the Dynamic Biplot.* International Statistical Review, **65**, 325 – 349.

Stewart D. K. and Love W. A. (1968). *A general canonical correlation index.* Psychological Bulletin **70**, 160 – 163.

Tenenhaus M. (1998). *La régression PLS.* Editions Technip, Paris.

Tenenhaus M. (1999). *L'approche PLS.* Revue de Statistique Appliquée **47**, 5 – 40.

Ter Braak C. J. F. (1990). *Interpreting canonical correlation analysis through biplots of structure correlations and weights.* Psychometrika **55**, 519 – 531.

Torre F. and Chessel D. (1995). *Co–structure de deux tableaux totalement appariés.* Revue de Statistique Appliquée **43**, 109 – 121.

Tucker L. R. (1958). *An inter-battery method of factor analysis.* Psychometrika **23**.

Van den Wollenberg A. L. (1977). *Redundancy analysis: an alternative for canonical analysis.* Psychometrika **42**, 207 – 219.

Verneaux A. L. (1973). *Cours d'eau de Franche–Comté (Massif du Jura). Recherches écologiques sur le réseau hydrographique du Doubs.* Essai de biotypologie. Thèse d'état, Besançon.

THE NON-SYMMETRICAL ANALYSIS OF MULTIATTRIBUTE PREFERENCE DATA

Giuseppe Giordano

University of Naples Federico II, Faculty of Economics, Department of Mathematics and Statistics, Via Cintia – Complesso Monte Sant'Angelo, I – 80 126 Naples, Italy; giordano@dms.unina.it

Abstract: Multiattribute preference data refer to judgements collected with respect to a set of stimuli described by relevant attributes. The elements involved in the analysis are a group of judges, a collection of stimuli and a set of attributes characterising the stimuli. In particular, this data structure is used in order to apply the full profile approach to Conjoint Analysis (Green and Rao (1971), Green and Srinivasan (1978), Green and Srinivasan (1978), Green and Srinivasan (1990)). A multidimensional approach to Conjoint Analysis by means of non-symmetrical factorial analysis will be discussed. In this framework, we show how to deal with the complexity of multiattribute data structure. The aim is to characterise, on a two-dimensional subspace, the relationships among the judges, the attribute levels and the stimuli. Each factorial axis is interpreted as a synthesis of the original judgements. For example, the results obtained can be useful in marketing for describing consumer preference towards new product features. A case study showing the results of Conjoint Analysis, carried out both at individual and aggregated level, allows to compare the different approaches.

1. Introduction to multiattribute preference data

Multiattribute preference data refer to preference judgements expressed with respect to a set of stimuli described by a short number of relevant attributes. Various objects are involved in this kind of data: a collection of stimuli, a group of judges and a set of attributes characterising the stimuli. Judges and attributes are statistical variables, stimuli are statistical units.

The stimuli can be related to real goods, services or to complex statistical observations (for example: business functions, commercial brands, etc.). These units are "decomposable", i.e. they can be described in terms of their simple parts. The attributes can be either set as nominal variables, or as continuous variables. In the following we deal with qualitative attributes. The description of the stimuli is included in a design matrix, where K levels of p attributes are retained in K dummy variables.

2000 *Mathematics Subject Classification.* 62-01, 62G09, 62H25, 62K15, 62P30, 62F40.

Key words and phrases. Conjoint analysis, design of experiments, preference scenario, principal component analysis.

Acknowledgement. The work was partially supported by the MURST (Italian Ministry for University and Scientific and Technological Research) 1999 – 2000 grant on *Multivariate Analyses for Total Quality: Methodologies, Computational Aspects and Applications.*

The required number of stimuli (observations) can be determined, for example, by using the ***design of experiments*** (DoE) theory and it is often based on a fraction of the whole number of stimuli obtained as a full combination of attributes and levels. Various schemes of fractional factorial designs are possible. In particular *orthogonal arrays*, suggested by Addelman (1962), represent the most parsimonious designs allowing the unconfounded estimation of all main effects, cf. Green (1974).

Let N be the number of stimuli submitted to judges. The N different stimuli will be defined as a combination of feasible alternatives of the attributes. They can be conjointly described with graphic designs, prototypes, etc.

The experiment we are interested in consists of submitting the descriptions of the stimuli to a judge and collecting his preferences. The experiment is conducted on different respondents according to a sampling design where we assume that the judges have to be representatives of a larger population. Otherwise, if the judges are opinion leaders we may be interested in their own perceptions.

Hereinafter, J is the number of preference judgements to be collected. Thus, we denote with $\boldsymbol{Y}$ the $(N \times J)$ preference data matrix and, with $\boldsymbol{Y}_j$ the column vector associated to the j-th preference response. These preferences can be either expressed as interval scaled rating data or, less restrictively, as ordinal scaled ranking data. The structure of the multiattribute preference data is represented as follows

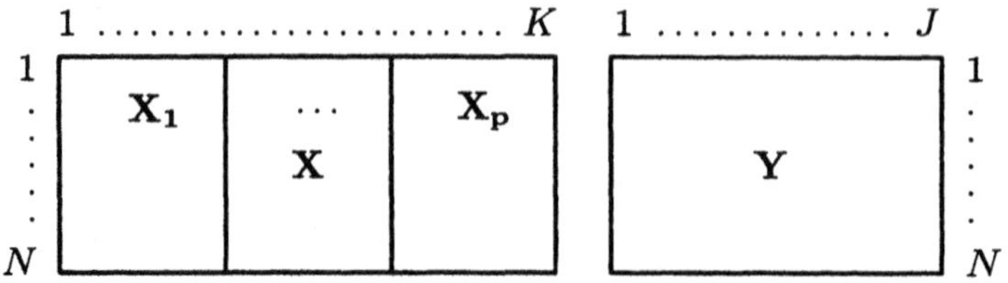

Each preference judgement $\boldsymbol{Y}_j$ represents a response variable allowing to compute the effects of the attribute levels $\boldsymbol{X}_k$ according to a compositional model. Similarly to a common experimental design, one may account for main effects, interactions and linear or higher order effects. The matrix $\boldsymbol{X}$ will be coded and shaped accordingly. As mentioned above, dealing with main effects, we assume that for each attribute one level is dropped (e.g. the last level) and, therefore, subtracted from the other columns. The result will be a full-rank design matrix of dimension $N \times (K - p)$.

2. Individual models and conjoint analysis

The data structure considered above can be applied to various fields of research related to the DoE, such as *Integrated Information Theory* by Louviere (1988) or *Conjoint Measurements* by Green and Rao (1971). In particular, we are interested in using the multiattribute preference data in order to perform a *Conjoint Analysis* (CA) following Green and Srinivasan (1978), Green and Srinivasan (1978) and Green and Srinivasan (1990).

Since its introduction CA has played an important role in the product positioning and new product development strategies. In this context, we can assume that the N stimuli refer to a new concept of a product and the attributes are its different features.

It is a common assumption that all J respondents share the same preference model. The current practice of CA has relaxed many assumptions of the psychometric tradition and many more simplifications are involved in a conjoint experiment. For instance, the assumed model is often the additive linear one and any potential meaningful interactions among levels are disregarded. Furthermore, it is also supposed that the collected dependent variables are interval scaled – e.g. continuous values from 1 to 100 – so that the estimation algorithm adopted can be the *Ordinary Least Squares* for multiple regression. Conversely, ordinal response variables lead to the use of non-linear estimating methods as is *monotone regression* of Tenenhaus (1988), or *Monanova* by Kruskal and Carmone (1969).

The $K-p$ dummy variables are used as predictors in the regression procedure. The response variable is the centred rating score expressed by each respondent. The least squares method applied on the data provides the estimates of the main effects for the utility model.

Since there are J response variables, one needs to perform J separate regressions, one for each respondent in the task.

The CA model for the i-th respondent is

$$\boldsymbol{Y}_i = b_{i1}\boldsymbol{X}_1 + \cdots + b_{ik}\boldsymbol{X}_k + \cdots + b_{iK}\boldsymbol{X}_K + \boldsymbol{E}_i, \qquad i = 1, \ldots, J, \qquad (2.1)$$

where $\boldsymbol{E}_i$ is the vector of the individual error terms. In the CA framework, the regression coefficient b_{ik} is defined as the individual *"part-worth coefficient"* for the generic k-th level. The model (2.1) allows to explain how the different attribute levels affect the development of the global preference. In *DoE* terminology, the part-worth coefficient is the effect of the level $\boldsymbol{X}_k$ on the individual response $\boldsymbol{Y}$.

The model (2.1) also explains how important each attribute is in the stated overall preference. The relative importance of each attribute is determined according to the range spanned by the set of coefficients. It is actually computed by dividing the range of coefficients, for each attribute, by the sum of all ranges settled on the attributes, and multiplying by 100.

A typical output of CA is extensive and contains useful information on the individual choice behaviour. The part-worth coefficients allow predicting the ideal product combination through a mere substitution in the compositive model. Furthermore, some new products can be *simulated* and some *"hold-out"* stimuli can be used to validate the estimated model. The "simulation" consists in defining a combination of attribute levels and estimating ex-post the derived total utility for that stimulus. The "hold-out" is a stimulus rated by a respondent but not used in the estimation procedure, so that the preference value estimated by the model can be compared with the real one given by the respondent.

The following example illustrates many steps involved in a classical study of CA.

Example (part 1) - conjoint analysis: tire pressure[1]
The following four attributes are related to purchasing automobile tires:

Expected Tread Mileage:	40,000	60,000	80,000
Brand Name:	TireMax	GoodTreads	RollsAhead
Purchase Price:	$45	$60	$75
Installation Cost:	$0	$7.5	—

Table 1. Attributes and levels.

Seven customers ranked (in order of preference) 18 out of 54 possible combinations of attribute levels. The goal of the analysis is to determine the importance of each attribute to the stated consumer preferences for a potential tire purchase.

The CA requires preference and attribute variables. Each row of the data set (see Table 2) corresponds to one of the evaluated products. Each column corresponds to either a product attribute level or a preference score. In this example, the preference variables are the ranks from the seven respondents, and the attribute variables are the four factors.

At first, we need to take into account how the values of each preference variable have to be handled. In the TIRES data set, a ranking 1 by a respondent indicates the most preferred stimulus, while a ranking 18 indicates the least preferred. Since the smallest ranking value corresponds to the preferred combination of attribute values, it is usual to *reflect* the preference variable (i.e. $y_{j(reflected)} = N - y_j + 1$).

In this example we should use the *nonmetric approach* to CA because we deal with ordinal preference responses. The nonmetric analysis allows a monotone transformation of the response variables, so that the estimated rank order violates the original as little as possible, Kruskal (1965). On the other hand, the metric CA requires the estimation of fewer parameters than the nonmetric one and is the most commonly used method. Therefore, in this application the metric approach is carried out too.

Consequently, we must define a measurement level for each attribute variable. In this example, all four attributes are regarded as qualitative rather than quantitative factors so that we will estimate as many part-worth coefficients as the number of attribute levels.

---

[1]This example has been adapted from the original one reported on the web site of the SAS Institute Inc.: `http://www.sas.com/`

ID	X1 Brand	X2 Install	X3 Price	X4 Mileage	Y1	Y2	Y3	Y4	Y5	Y6	Y7
1	TireMax	7.5	45	40,000	17	17	7	14	13	15	14
2	TireMax	0	75	40,000	18	15	17	17	17	18	17
3	TireMax	0	60	60,000	15	10	9	9	9	8	1
4	TireMax	7.5	75	60,000	16	11	15	12	14	16	6
5	TireMax	0	45	80,000	13	3	1	1	2	3	9
6	TireMax	7.5	60	80,000	14	4	8	4	5	9	8
7	GoodTreads	0	60	40,000	11	12	13	15	16	12	15
8	GoodTreads	7.5	75	40,000	12	14	18	18	18	17	18
9	GoodTreads	0	45	60,000	9	5	4	7	4	4	3
10	GoodTreads	7.5	60	60,000	10	7	10	10	11	6	2
11	GoodTreads	7.5	45	80,000	7	2	2	2	3	2	10
12	GoodTreads	0	75	80,000	8	6	12	5	6	7	11
13	RollsAhead	0	45	40,000	5	16	6	13	12	14	13
14	RollsAhead	7.5	60	40,000	6	13	11	16	15	13	16
15	RollsAhead	7.5	45	60,000	3	8	5	8	7	5	4
16	RollsAhead	0	75	60,000	4	9	14	11	10	11	5
17	RollsAhead	0	60	80,000	1	1	3	3	1	1	7
18	RollsAhead	7.5	75	80,000	2	18	16	6	8	10	12

Table 2. The TIRES data set

In order to obtain an aggregate model, the preference variables $\boldsymbol{Y}_j$ are often summarised by the unique response variable computed as the average of the whole set of variables. Table 3 displays a typical output of utility coefficients – for each level of the attributes – estimated for the average of the original seven preference responses. Furthermore, a box-and-whisker plot in Figure 1 shows the relative importance of each attribute among the seven respondents. In this example, MILEAGE is the most important attribute, with an average relative importance of 49%.

In the Figure 2 the BRAND utilities indicate that one respondent clearly prefers RollsAhead because that respondent's line shows a peak at RollsAhead, indicating high utility (preference). The other respondents only mildly prefer one brand to another. Indeed, the Table 4 indicates that BRAND is the most important attribute for the first respondent, and that RollsAhead is the most preferred brand, because it has the highest utility coefficient value. Thus, the first respondent preferred an 80,000-mile, $45, RollsAhead with no installation charge.

Although the standard ANOVA assumptions are not met in CA, some traditional statistics are displayed anyway. A low value of R-square may indicate either that there is an error in the data for that respondent or that the subject inconsistently rated the choices. Since only 18 of the 54 possible tire purchase combinations were presented to the respondents, we may want to predict the expected global preference of simulated combinations excluded from the set, e.g., in Table 4, the stimulus with the largest utility is the 80,000-mile, $45, RollsAhead with no installation charge.

The TRANSREG Procedure Hypothesis Tests for LINEAR(AVERAGE)

Source	DF	Sum of Squares	Mean Square	F Value	p
Model	7	304.892290	43.556041	54.615	0.0001
Error	10	7.975057	0.797506		
Total	17	312.867347			

Root MSE	0.8930317	R-square	0.97451
Dep Mean	0	Adj R-sq	0.95667
CV	.		

Utilities Table

Label	Utility	Importance (% Utility Range)
Intercept	0.000	
BRAND GoodTreads	0.357	12.986
BRAND RollsAhead	0.880	
BRAND TireMax	-1.238	
MILEAGE 40000	-4.642	49.319
MILEAGE 60000	1.238	
MILEAGE 80000	3.404	
PRICE 45	2.071	29.621
PRICE 60	0.690	
PRICE 75	-2.761	
INSTALL 0	0.658	8.074
INSTALL 7.5	-0.658	

Table 3. Anova table and utility coefficients for the average preference.

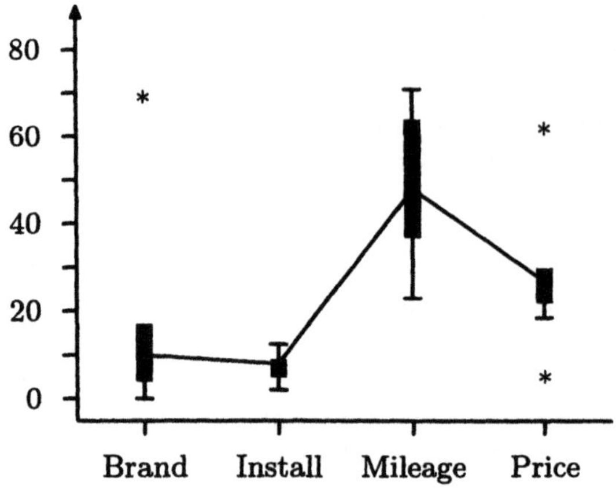

Figure 1. Relative importance of attributes

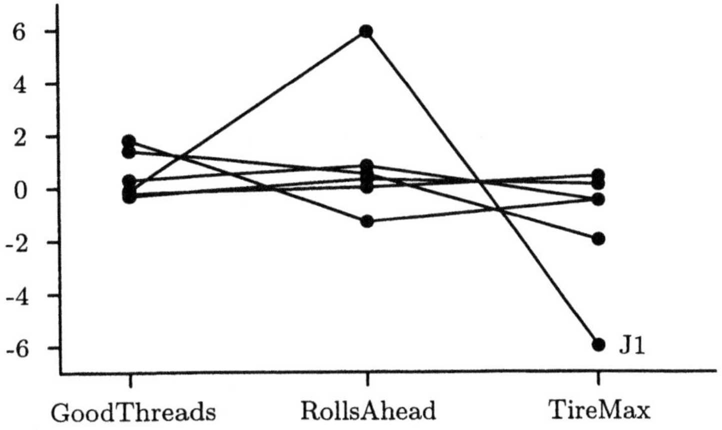

Figure 2. Utilities for brand values.

The TRANSREG Procedure Hypothesis Tests for LINEAR(J1)

Source	DF	Sum of Squares	Mean Square	F Value	p
Model	7	483.500000	69.071429	690.714	0.0001
Error	10	1.000000	0.100000		
Total	17	484.500000			

Root MSE	0.3162278	R-square	0.99794
Dep Mean	0	Adj R-sq	0.99649
CV	.		

Utilities Table

Label	Utility	Importance (% Utility Range)
Intercept	0.000	
BRAND GoodTrea	0.000	69.231
BRAND RollsAhe	6.000	
BRAND TireMax	-6.000	
MILEAGE 40000	-2.000	23.077
MILEAGE 60000	0.000	
MILEAGE 80000	2.000	
PRICE 45	0.500	5.769
PRICE 60	-0.000	
PRICE 75	-0.500	
INSTALL 0	0.166	1.923
INSTALL 7.5	-0.166	

Table 4. Anova table and utility coefficients for Judge1.

3. Aggregating models: The non-symmetrical factorial approach

The previous example has provided a first look at CA. In the following, some drawbacks will be highlighted and new interpretative tools will be discussed in the framework of multidimensional exploratory data analysis following Giordano (1997). In particular, we discuss an explorative strategy of data analysis which allows to take into account the peculiarity of the multiattribute data structure.

In the marketing field it may be of interest to describe whether a particular market may be segmented according to some preferences or to some consumer behaviour similarities. This is also the case for *Customer Satisfaction* requirements, wishing to define the ideal product for a given consumer segment. A set of respondents (consumers) can be defined homogeneous either according to some, a priori known, demographic attributes or, ex post, by investigating the similarity of preference responses. This last goal can be achieved by analysing the part-worth coefficients and looking for similar patterns of utilities coefficients.

For instance, we consider the non-symmetrical role played by the two sets of variables (preferences and attributes) involved in multiattribute preference data. Note that, in multidimensional data analysis, *non-symmetry* refers to the different role played by two or more set of variables when we observe a particular phenomenon. In our context, we highlight the dependence relation between the set of the J preference response variables and the set of the $K-p$ attribute levels described in the design matrix.

For our purpose, the model (2.1) will be restated in matrix notation. Thus, the CA model is written as the following multivariate multiple regression model

$$\boldsymbol{Y} = \boldsymbol{X}\boldsymbol{B} + \boldsymbol{E}, \qquad (3.1)$$

where $\boldsymbol{Y}$ is $N \times J$ matrix of preferences, $\boldsymbol{X}$ is the $N \times (K-p)$ full rank design matrix, $\boldsymbol{B}$ is the $(K-p) \times J$ matrix of individual part-worth coefficients and $\boldsymbol{E}$ is the $(N \times J)$ matrix of error terms for the set of J regressions.

Indeed, the simultaneous estimation of the elements of the coefficient matrix $\boldsymbol{B}$ yields the same results as a set of J separate multiple regressions, since the relations within the multiple responses are not involved in the least squares estimation method.

Different techniques have been proposed in order to take into account the dimension reduction aspect of multivariate regression, particularly when they derive from the interrelationships among the multiple responses. Among others, we mention the *Reduced-Rank Regression Model* (Anderson (1951), Izenman (1975)), the *Principal Component of Instrumental Variables* (Rao (1964)), the *Simultaneous Linear Prediction Modelling* (Fortier (1966)), the *Redundancy Analysis* (van den Wollenberg (1977)) and the *Principal Component Analysis on a Reference Subspace* (D'Ambra and Lauro (1982)).

The peculiarity of all these techniques is the connection which arises between the regression coefficients and the descriptive tools of principal components and canonical variates. This is just the reason why we can think about a suitable aggregation of preference responses defined in a common factorial subspace.

The optimal synthesis is thus achieved through a factorial technique; in particular, we refer to *Principal Component Analysis on a Reference Subspace* (PCAR). This technique allows to summarise the multivariate set of preference response variables by performing a *Principal Component Analysis* (PCA) of the matrix stated in (3.2) to get

$$\widehat{\boldsymbol{Y}} \equiv \boldsymbol{X}\widehat{\boldsymbol{B}} = \boldsymbol{X}(\boldsymbol{X}'\boldsymbol{X})^{-1}\boldsymbol{X}'\boldsymbol{Y}. \tag{3.2}$$

In order to preserve the magnitude of part-worth coefficients, the normalising factor $1/N$, appropriate in PCA, will not be applied here.

In this way, the individual part-worth coefficients will be aggregated by means of a suitable weighting system (the principal component coefficients) reflecting the judges' preference heterogeneity.

It is worth noting, by comparing expressions (3.1) and (3.2), that the criterion optimised with PCAR (i.e. variance accounted by attribute levels) is fully consistent with CA. The data structure is the same as in CA and the estimated part-worth coefficients are just related to the PCAR eigen-equation as it will be clearer in the following.

In order to enhance the CA results and their interpretation, the multidimensional factorial approach considers the representation of the relationships, existing between attribute levels and individual preference judgements, on a two dimensional subspace. We refer to such representations as *Preference Scenarios*.

The PCAR geometrical interpretation allows to enrich even more CA by joint plots of attribute levels, judges and stimuli. Some extra information on consumers (e.g. a priori cluster or social-demographic characteristics) can be also added on the maps.

In the next, traditional interpreting tools of CA will be read in the context of multidimensional data analysis; for example, the relative importance of each attribute (see Figure 1) can be argued by looking at the range of the level-point coordinates on each factorial axis.

Considering the expression (3.2), the principal axes of inertia are obtained as solution of the following characteristic equation under orthonormality constraints

$$\widehat{\boldsymbol{Y}}'\widehat{\boldsymbol{Y}}u_\alpha = \lambda_\alpha u_\alpha, \tag{3.3}$$

that is a PCA of the matrix $\widehat{\boldsymbol{Y}}$, noting that

$$\widehat{\boldsymbol{Y}}'\widehat{\boldsymbol{Y}} = \boldsymbol{Y}'\boldsymbol{X}(\boldsymbol{X}'\boldsymbol{X})^{-1}\boldsymbol{X}'\boldsymbol{X}(\boldsymbol{X}'\boldsymbol{X})^{-1}\boldsymbol{X}'\boldsymbol{Y}, \tag{3.4}$$

which leads to

$$\boldsymbol{Y}'\boldsymbol{X}(\boldsymbol{X}'\boldsymbol{X})^{-1}\boldsymbol{X}'\boldsymbol{Y}u_\alpha = \lambda_\alpha u_\alpha \tag{3.5}$$

with $u'_\alpha u_\alpha = 1$ and $u'_\alpha u_{\widetilde{\alpha}} = 0,\ \forall \alpha \neq \widetilde{\alpha}$.

For sake of computational efficiency, the (3.5) can be differently stated by looking for the solution of the eigen-equation

$$(\boldsymbol{X}'\boldsymbol{X})^{-1/2}\boldsymbol{X}'\boldsymbol{Y}\boldsymbol{Y}'\boldsymbol{X}(\boldsymbol{X}'\boldsymbol{X})^{-1/2}w_\alpha = \lambda_\alpha w_\alpha \tag{3.6}$$

with $w'_\alpha w_\alpha = 1$ and $w'_\alpha w_{\widetilde{\alpha}} = 0,\ \forall \alpha \neq \widetilde{\alpha}$, which is equivalent to the two-sided eigenvalue equation

$$\boldsymbol{X}'\boldsymbol{Y}\boldsymbol{Y}'\boldsymbol{X}v_\alpha = \lambda_\alpha(\boldsymbol{X}'\boldsymbol{X})v_\alpha \tag{3.7}$$

with $v'_\alpha \boldsymbol{X}'\boldsymbol{X}v_\alpha = 1$ and $v'_\alpha \boldsymbol{X}'\boldsymbol{X}v_{\widetilde{\alpha}} = 0,\ \forall \alpha \neq \widetilde{\alpha}$.

The $v_\alpha = (\boldsymbol{X}'\boldsymbol{X})^{-1/2}w_\alpha$ are orthonormal column eigenvectors; the λ_α in (3.6) and (3.7) are the eigenvalues corresponding to the same non-zero positive values of the (3.5) arranged in non-increasing order. The (3.7) gives the spectral decomposition of the matrix $\boldsymbol{X}'\boldsymbol{Y}$ in the metric $\boldsymbol{X}'\boldsymbol{X}$.

The transition formulas are given by

$$u_\alpha = \frac{1}{\sqrt{\lambda_\alpha}}\boldsymbol{Y}'\boldsymbol{X}(\boldsymbol{X}'\boldsymbol{X})^{-1/2}w_\alpha = \frac{1}{\sqrt{\lambda_\alpha}}\boldsymbol{Y}'\boldsymbol{X}v_\alpha \tag{3.8}$$

$$w_\alpha = \frac{1}{\sqrt{\lambda_\alpha}}(\boldsymbol{X}'\boldsymbol{X})^{-1/2}\boldsymbol{X}'\boldsymbol{Y}u_\alpha \tag{3.9}$$

$$v_\alpha = \frac{1}{\sqrt{\lambda_\alpha}}(\boldsymbol{X}'\boldsymbol{X})^{-1}\boldsymbol{X}'\boldsymbol{Y}u_\alpha \tag{3.10}$$

The eigenvectors u_α are the weights for the J respondents in the aggregated preference model

$$\widehat{\overline{\boldsymbol{Y}}}_\alpha = \boldsymbol{X}(\boldsymbol{X}'\boldsymbol{X})^{-1}\boldsymbol{X}'\boldsymbol{Y}u_\alpha \equiv \boldsymbol{X}\widehat{\boldsymbol{B}}u_\alpha. \tag{3.11}$$

Since there are at most $\alpha = 1, \ldots, K - p$ different weighting systems with decreasing order of importance, we refer to $\alpha = 1, 2$ as the *principal judgement system* and define the first factorial plan as a *Preference Scenario.*

The (3.11) is used in computing the coordinates of the N stimuli. The simulated stimuli can be represented as supplementary points

$$\boldsymbol{X}^*\widehat{\boldsymbol{B}}u_\alpha, \tag{3.12}$$

where $\boldsymbol{X}^*$ consists of the attribute levels combinations describing any new product.

The coordinates of the attribute levels are

$$Coor(X_k) = \sqrt{\lambda_\alpha}w_\alpha \equiv \sqrt{\lambda_\alpha}(\boldsymbol{X}'\boldsymbol{X})^{1/2}v_\alpha \equiv (\boldsymbol{X}'\boldsymbol{X})^{-1/2}\boldsymbol{X}'\boldsymbol{Y}u_\alpha. \tag{3.13}$$

The p remaining coefficients are determined by the constraint of summing up to zero within each attribute.

According to (3.5) – (3.7) the coordinates of the J respondents are

$$Coor(Y_j) = \sqrt{\lambda_\alpha}u_\alpha \equiv \boldsymbol{Y}'\boldsymbol{X}(\boldsymbol{X}'\boldsymbol{X})^{-1/2}w_\alpha \equiv \boldsymbol{Y}'\boldsymbol{X}v_\alpha, \tag{3.14}$$

which give the directions along which to read the individual preference models.

If social-demographic information about respondents are known, they can be represented on the factorial plan in order to highlight the preference of *a-priori* judges typology. Let $\boldsymbol{Z}$ be the $J \times C$ matrix of dummy variables associated to C exogenous characteristics and to the J respondents, the representation of the C characters' *centre of gravity* is obtained from the (3.13) as

$$\boldsymbol{\Delta}_Z^{-1}\boldsymbol{Z}'\boldsymbol{Y}'\boldsymbol{X}(\boldsymbol{X}'\boldsymbol{X})^{-1}w_\alpha, \tag{3.15}$$

where the matrix $\boldsymbol{\Delta}_Z^{-1}$ is the inverse of the matrix whose elements are the diagonal terms of $\boldsymbol{Z}'\boldsymbol{Z}$. The coordinates in (3.15) are obtained as illustrative points of the exogenous characteristics: such information does not influence the spanning of the principal axes. Giordano and Scepi (2001) show how to obtain a suitable representation where both the information on respondents and the information on stimuli (the attribute levels) play an active role.

4. Interpreting results: Preference scenarios

The factorial approach to CA by means of PCAR allows to represent on a two dimensional space the relationships among the judges, the attribute levels and a set of stimuli.

In order to correctly interpret the position of judges, levels and stimuli on each factorial axis, the quality of their representation can be evaluated according to the classical formulas used in PCA (i.e. relative and absolute contributions).

Each factorial axis is a synthesis of the preference variables. They describe the preference of a homogenous subset of respondents towards the attribute levels. The first factorial axis determine the maximum agreement system within judges while the successive ones establish alternative preference patterns of judges' subsets.

The interpretation of the maps is obtained by looking at

(1) *the direction of judge-vectors;*
(2) *the direction of level-vectors;*
(3) *the position of stimuli-points;*
(4) *the position of illustrative characteristics of the respondents.*

The representation of each attribute level explains the preference of the whole set of judges: the closer is a judge-vector to a level-vector, the higher is the preference of this judge for that level.

Since the level coordinates (3.13) are computed as a linear combination of the individual part-worth coefficients, and assuming the relation between a judge and a factorial axis as weighting system (i.e. the vector u_α), we can state that they represent different syntheses of the estimated part-worth coefficients. In this way, we obtain different syntheses of the individual estimates instead of the unique average which is traditionally considered.

Consumers which show the same preferences are represented by judge-vectors pointing in the same direction, e.g. they delineate market segments.

Such groups of judges can be illustrated by the position of social-demographic indicators (the auxiliary dummy variables in the matrix $\boldsymbol{Z}$) according to the (3.15).

The *Preference Scenario* allows to analyse the importance of each attribute levels and, at the same time, to recognise latent market segments.

The position of the stimuli points is a function of their own attribute levels. Supplementary stimuli (e.g. simulated scenarios) can be easily shown on the map. They are useful to illustrate the preference of respondents for either new or actual products.

It is worth noting that when the whole group of respondents is very homogenous with respect to their preferences, the first factorial axis is very close to the average aggregate model.

When manifold subsets are present in the data the successive axis will constitute a synthesis obtained as a weighted average of the part-worth coefficients; the weights are the importance of each preference vector in determining the direction of inertia. For this reason, we can state that each obtained component represents a synthesis of decreasing importance in explaining the preference structure.

On each axis, the relative importance of an attribute can be evaluated by looking at the widespread of the level coordinates. Analogously, it can be calculated by dividing the range of coordinates for each attribute by the sum of the level coordinates ranges for all attributes. In such a way, we obtain a different attribute importance for each direction.

If we are interested in a vector-judgement which lies in a particular direction, we can read, on that direction, the position and the range of the attribute levels coordinates and the position of the stimuli. Furthermore, in order to enhance the interpretation of a map, the configuration of points can be rotated so that the horizontal axis shows the leading direction we wish to describe.

As an illustrative example, we show the results of the non-symmetrical approach to multiattribute preference data on the *Tires* data set presented in Section 2.

Example (part 2) – multidimensional conjoint analysis: tire pressure

The data set used in this example has been shown in Table 2. The output produced by performing the multidimensional approach to CA has been realised by using the *Interactive Matrix Language* (IML) of the SAS System. In this section the most significative results will be highlighted. Table 5 shows the non zero eigenvalues of the eigenequations (3.5) – (3.7). Note that the first two dimensions account for the 84% of the whole information.

The *Preference Scenario* in Figure 3 shows the placement of attribute levels and judges. Six respondents seem to have quite homogenous preferences, while the judge 1 shows a different pattern of preference. In fact, on the first axis the most important attributes are MILEAGE and PRICE; most judges have preferred the high levels of MILEAGE (*Mile80, Mile60*) and the low levels of PRICE (*Price45, Price60*). For each axis it is possible to determine the importance of the attributes according to the widespread of the relative level coordinates. For instance, the attributes importance on the first two axes are shown in Table 6.

Eigenvalue	**Percent**	**Cumulated**
2228.736	69.790	69.79
442.935	13.870	83.66
291.104	9.116	92.77
171.180	5.360	98.13
57.324	1.795	99.93
1.965	0.062	99.99
0.252	0.008	100.00

Table 5. Eigenvalues and their relative importance.

Attribute	**Level**	**Label**	**Axis 1**	**Axis 2**	**Attrib. Imp. Axis 1**	**Attrib. Imp. Axis 2**
BRAND	TireMax	TireMax	-5.879	-13.866	8.57%	68.00%
	GoodTread	GoodTre	2.580	-0.274		
	RollsAhead	RollsAh	3.299	14.140		
INSTALL	INST free	INSfree	5.453	-0.588	10.19%	2.85%
	INST 7.5	INS7.5	-5.453	0.588		
PRICE	Price 45$	Price45	14.032	-0.657	30.39%	10.03%
	Price 60$	Price60	4.449	-1.737		
	Price 75$	Price75	-18.482	2.394		
MILEAGE	Mileage 40	Mile40	-30.846	2.655	50.83%	19.10%
	Mileage 60	Mile60	7.311	-5.212		
	Mileage 80	Mile80	23.535	2.557		

Table 6. Coordinates of levels and attributes importance.

The attribute INSTALL shows very poor widespread, it is quite indifferent in discriminating about stimuli preference. Nevertheless, the Free Install (*INSfree*) level is preferred over an Install Cost of $7.5 (*INS7.5*). Finally, the BRAND attribute is much more important on the second axis. It explains the difference between the *Judge 1* and the other consumers. These outcomes can be compared with the results shown in the first part of the example (Figures 1–2; Tables 2–3).

The Judges coordinates and the cosines of the angles spawned by each vector-judge with the first two factorial axes are reported in Table 7.

As a general rule, we look at the direction spanned by a judge-vector. The attribute levels lying in the same direction explain how the judge composes his preferences. For the *Judge 1* the leading direction is spanned by the attribute BRAND. It is quite indifferent to attributes MILEAGE, PRICE and INSTALL.

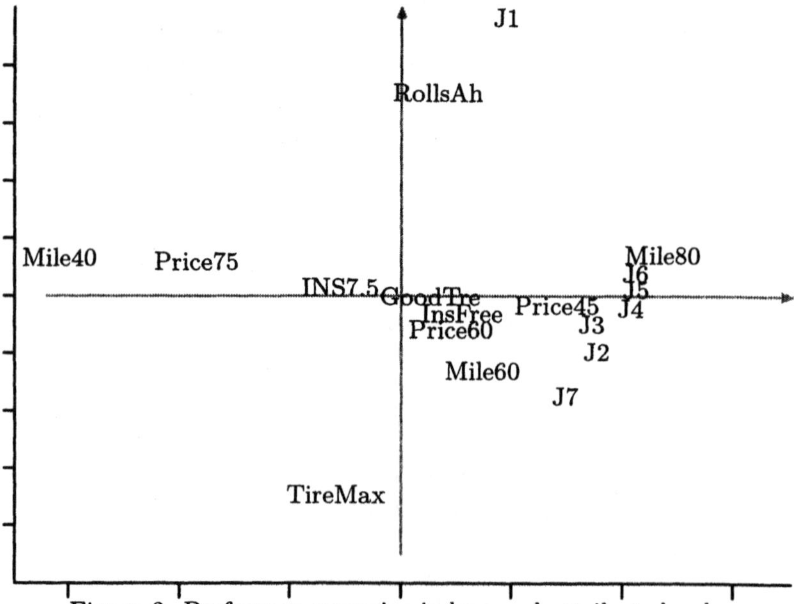

Figure 3. Preference scenario: judges and attribute levels ($\lambda_1 = 69.79\%; \lambda_2 = 13.87\%$).

	Coordinates		Cosines	
	Axis 1	*Axis 2*	*Axis 1*	*Axis 2*
Judge 1	9.45	19.32	0.44	0.89
Judge 2	17.62	-3.87	0.97	-0.21
Judge 3	17.15	-1.73	0.99	-0.10
Judge 4	20.65	-0.86	0.99	-0.04
Judge 5	21.10	0.44	0.99	0.02
Judge 6	21.06	1.28	0.99	0.06
Judge 7	14.79	-6.98	0.90	-0.42

Table 7. Coordinates of judges and cosines.

Conversely, the group of respondents prefers the high levels of attribute MILEAGE (*Mile80* and *Mile60*) and the low levels of PRICE (*Price45* and *Price60*). Most judges seem indifferent to attribute INSTALL whose coordinates are very close to the origin.

Finally, we can state that the key-factors are PRICE and MILEAGE; they are "*size*" attributes, whereas the BRAND factor can discriminate within the consumer group. It is a "*shape*" attribute.

In Figure 4 the configuration of stimuli points is shown. This configuration reproduces the original global preferences expressed by respondents. For example, in Figure 4, the estimated preference ranking for *Judge 1* is reproduced.

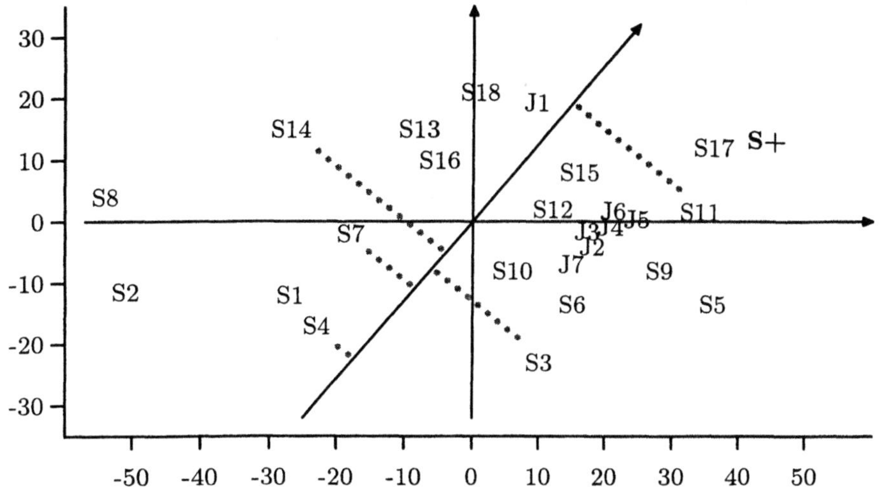

Figure 4. Preference scenario: the preferences of Judge 1 ($\lambda_1 = 69.79\%; \lambda_2 = 13.87\%$).

Nevertheless, we can reproduce all the individuals preference rankings, but our aim is to look for a suitable aggregated preference. The main candidate appears to be the first factorial axis. Different stimuli seem to encounter the favour of the judges, they are collocated at the far right on the map (for sake of interpretability, the positive sides of the axes show positive utility). The positive side of the first axis indicates the stimuli *S17, S5, S11* as the preferred ones for the set of judges as a whole. Indeed, they show lack of preference for stimuli *S2* and *S8.*

Finally, supplementary stimuli points can be plotted on the map in order to simulate the preference obtained by particular compositions of attribute levels.

For instance, according to the results of the individual models we may want to represent the stimulus preferred by the group as a whole (see Table 3) and by the *Judge 1* (see Table 4). As noted in Section 2, this optimal configuration is realised by combining the levels *Rolls Ahead, Mileage 80000, Price 45* and *Install 0*. This stimulus is represented in Figure 4 with the label "**S+**".

References

Addelman S. (1962). *Orthogonal main-effects plans for asymmetrical factorial experiments.* Technometrics **4**, 21 – 46.

Anderson T. W. (1951). *Estimating linear restrictions on regression coefficients for multivariate normal distributions.* Annals of Mathematical Statistics **22**, 327 – 351.

Bouroche J. M., Saporta G. and Tenenhaus M. (1977). *Some methods of qualitative data analysis.* In: Recent developments in statistics, J. R. Barra et al., eds., North Holland, Amsterdam.

D'Ambra L. and Lauro N. C. (1982). *Analisi in componenti principali in rapporto ad un sottospazio di riferimento.* Rivista di Statistica Applicata **1**, 51 – 67.

Fortier J. J. (1966). *Simultaneous linear prediction.* Psychometrika **31**, 369 – 381.

Giordano G. (1997). *Multidimensional Preference Data Analysis: An Exploratory Strategy to Conjoint Analysis.* Ph.D. Thesis, University "Federico II", Naples. (In Italian).

Giordano G. and Scepi G. (1998). *A strategy for dealing with nonresponses in marketing research.* In: Proceedings of the International Seminar on New Techniques & Technologies for Statistics, Garonna L. et al. eds., Sorrento (Napoli - Italy), 199 – 204.

Giordano G. and Scepi G. (2001). *Different informative structures for quality design.* J. of Italian Statistical Society. (in press).

Green P. E. (1974). *On the design of choice experiments involving multifactor alternatives.* J. of Consumer Research **1**, 71 – 74.

Green P. E. and Krieger P. (1989). *Recent contributions to optimal product positioning and buyer segmentation.* European J. of Operational Research **41**, 127 – 141.

Green P. E. and Rao V. R. (1971). *Conjoint measurement for quantifying judgmental data.* J. of Marketing Research **8**, 355 – 363.

Green P. E. and Srinivasan V. (1978). *Conjoint Analysis in consumer research: issues and outlook.* J. of Consumer Research **5**, 103 –123.

Green P. E. and Srinivasan V. (1990). *Conjoint Analysis in marketing: New developments with implications for research and practice.* J. of Marketing **54**, 3 – 19.

Green P. E. and Wind Y. (1975). *New way to measure consumer's judgements.* Harvard Business Review **53**, 107 – 117.

Hagerty M. R. (1985). *Improving the predictive power of conjoint analysis: The use of factor analysis and cluster analysis.* J. of Marketing Research **22**, 168 – 184.

Israels A. Z. (1984). *Redundancy analysis for qualitative variables.* Psychometrika **49**, 331 – 346.

Izenman A. J. (1975). *Reduced rank regression for the multivariate linear model.* J. of Multivariate Analysis **5**, 248 – 264.

Kruskal J. B. (1965). *Analysis of factorial experiments by estimating monotone transformation of the data.* J. of Royal Statistical Society B **27**, 251 – 263.

Kruskal J. B. and Carmone F. J. (1969). *MONANOVA: A Fortran IV program for monotone analysis of variance.* J. of Marketing Research **6**, 497.

Lauro N. C., Giordano G. and Verde R. (1998). *A multidimensional approach to conjoint analysis.* Applied Stochastic Model and Data Analysis **14**, 265 – 274.

Louviere J. J. (1988). *Analyzing Decision Making – Metric Conjoint Analysis.* Sage University Papers.

Rao C. R. (1964). *The use and the interpretation of principal component analysis in applied research.* Sankhya **A 26**, 329 – 358.

Shocker A. D. and Srinivasan V. (1979). *Multiattribute approaches for product concept evaluation and generation: A critical review.* J. of Marketing Research **16**, 158 – 180.

Tenenhaus M. (1988). *Canonical analysis of two polyhedral convex cones and applications.* Psychometrika **53**, 503 – 524.

van den Wollenberg A. L. (1977). *Redundancy analysis: An alternative for canonical correlation analysis.* Psychometrika **42**, 207 – 219.

van der Lans I. A. (1995). *Nonlinear Multivariate Analysis for Multiattribute Preference Data.* DSWO Press, Amsterdam.

Young F. W., De Leeuw J. and Takane Y. (1976). *Regression with qualitative and quantitative variables: An alternating least squares approach with optimal scaling features.* Psychometrika **41**, 505 – 529.